Use R!

Series Editors:
Robert Gentleman   Kurt Hornik   Giovanni Parmigiani

Use R!

Julien Claude

Morphometrics with R

 Springer

Julien Claude
Université de Montpellier II
ISEM, UMR 5554 CNRS
Laboratoire de Morphométrie
2 place Eugène Bataillon
34095 Montpellier
France
julien.claude@univ-montp2.fr

Series Editors:
Robert Gentleman
Program in Computational Biology
Division of Public Health Sciences
Fred Hutchinson Cancer Research Center
1100 Fairview Avenue, N. M2-B876
Seattle, Washington 98109
USA

Kurt Hornik
Department of Statistik and Mathematik
Wirtschaftsuniversität Wien Augasse 2-6
A-1090 Wien
Austria

Giovanni Parmigiani
The Sidney Kimmel Comprehensive
Cancer Center at Johns Hopkins University
550 North Broadway
Baltimore, MD 21205-2011
USA

ISBN 978-0-387-77789-4   e-ISBN 978-0-387-77790-0
DOI: 10.1007/978-0-387-77790-0

Library of Congress Control Number: 2008927603

Printed on acid-free paper.

9 8 7 6 5 4 3 2 1

springer.com

To my parents

Preface

This book aims to explain how to use R to perform morphometrics. Morphometric analysis is the study of shape and size variations and covariations and their covariations with other variables. Morphometrics is thus deeply rooted within statistical sciences. While most applications concern biology, morphometrics is becoming common tools used in archeological, palaeontological, geographical, or medicine disciplines. Since the recent formalizations of some of the ideas of predecessors, such as D'arcy Thompson, and thanks to the development of computer technologies and new ways for appraising shape changes and variation, morphometrics have undergone, and are still undergoing, a revolution. Most techniques dealing with statistical shape analysis have been developed in the last three decades, and the number of publications using morphometrics is increasing rapidly. However, the majority of these methods cannot be implemented in available software and therefore prospective students often need to acquire detailed knowledge in informatics and statistics before applying them to their data. With acceleration in the accumulation of methods accompanying the emerging science of statistical shape analysis, it is becoming important to use tools that allow some autonomy. R easily helps fulfill this need.

R is a language and environment for statistical computing and graphics. Although there is an increasing number of computer applications that perform morphometrics, using R has several advantages that confer to users considerable power and possible new horizons in a world that requires rapid adaptability. Indeed, the R language and environment is suitable for both basic users and developers, and can run on most operating systems (Windows, Linux, or Apple OS). With one single environment, morphometric analysis can be performed from data acquisition to data analysis, and results can be presented in the form of graphs, both accurate and esthetic. Results can also be included in further tests with independent data, using the wide range of functions and packages available with R. R gathers the achievements of the *R core development team*, numerous contributors of online available packages, and possibly your own interaction. The advanced user can develop and modify his/her own or programmed functions. R is highly evolvable and offers a single and integrative environment to perform a wide range of statistical analyses; all characteristics make this software suitable for beginners. In addition, R is taught more and more in universities and increasingly used worldwide. Newcomers can easily get specific advice about both practical and scientific questions with the international assistance provided by the R-help web list and many online and inexpensive manuals. Finally, R is freely distributed over the Internet.

However, there was obviously a need for bringing the R and morphometric approaches together. The book is a guide for performing and for developing modern morphometrics with R. Exercises and examples can be used as tutorials, and the combination of the book with R can be used as a teaching instrument. Besides the need for supplying knowledge about R and morphometrics, the book expresses the need for reducing the gaps between theoreticians, developers, and users. For this goal, I deliberately favoured an approach involving customized functions directly written by the user, rather than the explanation of a selected package. Functions from specific "morphometric" packages are nevertheless briefly presented for faster computation. I hope that this book will not be only a guide for using R code for performing morphometrics, but also a helpful document for helping users to acquire autonomy and to develop programs for their own scientific needs. For the book to be fully useful, I strongly encourage you to work with R. It only requests you to download and install R on your computer: something that will take a few minutes.

The first chapter of the book deals with general considerations and an introduction to R. It directly brings users into contact with the R language and environment. The second chapter explains how to gather and capture data with R; a section of this chapter is concerned with the organization of files. Since image analysis is often a prerequisite for morphometric measurement, some applications and development of R applied to basic image analysis is provided as well. The third chapter is designed to guide you within the field of traditional and multivariate morphometrics. The fourth chapter deals with statistical analysis of shape using landmark data, and the fifth chapter presents R applications and developments for the statistical analysis of outlines. The sixth chapter presents statistical analysis considering the specifics of morphometric data (especially those based on landmarks and outlines). This chapter mainly relies on biological applications, but readers will find examples of code that can be applied to other fields. Finally, the last chapter explains how to progress further with R to perform simulation, and to hybridize R with other software for more specialized issues. For some examples, I used original data files (image files, codes, and datasets); these are provided online.[1]

I am grateful to Emmanuel Paradis, Michel Baylac, and Tristan Stayton who have contributed to parts of Chapters 3 and 7. In addition to their chapter contributions, Michel Baylac allowed me to explore the 2006 pre-releases of the package Rmorph and offered some basic code for Fourier analysis useful for developing functions provided in Chapter 5, Emmanuel Paradis supplied very useful comments on the first chapters and for improving and shortening some code, and Tristan Stayton checked Chapter 3 in integrality. The french research group *GDR 2474 CNRS Morphométrie*

---

[1] http://www.springer.com/statistics/stats+life+sci/book/
978-0-387-77789-4
and http://www.isem.cnrs.fr/spip.php?article840

and Évolution des Formes has been very stimulating for listening to the needs of the morphometric user's community. Some of my students (Piyatida Pimwichai, Saksiri Soonchan, Cédric Delsol, Guillaume Chatain) and colleagues (Ylenia Chiari, Lionel Hautier, Renaud Lebrun) have tested some previous versions of code, functions, and pedagogical issues of the book, and I thank them for their comments and support. Vincent Lazzari kindly supplies one dataset for outline analysis used in the fifth chapter, and Sylvie Agret digitized one landmark dataset and her help was appreciated for selecting images that are used as examples. I am grateful to Nicolas Navarro, Leandro Monteiro, Chris Klingenberg, Vincent Debat, Jean-Baptiste Ferdy, Florent Détroit, Jean-Christophe Auffray, and Sylvain Adnet for helpful short or long discussions and corrections. The very active MORPHMET list forum[2] as well as the R mailing lists[3] have been very important resources for developing techniques and functions, and understanding codes and algorithms. I thank the reviewers for their patience reading and correcting first manuscripts. I thank the editorial board who enriched the book with many suggestions and who improved the English of earlier versions. I would finally like to express my admiration to the *R Core Development Team* for the remarkable efforts they have made in providing this wonderful tool for all. This book is publication ISEM 165-2007.

Montpellier   *Julien Claude*
March 2008

---

[2] http://www.morphometrics.org/morphmet.html
[3] http://tolstoy.newcastle.edu.au/R

Contributors

Addresses of the Author and Contributors

- **Julien Claude**
  Institut des Sciences de l'Evolution de Montpellier
  UMR 5554 CNRS
  Université de Montpellier 2
  2, place Eugène Bataillon
  34095 Montpellier cedex 5, France

- **Michel Baylac**
  Origine, Structure et Evolution de la Biodiversité
  UMR 2695 CNRS
  Muséum national d'Histoire naturelle de Paris
  75005 Paris, France

- **Emmanuel Paradis**
  Institut de Recherche pour le Développement
  UR175 CAVIAR, GAMET
  BP 5095, 361 rue Jean François Breton
  34196 Montpellier cedex 5, France

- **Tristan Stayton**
  Bucknell University
  701 Moore Avenue
  Lewisburg, PA 17837
  United States of America

Abbreviations and Notations

k number of dimensions of the object or of the configuration
p number of landmarks
n number of objects or observations
M configuration matrix
A array of matrix configurations
m vectorized configuration matrice
A data frame or array containing a set of configurations
M' transpose matrix of M
d scalar distance
M^{-1} inverse matrix of M
PCA   principal component analysis
R_v R_v correlation coefficient of Escoufier (1973)
T^2_{HL} Hotelling Lawley trace
t^2 Hotelling t square
d_m Mahalanobis distance
sc_M centroid size
M_c vector of centroid coordinates
Ms configuration matrix scaled to unit centroid size
Mb Bookstein shape coordinates matrix
Mk Kendall shape coordinates matrix
$\mathbf{\Gamma}$ rotation matrix
d_F full Procrustes distance
d_P partial Procrustes distance
β scale
α translation vector
$\mathbf{H}_1$ lower right $(p-2) \times (p-2)$ partition matrix of the Helmert matrix
Mh Helmertized configuration
Z preshape matrix
X centered configuration
Zc centred preshape configuration matrix

$\mathbf{Be}$ bending energy matrix
M_p Procrustes superimposed configuration
GPA   generalized Procrustes analysis
EDMA   Euclidean distance matrix analysis
$\mathbf{FM}$ Euclidean distance matrix form
fm vectorized Euclidean distance matrix form
$\mathbf{FDM}$ form difference matix
ρ Procrustes trigonometric distance
TPS   Thin-Plate Splines
T period
ω pulse

Contents

1

Introduction

This chapter is a short introduction to geometric morphometrics and R. It explains the aims and applications of the first, while providing minimal requirements for using R for morphometrics. If you are familiar with one or both of these topics, you can skip those respective sections and proceed directly to the following chapters.

1.1 Morphometrics Today

Morphometric analysis is the statistical study of shape and size and their covariations with other variables. Shape is commonly described as the property of an object invariant under scaling, rotation, or translation. Size is a scalar, based on distances or coordinates of points specified on the object. An object can have other attributes that are close to the above-mentioned definition of shape but that are not shape properties, such as colors or texture. Although one can estimate these two attributes quantitatively, I will not consider their treatment in this book. We will consider shape as *the geometric property of an object invariant under rotation, scale, or translation.*

We regularly estimate the size and shape parameters of objects in our daily lives. For instance, every time we go to the market, we select the ingredients for our meals according to their size, volume, mass, and shape. Moreover, the study of shape and size variation or difference occupy, and will occupy, a more and more prominent place in our life with the development of technologies linked to the needs of our societies: fingerprints or calligraphic recognition, calibration of food and industrial products, appraising quantitatively developmental effects of some defective gene, etc. Morphometrics is an important tool in biological, agricultural, archeological, geological, geographical, and forensic sciences.

Until recently, the shape of objects and their variation were estimated using ad-hoc distances, angles, or proportions. Shape analysis received many improvements during the 20th Century with the development of concepts and methods aiming to describe shape parameters objectively and quantitatively. These developments became more numerous with the introduction of computers, and with the development and maturing of multivariate statistics, which took into account the ensemble of the attributes (variables) of objects used for describing shape and size. Things changed at the end of the 20th Century, when studies of shape variation became more than the application of multivariate methods to a series of ad-hoc measurements. Indeed, multivariate methods were just a small step toward what is today a mature discipline and permits to extract and assess reliably the variation of the geometric properties of objects. Morphometrics today considers the shape of an object as a whole and the interdependence of its parts. Although multivariate techniques could have resolved this dependence, modern statistical shape analysis goes further, and supply methods for easily describing shape variation or shape change in both qualitative and quantitative terms, with statistics appropriate for the task.

Modern shape analysis was initiated with the seminal work of D'Arcy Thompson [117], that offered new ways of understanding shape variation. The idea to map a shape to a grid and to deform this grid to fit a target shape changed the perception of shape changes in terms of mathematical transformations. Indeed, dramatic changes between comparable biological objects can be achieved by changing a few properties of objects. One can interpret the changes as resulting from distending, flattening, or shearing forces applied to the objects themselves. The expression of these forces can be visualized with deformation grids. Although the intuitive idea of using map functions between two configurations was first developed by artists such as Leonardo di Vinci or Dürer for aesthetic purposes and drawing constructions, it was D'Arcy Thompson who developed these grids with the scientific purpose of describing morphological changes (Fig. 1.1). It was only several decades later, at the end of the 20th Century, and after several attempts, that mathematical frameworks were developed for constructing these grids. This achievement was part of the emerging shape statistics that proposed several methods for analyzing the parameters and variation of shape and form of objects: superimposition methods (Section 4.2), application of Fourier analysis to outlines (Section 5.2), Thin-Plate Splines (Section 4.3), and Euclidean Distances Matrix Analysis (Section 4.4), etc. Some other methods were applied to shape and image recognition; among them, the Hough transform played an important role for image recognition. Most of these methods are based on multivariate techniques and are now the core of an emergent discipline at the interface of statistics and other sciences. The important developments in morphometrics at the end of the last century led several authors to write about a "revolution in morphometrics." [1, 106]

Development of geometric morphometrics has been driven by the increasing demands of other disciplines. Morphometrics changed within only one century from univariate, to bivariate, and to multivariate analyses, and saw the inventing of its own statistical toolkit before the beginning of this century. Today, morphometrics is not only an application of univariate or multivariate statistics to shape and size variables; it is a new emerging discipline with its own descriptors of shape.

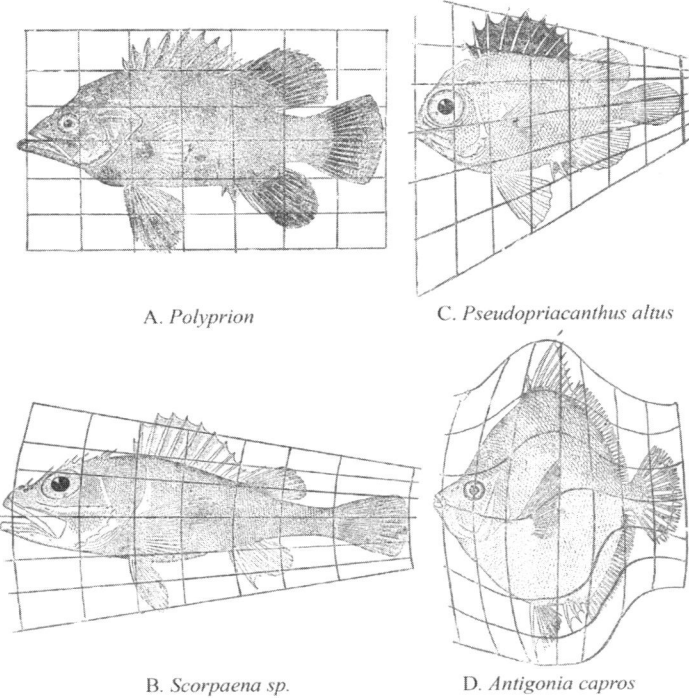

A. *Polyprion*

C. *Pseudopriacanthus altus*

B. *Scorpaena sp.*

D. *Antigonia capros*

Fig. 1.1. Biorthogonal grids monitoring shape changes between different species of fish (from d'Arcy Thompson [117])

1.2 Shapes and Configurations

Objects analyzed by morphometricians are mostly two dimensional (2D) or three dimensional (3D). In this book k will refer to the dimensions. For one-dimensional objects (when $k = 1$), form, size, and shape overlap, and traditional statistical methodology (for example, analysis of variance) is easily applied. When $k = 2$ or 3, the geometric characterization of an object includes volumes, surfaces, outlines, locations of segments, positions of peculiar points, and the decomposition of the form into size and shape ensues. Extracting shape and size information is then less obvious, and applying statistics becomes more complex. The full surface or outline of the object corresponds to the location of infinitely many points in 2D or 3D cases. Users can define a sampling procedure to obtain the coordinates of points belonging to the object (for example, equally spaced points on the outline). However, more information will be retained if, among the sampled points, the positions of peculiar structures that are recognizable for every object of the set are included.

To correctly compare shapes, one must first define some of the structural properties of objects (bottom, left side, anterior part, etc). This is particularly true for biological objects, where structures that are recognizable and comparable among specimens are said to be homologous (here homology is taken in its widest biological acceptance). We often use some kind of homology for comparing shape of objects other than biological objects as well. For example, to compare two bottles of milk with some reliability, it is necessary to define the parts that will be further compared together. If you explore the variation of shape between bottles, perhaps it is not useful to compare the bottleneck of one bottle with the bottom of the other. The interbottleneck, inter-bottom comparisons and differences in the structural relationships between bottleneck and bottom are more reliable for the purpose of bottle comparison. In morphometrics, the statement of homology is often a statement of topological features, rather than a statement based on a proven biological and historical basis.

A peculiar point for which position is comparable among objects is a landmark. For biological objects, a point that correspond to an anatomical position comparable among objects is defined as an anatomical landmark. Anatomical landmarks are said to be homological if one can state a similar embryological, anatomical, or historical origin. Outlines or surfaces depicting the same anatomical structure between individual organisms can be homological as well. It is often necessary to take into account more general features or homologies that objects can share together (for instance, their antero-posterior or dorso-ventral axes) in order to compare them.

One can rely on the relative position of landmarks invariant to rotation, scaling, or translation to make shape comparisons, as we will see later. Most morphometric methods use the locations of landmarks as first input. Indeed, even when you measure distances on a series of objects with a caliper, you often collect interlandmark measurements. Anatomical landmarks are of several types. Bookstein [10] provided a classification of landmarks and recognized three categories (Fig. 1.2). A landmark of type I corresponds to discrete juxtaposition of tissues, or sufficiently small features to be defined by a single point (for example, a small foramen). This kind of landmark is probably the only one to have a true biological homology origin. A type II landmark corresponds to maximum of curvature; it may correspond to similar developmental features (for example, a meristem), but its homological basis may have weaker biological grounds. A type III landmark is an extremal point that can correspond to an end-points of diameters, a centroid, or an intersection between interlandmark segments; it is constructed geometrically. I do not distinguish the latter type from pseudolandmarks that are points defined by construction, too: this may be points regularly sampled on an outline or on a surface, or points at the intersection between the outline of the structure and a segment defined by two anatomical landmarks. Although the last two types do not carry as much biological information in terms of homology as type I (Bookstein [10] calls them "deficient"), they can be useful for including geometric information in regions of the object where digitized points are under-sampled, or to extract geometric properties of outlines or surfaces. Although an object may not have recognizable points, sometimes a small surface or characteristic segment, assumed to be homologous among specimens, enables one to record an important geometric feature comparable between objects. The exact location of this feature can only be approximately digitized on the structure (for example, a point located approximately at the middle of the surface), and is thus named a "fuzzy landmark." [118] This kind of landmark location is common in biological specimens (structures like bulges or bosses that have no clear boundaries such as articulation points between segments). Finally, even when ignoring the biological basis for a landmark definition, one can still define mathematical landmarks that are points defined by some mathematical property (i.e., high curvature).

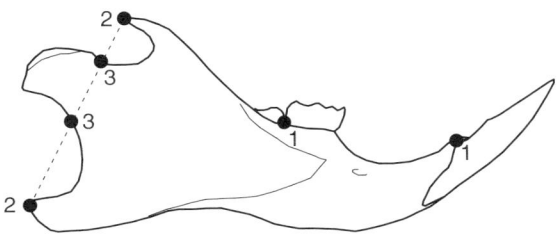

Fig. 1.2. The three types of landmarks illustrated on the jaw of a mouse

As long as it is possible to identify landmarks and measure their coordinates, it is rather easy to calculate distances, angles, etc. Further statistical analyses can deal with the coordinates themselves or with distances and angles calculated from them. One can also deduce coordinates of landmarks (or the relative positions between landmarks), but only if enough interlandmark distances or angles are measured following a given procedure (EDMA, triangulation, or the truss methods [116]). Furthermore, visualizing shape differences is easier to interpret by directly examining the relative positions of landmarks rather than examining a table of varying distances or angles. The coordinates of landmarks thus contain the richest geometric information available in any object. A configuration is defined by the collection of landmark coordinates in one object. The more landmarks are collected for a given object, the more shape information is gathered in the configuration depicting the object, and the better its morphological features are appraised quantitatively.

Most morphometric methods described here use either coordinates, angles, distances, or any other property of objects that one can obtain using landmark or pseudolandmark coordinates. Chapter 2 explains how to organize or obtain measurements using coordinates of landmarks with R, while following chapters explain how to implement statistical methods using R for analyzing these raw or transformed data.

1.3 An R Approach to Morphometrics

Let's first see why R is appropriate to deal with morphometric data. A useful strategy for a morphometric analysis requires:

- Collecting data from the object and transferring to computer files as directly as possible, such as digitization of point coordinates in a numerical image.
- An interface allowing a simple manipulation, including visualization or modification of the raw data.
- An interface between the user and the data that can perform several operations or apply several tests to the data (usually sets of coordinates in our case), and that present results in diverse ways (graphs, tables, etc).
- To avoid multiplication of software applications for obtaining data used in statistical analyses.
- For 3D data, the graphic devices should allow some interactivity. Since the screen is flat, one needs to observe and rotate 3D data in a convivial way.
- To obtain results and store these results under the form of graphs or tables.
- To adapt and to develop customized functions that are less common in traditional statistics.
- To convert files that have been treated with other software and export files for users that may be still reluctant to use R as well.

R is a language and environment that allows diverse statistical treatments of objects, and even more. R is well designed for our needs. R can be adapted to become an environment for performing any morphometric analysis. More than this, R has many other qualities that can be seriously attractive and can confer it some superiority above other software. I am quite sure that you will learn that "R is very much a vehicle for newly developing methods of interactive data analysis" for quoting the online manual "An Introduction to R." [120]

- R runs on all the common operating systems (Unix, Linux, Apple OS, or Windows).
- R is free.
- R is a language rather easy to understand and to learn.
- There are multiple ways to get help: books, online help, help forums, online manuals, articles, courses, and even meetings.
- Using R for morphometrics will considerably reduce the number of software applications that one needs to perform analyses.
- If you save what you did on a file (for example, a text file), you can re-run the complete analysis, doing adjustments, corrections, and adapt the code to one new similar treatment.
- R is evolvable and adaptable, the user does not have to adapt to R for his/her needs but adapt R to his/her own requests.
- R can do repetitive things with a minor contribution of the user.
- R is designed for efficient scientists who do not wish to waste their time in repetitive and useless manipulations, but who are ready to waste their time in finding syntax error in R programs.
- A lot of packages that perform an exponentially number of analyses have been developed around the basis of R.
- The development of R is accessible for experienced users and for novices.
- Codes are available.

- R is not wysiwyg (what you see is what you get). Users have to think before trying pseudo-treatments and pseudo-exploration with interactive menus and windows. With R, what you get is what you want.
- R has already been adopted as a computational environment for data analysis in many fields (e.g., phylogenetics).[1]

We are starting with some basics before going more deeply into statistics. The first step is to install R on your computer. R is freely available on the website of the *Comprehensive R Archive Network (CRAN)*[2] from where you can download the files needed for your operating system (Linux, Unix, Windows, or Apple OS). Then, just follow the available instructions for installation.

Once R is installed on your computer, you can launch it using the corresponding executable. After the user interface appears on the screen, a short text appears, and the sign ">" indicates that R is waiting for your command. Under some operating systems (for example, Windows), this interface consists of pull-down menus and a toolbar that can be used for some commands (preferences, access to the online help, loading and installing packages). Since R is not "wysiwyg," it is essential to spend a few minutes to acquire basic information. Once done, using R rapidly becomes very intuitive, although the new user may feel anxious with this new and seemingly very simple interface. This has, at least, the enormous advantage that you are not proceeding with any analysis without thinking a bit about what you are going to do.

Finding assistance with R is easy: first, you should read the manual "An Introduction to R" that one can freely download from the *CRAN* family of Internet sites.[3] Other manuals are available. For a shorter text, you can read R for Beginners from Paradis [80] which provides most of what you need for starting; it is freely available on the web.[4] Other sources of help are available through an increasing number of books (such as [120]). You can directly use the `help(topic)` or `?(topic)` commands from the user interface of R to display online help for any R-topics. If your problem persists, it is possible to access the archives of the "r-help" general discussion list.[5] The discussion list is useful because it reports unknown bugs, statistical questions, or user problems. If your problem is new, you can subscribe to the "r-help" mailing list. Before doing that, check whether your problem is really new using a web-search engine or by reading the archives of the online mailing list.[6] Finally, some of your colleagues, teachers, or friends may be R users and can guide you in your first steps and answer your questions about R.

I first introduce some of the graphical possibilities of R to show how to efficiently use the command line later. We need graphs to analyze and communicate our results.

---

[1] http://cran.r-project.org/src/contrib/Views/
[2] http://www.cran.r-project.org/
[3] http://CRAN.R-project.org
[4] http://cran.r-project.org/doc/contrib/Paradis-rdebuts_en.pdf
[5] http://finzi.psych.upenn.edu/search.html
[6] http://www.R-project.org/mail.html

You can examine some of the possibilities offered by the basic graphs in R in typing the two command lines:[7]

```
> par(ask=TRUE)
> demo(graphics)
```

The command `demo(graphics)` opens a graphical interface and displays successive examples of graphs. Actually, **graphics** is a package of R. On the command line interface (CLI), you can read the code and the commands that R types corresponding to the graphs. Since there are successive graphs, we first need to type the command `par(ask=TRUE)`, to be asked for typing the "Enter" key, before the next figure to be drawn and the new corresponding code to be typed by R. Other graphical demonstrations are displayed for the function `image`.

```
> par(ask=TRUE)
> demo(image)
```

You can visit the `addictedtor`[8] webpage for discovering the remarkable possibilities that R offers for graphical visualisation. In addition, the corresponding code is available for producing your own graphs as a tutorial.

The function `example` is a common way to obtain a tutorial and ways to use R functions. Writing the command `example(topic)` runs all the R code from the `Examples` part of R's online help topic.

```
> example(persp)
```

The function `persp` draws graphs of 3D data. There are several examples for this function (use `help(persp)` to see them). Notice that you can copy selected command lines of interest (one or several) from the example section, and then paste them on the CLI. It is a way to make a selection among the possible examples of this kind of help file. This demonstrates a very useful property of the R environment: indeed, you can write successive commands (separated either by a new line or a semicolon) on a text file that you save somewhere, and paste them later on R. This has three advantages:

1. It is possible to correct isolated mistakes in command or data, and simply copy and to paste the new text file on the R command line. R will rerun the full analysis for you. This avoids wasting time in rethinking and reclicking a complete analysis as it is the case with software applications providing menus and other right-left click button.
2. You can use and transform your text file to adapt for new datasets.
3. You can keep an archive of what you have already done.

---

[7] except prompt signs, and R results, the code typed on R will be always be typed in small typewriter font, when needed basic explanation of the code will be written in small usual fonts

[8] http://addictedtor.free.fr/graphiques/

Note that on some operating systems (Unix, Windows), the vertical-arrow key of the keyboard can scroll forward and backward through a command history. This is another simple way to do corrections or to modify a command.

For interacting with 3D data, you can download and install the rgl package. Once done, type:

```
> library(rgl)
> par(ask=TRUE)
> demo(rgl)
```

A second interface is displayed on the screen. This is the rgl device; you can interact with the device using the mouse and positioning or scaling the graph as you wish.

All R functions and datasets are stored in packages. Aside from the base, utils, stats, graphics, datasets, grDevices packages that are automatically available and loaded in any R installation, contributed packages must be downloaded from the *CRAN* website or one of its mirror sites (this can be done using the menu "package" on some operating systems if you have an Internet connection). A package is loaded using the function library(package). Listing available functions, datasets, and summary information is obtained as follows:

```
> library(help=rgl)
```

Finally, it is sometimes useful to interact with other programs: R can invoke a system command, using a shell. The functions system or shell can interact with flexible programs. They become useful if you want to do everything from R. We will see this for converting image files in Chapter 2. For instance, if notepad is on your computer, you can type shell(notepad). Respect the syntax of your operating system for writing the path: be careful with "/" or "\" that do not have similar meaning under Linux or Windows.

1.4 Starting with R

1.4.1 Expression, Assignment and Other Basics

There are two elementary commands with R: assignment and expression. When you write an expression, R is evaluating the command and prints the value, while an assignment stores the value of this command to a variable.

```
> 1 + 1
[1] 2
```

This command is an expression that uses the sum operator between two vectors of value 1 and of unitary length. R prints the result, preceded by the digit 1 between brackets. This digit indicates that the displayed line start with the first element.

```
> a<-1+1
> a
[1] 2
> A
Error: Object "A" not found
```

The first command is an assignment. For assigning a value to a variable, we use the "<−" operator. Then it is followed by two expressions; the first returns the value of a while the second return an error message. Notice that R is case sensitive. Many errors arrive because of syntax mistakes. The name of a variable must start with a letter (A-Z and a-z) and can include letters, digits (0-9), and dots (.). We call these variables "objects" in the text for avoiding confusion with mathematical variables.

```
> a+
+
+ 1
[1] 2
```

If a command is not complete at the end of a line, the prompt will change to become the continuation symbol +, indicating that R is waiting for the user to write the end of the command. Finally, it is possible to use a function rather than an operator.

```
> sum(1, a)
[1] 2
```

Functions are sometimes preferred over operators, since one can enter other arguments into them. For example, the function sum allows the handling of missing data in different ways.

R can print the value of some expressions on a graphical interface.

```
> plot(1,1)
```

This command plots a point of coordinates $x = 1$ and $y = 1$.

1.4.2 Objects

Generalities

Aside from their names and values, objects are characterized by their attributes. The mode and length are attributes of all objects in R. The mode is the basic type of the elements of the object and is returned with the function mode. If elements are data, they can have four different modes: numeric, character, complex, and logical (FALSE or TRUE, alternatively typed as F or T). The length is the number of elements of the object and is returned by typing length(object). Finally, the function str displays the internal structure of a R object.

```
> b<-1+2i; d<-"ab"; e<-FALSE
> mode(a); mode(b); mode(d); mode(e)
[1] "numeric"
```

```
[1] "complex"
[1] "character"
[1] "logical"
> length(a); length(b); length(d); length(e)
[1] 1
[1] 1
[1] 1
[1] 1
> str(b)
 cplx 1+2i
> e<-c(a, 2)
> e
[1] 2 2
> mode(a)
[1] "numeric"
> length(e)
[1] 2
```

Commands are separated by semicolons, while outputs are printed on new lines. The syntax of characters and strings uses double quotation marks ("). The syntax of the imaginary part of complex numbers uses `i` without the multiplication operator. The function `c(object1, object2, object...)` combines objects to form a vector of content and length equal to the total of the combined objects. It is possible to check and coerce the mode of objects using the respective functions: `is.numeric`, `is.complex`, `is.character`, `is.logical` and `as.numeric`, `as.complex`, `as.character`, `as.logical`.

```
> is.complex(b)
[1] TRUE
> is.logical(b)
[1] FALSE
> as.complex(a)
[1] 2+0i
> as.character(b)
[1] "1+2i"
> as.numeric(b)
[1] 1
Warning messages:
1: imaginary parts discarded in coercion
2: out-of-range values treated as 0 in coercion to raw
```

Note the warning message. You have to be careful, as assigning a coerced object may yield undesirable results.

All objects in R have a class attribute, reported by the `class` function. The definition of the class contains the names of all slots directly and indirectly defined. Each slot has a name and an associated class. One extracts a slot with the @ operator. For simple vectors, the class corresponds to the mode (`numeric`, `logical`, `character` or `list`). However, there are many other classes for R objects.

matrix, array, factor, data.frame, formula, and function are among
the most common and most useful classes of objects. Some functions are sensitive
to the class of the object and can display or yield different results accordingly. These
functions are "generic" because they perform different tasks depending on the class
of the object. We will use objects of the matrix and factor classes frequently
in this book because we will perform matrix operations and categorize our observations. We will frequently use vectors as well. As for the mode attribute, one can
sometimes coerce the class of an object (for example, the as.matrix(object)
command can coerce an object of the vector or data.frame classes to an object
of the matrix class). As for the mode, coercion should be used carefully; usually
R yields a warning message or an error message if coercion is problematic. We will
examine how to handle these classes of objects.

Objects of the vector Class

All elements of a vector have a similar mode. Traditional operations can be applied
to vectors of numeric mode.

```
> a<-1:5
> a
[1] 1 2 3 4 5
> b<-rep(1,5)
> b
[1] 1 1 1 1 1
> a+b
[1] 2 3 4 5 6
```

The ":" operator generates a regular sequence of numbers in step 1, while the function rep(x, n) repeats the x vector, n times. The arithmetic operators operate on
each element of the two objects, and thus return an object of the same size.

```
> a<-c(1, 6); b<-c(2,3,5); d<-c(2, 3)
> d[2]
[1] 3
```

For accessing an element of an object, we use an index between underbraces. The
index is generally a numeric vector of indices. We can select indices by writing an
expression between underbraces as well.

```
> b[2:3]
[1] 3 5
> b[c(1,3)]
[1] 2 5
> b[-3]
[1] 2 3
```

The last command returns a vector with the third element removed.

```
> a+d
[1] 3 9
> a*d
[1]  2 18
> e<-10
> d*e
[1] 20 30
> a+b
[1] 3 9 6
Warning message:
longer object length is not a multiple of shorter
   object length in: a + b
```

If the vectors are not of the same size, R is recycling the shortest one. If the shortest vector does not have a multiple length of the longest one, R proceed with the operation, returns a vector of equal size to the longest initial vector, and prints a warning message.

R offers the possibility to work on complex vectors as well.

```
> a<-1+1i
> Re(a); Im(a)
[1] 1
[1] 1
> Conj(a)
[1] 1-1i
> Arg(a)
[1] 0.7853982
> Mod(a)
[1] 1.414214
```

The above functions successively return the real part, imaginary part, conjugate, argument and modulus of a complex vector.

One can apply diverse functions and operations to vectors of `character` mode. Among them, the `paste` function concatenates vectors of `character` mode.

```
> a<-c("a", "b")
> b<-c("c", "d")
> paste(a, b)
[1] "a c" "b d"
> paste(a, b, sep="")
[1] "ac" "bd"
>  paste(a,b, sep = "", collapse = "")
[1] "acbd"
```

The third argument entered through `sep=""` specifies nothing between strings of each vector, and its default value corresponds to a space. `collapse` is another argument of the function paste. If a value is specified for `collapse`, the elements are concatenated into a single string and separated in that string by the value of `collapse` mentioned.

Objects of the `factor` Class

An object of the `factor` class specifies a grouping (or categorization). Factors can be ordered or unordered. A `factor` object contains the values of the categorical variable (usually a numeric or a character), and the different levels of that variable. The `levels` function returns the levels of a given factor.

```
> factor(c(1, 1, 2, 2, "a", "a"))
[1] 1 1 2 2 a a
Levels: 1 2 a

> gl(3, 2)
[1] 1 1 2 2 3 3
Levels: 1 2 3
```

One can use the `gl` function for specifying groups of equal size.

```
> a<-gl(3, 2)
> b<-gl(2, 3)
> a:b
[1] 1:1 1:1 2:1 2:2 3:2 3:2
Levels: 1:1 1:2 2:1 2:2 3:1 3:2
```

Specifying interacting factors is achieved using the ":" operator; this operator is useful for entering elements of the `formula` used as an argument of some functions (see Chapter 3).

Objects of the `matrix` Class

A matrix is similar to a collection of scalar organized in r rows and c columns. It is also a collection of r vectors of same mode and same length (i.e., c). It can be considered as a vector where components are subscripted by two entries indicating rows and columns.

For generating a matrix, we use the `matrix` function, with two arguments specifying the number of columns and rows. By default, matrices are filled by columns, but can be filled by rows specifying the value of a fourth argument. A matrix of p rows and k columns can describe a configuration of p landmarks with k dimensions.

```
> matrix(1:6, 3, 2)
     [,1] [,2]
[1,]    1    4
[2,]    2    5
[3,]    3    6

> matrix(1:6, 3, 2, byrow=T)
     [,1] [,2]
[1,]    1    2
[2,]    3    4
[3,]    5    6
```

Objects of the `matrix` class have a `dim` attribute that corresponds to the number of rows and columns. It is possible to extract any element of a matrix using a vectorized indexing, or a matrix indexing, where two indices between underbraces indicate the row and column position.

```
> a<-matrix(1:6, 3, 2)
> dim(a)
[1] 3 2
> a[2,2]
[1] 5
> a[5]
[1] 5
> a[-1,]
     [,1] [,2]
[1,]    2    5
[2,]    3    6
```

We use the function `t` to obtain the transpose of a matrix.

```
> t(a)
     [,1] [,2] [,3]
[1,]    1    2    3
[2,]    4    5    6
> rbind(a[1:2,], a)
     [,1] [,2]
[1,]    1    4
[2,]    2    5
[3,]    1    4
[4,]    2    5
[5,]    3    6
> cbind(a, a)
     [,1] [,2] [,3] [,4]
[1,]    1    4    1    4
[2,]    2    5    2    5
[3,]    3    6    3    6
```

The two functions `rbind` and `cbind` combine `matrix`, `data.frame`, or `vector` objects by columns or rows, respectively. In addition, usual operators work on matrices as they work on vectors, and R allows matrix operation using matrix operators and functions (see Sections 1.4.3 and 1.4.4).

Objects of the `array` Class

Objects of the `array` class are expanded matrices with dimensions > 2 and are indexed in the same way as matrices. A matrix is a special case of an array. Morphometric analyses make extended use of matrix operations, and array are a convenient way to store datasets of configurations, as we will see in Chapter 2.

```
> is.matrix(a)
[1] TRUE
> is.array(a)
[1] TRUE
> array(1:4, c(2, 3, 2))
, , 1

     [,1] [,2] [,3]
[1,]    1    3    1
[2,]    2    4    2

, , 2

     [,1] [,2] [,3]
[1,]    3    1    3
[2,]    4    2    4

> b<-array(1:4, c(2, 3, 2))
> b[1,,]
     [,1] [,2]
[1,]    1    3
[2,]    3    1
[3,]    1    3
```

Note that data are recycled in this case, since the size of the array is greater than the length of the data.

Objects of the `data.frame` Class

A `data.frame` object is organized like a `matrix`, but its elements can have different modes. Indexing and combinations are similar with `matrix` objects. As for matrices, it is possible to assign a name to the rows and columns, using the functions `rownames` and `colnames`. A `data.frame` is an object designed for performing tests because categories, individual labels, and variables can be stored in the same object.

```
> a<-as.data.frame(a)
> colnames(a)<-c("length", "weight")
> rownames(a)<-paste("ind", 1:3, sep="")
> a
     length weight
ind1      1      4
ind2      2      5
ind3      3      6
```

Objects of the `list` Class

R can handle objects called lists, which are not only a class but are also a specific mode. These are combinations of objects, which individually can be of any mode and length.

```
> j<-list(a, b, c, e)
> mode(j)
[1] "list"
> length(j)
[1] 4
> j
[[1]]
[1] 2

[[2]]
[1] 1+2i

[[3]]
[1] "ab"

[[4]]
[1] 2 2

> j[[2]]
[1] 1+2i
```

The length of a `list` object corresponds to the number of objects it contains. It is possible to access an object in the list using double brackets and indicating the index of the object. It is also possible to assign a name to each object.

```
> j<-list(A=a, B=b, C=d, K=e)
> j$K
[1] 2 2
```

Note the use of `object$name` to access a nominated object of a `list`.

1.4.3 Functions

Functions are objects of the `function` class. A help file is available for every function in R using `help(function)`. Examples are usually available and can be used as a tutorial by typing `example(function)`. In addition to a short description and example section, the help file contains "usage," "arguments," "details," "values," "references," and "see also" sections. "Usage" specifies which and how arguments are passed to the function; it gives default values (if there are any). The "argument" section details the class and mode of the objects passed as arguments. "Details" provides additional description. "Values" explains what objects are returned by the function, and "see also" refers to similar or related functions.

As an example, we look at the help file for the square root function `sqrt`:

```
> help(sqrt)

abs                       package:base                    R Documentation

Miscellaneous Mathematical Functions

Description:

     These functions compute miscellaneous mathematical functions. The
     naming follows the standard for computer languages such as C or
     Fortran.

Usage:

     abs(x)
     sqrt(x)

Arguments:

      x: a numeric or 'complex' vector or array.

Details:

     These are generic functions: methods can be defined for them
     individually or via the 'Math' group generic.  For complex
     arguments (and the default method), 'z', 'abs(z) == Mod(z)' and
     'sqrt(z) == z^0.5'.

References:

     Becker, R. A., Chambers, J. M. and Wilks, A. R. (1988) _The New S
     Language_. Wadsworth & Brooks/Cole.

See Also:

     'Arithmetic' for simple, 'log' for logarithmic, 'sin' for
     trigonometric, and 'Special' for special mathematical functions.

Examples:

     require(stats) # for spline
     xx <- -9:9
     plot(xx, sqrt(abs(xx)),   col = "red")
     lines(spline(xx, sqrt(abs(xx)), n=101), col = "pink")
```

If the function is written in R code, it is possible to see the content of that function. Sometimes parts of a function or the whole function call an internal code built into the R interpreter. Modifying this code is mainly the affair of R architecture developers. For example, let's see the function $ginv$ that computes the Moore-Penrose generalized inverse of a matrix that we will use in Chapters 5 and 6. This function is in the package **MASS** that should be first loaded (MASS is a recommended package, downloaded on your computer once R is installed, but you have to load it).

```
> library(MASS)
> ginv
function (X, tol = sqrt(.Machine$double.eps))
{
    if (length(dim(X)) > 2 || !(is.numeric(X) || is.complex(X)))
        stop("X must be a numeric or complex matrix")
    if (!is.matrix(X))
        X <- as.matrix(X)
    Xsvd <- svd(X)
    if (is.complex(X))
        Xsvd$u <- Conj(Xsvd$u)
    Positive <- Xsvd$d > max(tol * Xsvd$d[1], 0)
    if (all(Positive))
        Xsvd$v %*% (1/Xsvd$d * t(Xsvd$u))
    else if (!any(Positive))
        array(0, dim(X)[2:1])
    else Xsvd$v[, Positive, drop = FALSE] %*% ((1/Xsvd$d[Positive]) *
        t(Xsvd$u[, Positive, drop = FALSE]))
}
<environment: namespace:MASS>
```

This provides a tutorial for understanding the R syntax of a function. For example, consider that we write a function that computes the sum of squares of a vector, and that we call it myfun.

Function 1.1. myfun

Argument:
    vec: *A numeric vector.*
Value:
    Sum of squared elements.

```
1  myfun<-function(vec)
2    { sum(vec*vec) }
```

```
> myfun(c(1:3))
[1] 14
```

Note that we use a name for declaring what arguments will be entered in our function. We will be free later to use any name for objects that will be entered as arguments of our function. It is necessary to call the function we have programmed, if we want it to work for a given work session. One can directly paste the function to the prompt. It is often easier to save all your functions in a text file, and then copy them on the R prompt when you want to use them. If the function has been saved in an ASCII file, one can load it with the function source() like any other program. Most functions that we will develop to perform our analyses are customized and not accessible in any R package. As an exercise, I suggest you copy or modify the code (and maybe improve it) of functions that are supplied throughout the text. It can serve as the first architecture for your R toolbox or for a future package in morphometrics.

In the book, we will write several functions for performing morphometric operations. The function code will be enclosed in boxes using Courier font; useful comments will appear in slanted font for functions and code examples. For functions, lines of code will be numbered from 1 to n.

Table 1.1 summarizes some of the most frequently used functions in the book that are necessary for morphometric analyses. There are many other ones, and probably others to be implemented in the future, but the following ones are a foundation for constructing more elaborate functions.

Table 1.1. Commonly used functions in morphometric analysis

Function	Package	Short Description
data	utils	load existing data-sets
names	base	obtain the names of an object
strsplit	base	split string into substrings
sub	base	replacement of matches determined by regular expression matching
max	base	maximal value of a numeric object
min	base	minimal value of a numeric object
abs	base	absolute value
round	base	rounding of decimal number
cos, acos	base	cosine and arc-cosine
sin, asin	base	sine and arc-sine
tan, atan	base	tang and arc-tang
sqrt	base	square root
sort	base	sort a vector or a factor
which	base	extract indices of a logical vector
diag	base	extract the diagonal of a square matrix
apply	base	apply a function to margins of an array
mean	stats	arithmetic mean
cor	stats	correlation computation
var	stats	variance and covariance computation
lm	stats	linear models
aov	stats	analysis of variance
svd	stats	singular-value decomposition
dist	stats	compute distances matrices
hclust	stats	hierarchical clustering
plot	graphics	2D plot
points	graphics	points for 2D plot
text	graphics	text for 2D plot
abline	graphics	abline for 2D plot
segments	graphics	segments for 2D plot
persp	graphics	3D perspective plot
rgl.points	rgl	interactive 3D plot

1.4.4 Operators

R provides several kinds of operators (Table 1.2): arithmetic, relational, logical... and some others: we already know "<-" for assignments, ";" for separating commands, ":" for generating regular sequences, and others that are used for indexing objects ($, [], [[]]).

Table 1.2. Main operators

	Arithmetic		Relational		Boolean
+	addition	<	lesser than	!	NOT
−	substraction	>	greater than	&	AND
*	multiplication	<=	lesser than or equal	&&	AND
/	division	>=	greater than or equal	|	OR
^	power	==	equal	||	OR
%%	modulo	!=	different	xor	exclusive OR
%/%	integer division				

Modern morphometrics and associate multivariate statistics make intensive use of matrix operations. The +, −, %*% operators perform matrix addition, substraction, and multiplication respectively. Other matrix operations are obtained with the t, solve, ginv functions that work with a matrix as unique argument, and return the matrix transpose, matrix inverse, and Moore penrose generalized matrix inverse respectively. The ginv function belongs to **MASS** which should be loaded before to call the function. The svd and eigen functions returns respectively the matrices resulting of a singular-value decomposition and of a spectral decomposition.

"%in%" is an operator that returns a logical vector indicating whether there is a match or not for its left operand. It will be used in some of our programs. In addition, we will use the "#" operator for writing comment lines (right to this operator, R ignores the script), and "~" for specifying objects of the formula class.

```
> lm(x~y)
```

This command orders a linear model where x (left to ~) is the response and y (right to ~) is the predictor. The x~y object is of the formula class.

1.4.5 Generating Data

R can generate regular or random sequences of data with functions and operators. Generating sequences is useful for loops, indexing, simulation, or whatever.

You can use the function scan without arguments to enter real numbers directly using the keyboard. You can specify the mode of data you want to enter through the what argument.

```
> scan()
1: 1 5 8 5.9
5:
Read 4 items
[1] 1.0 5.0 8.0 5.9
> a<-scan(what="character")
1: 1:6 4 7 9 juju
6:
Read 5 items
> a
[1] "1:6"   "4"      "7"      "9"      "juju"
```

The function `seq` generates sequences of real numbers.

```
> seq(1, 5, by=2)
[1] 1 3 5
> seq(1, 5, length=9)
[1] 1.0 1.5 2.0 2.5 3.0 3.5 4.0 4.5 5.0
```

The function `rep` replicates the value of a vector a specified number of times.

```
> rep("a", 4)
[1] "a" "a" "a" "a"
> rep(1:3*2-1, 2)
[1] 1 3 5 1 3 5
```

Many statistical tests work with randomly distributed data for obtaining a null distribution. In addition, generating random data is necessary for some simulations. Several functions of R generate such data following several kinds of probability density functions. These functions are of the form `rfunc(n, p1, p2,...)`, where `func` indicates the probability law, n the number of data generated, and p1, p2,... are the values of the parameters of the law. In replacing the letter `r` with d, p, or q, it is possible to obtain the probability density, the cumulative probability density, and the value of quantile (`qfunc(p, ...)`), respectively . These functions will be useful for returning the p-values of tests. "rnorm", "rbinom", "rf", and "runif" generate randomly distributed data following normal, binomial, Fisher-Snedecor, and uniform laws.

```
> rnorm(10, 0, 5)
[1] -1.5015661 -3.6697118  2.9320062 -6.9773504 -1.6757256
[6]  4.6287676 -0.2262281 -0.3837671 -1.3207831 -0.5380926
> round(runif(10, 0, 5))
[1] 6 6 6 4 3 3 3 1 2 4
```

The first argument entered in these two functions specifies the number of generated data. In the first command, the second argument is the mean and the third is the standard deviation; for the second command, the second and third arguments correspond to minimal and maximal values, respectively. We can randomly extract components of objects using integer random vectors.

```
> palette(rainbow(6))[round(runif(10, 0, 5)+1)]
[1] "blue"    "magenta" "yellow"  "green"   "blue"
[6] "green"   "magenta" "red"     "red"     "green"
> letters[round(runif(10, 0, 25)+1)]
[1] "o" "r" "b" "v" "q" "j" "n" "n" "x" "l"
> sample(letters)[1:10]
[1] "q" "e" "h" "k" "n" "m" "b" "y" "z" "j"
```

The `letters` vector of R contains the 26 lower-case letters of the Roman alphabet. "`palette(rainbow(6))`" is a vector of six colors of the rainbow palette. The function `sample` can act similarly.

```
> sample(letters)[1:10]
[1] "q" "e" "h" "k" "n" "m" "b" "y" "z" "j"
```

We will see later how useful generating random sequences and resampling is to performing permutation tests and simulations (see Chapters 3 and 7).

1.4.6 Loops

R can do repetitive commands, which is often not the case with a graphical user interface (GUI). The language for writing loops and indexing is intuitive. Hereafter are some simple examples:

```
> x<- c("black", "red", "green")
> y<- matrix(round(runif(15, 0, 3)), 3, 5)
> y
     [,1] [,2] [,3] [,4] [,5]
[1,]    2    3    1    2    0
[2,]    1    2    2    1    1
[3,]    1    1    1    1    1
> z<-matrix(NA, 5, 3)
> for (i in 1:length(x)){z[,i]<-paste(x[i],y[i,],sep="")}
> z
     [,1]      [,2]     [,3]
[1,] "black2" "red1" "green1"
[2,] "black3" "red2" "green1"
[3,] "black1" "red2" "green1"
[4,] "black2" "red1" "green1"
[5,] "black0" "red1" "green1"
```

Note the use of `NA`, which is a constant of unitary length and indicates missing value. Here it has been used to fill z with missing values (empty matrix). `NA`, as a single assignment, can specify an empty vector or matrix. Conditional control flows are constructed with `if (expression is TRUE) {expression1}` `else {expression2}`. Another possible situation is to execute an instruction as long as a condition is true. The syntax is `while (expression is TRUE) {expression1}`. Here are some examples:

```
> x<-1:10
> for (i in 1:10)
+        {if(x[i]%%2==0){x[i]<-2*x[i]}
+                else {x[i]<-x[i]/2}}
> x
 [1]  0.5  4.0  1.5  8.0  2.5 12.0  3.5 16.0  4.5 20.0
> y<-2;    x<-0
> while(y<8) {y<-2*y ; x<-x+1}
> x
[1] 2
```

Most of the time, using function or logical indexing allows writing loops to be avoided. Logical indexing is a feature of R that often improves computation time.

```
> x<-1:10
> x[x %% 2 == 0]<- x[x %% 2 == 0] * 2
> x[x %% 2 != 0]<- x[x %% 2 != 0] / 2
> x
 [1]  0.5  4.0  1.5  8.0  2.5 12.0  3.5 16.0  4.5 20.0
```

Problems

1.1. Organizing a `data.frame` object
Define a hypothetical data frame containing five measurements normally distributed (size, head perimeter, pectoral width, area, and weight) for four individuals that you will name ind1, ind2, ind3, ind4. Use the function `paste` and a regular sequence function to create individual names. Add a column corresponding to the factor sex, with the first two individuals being males, and the last two ones being females.

1.2. Manipulating `array` and `matrix` objects
Fill a four-dimensional array of dimensions 2, 4, 3, 3 with a regular sequence. Using the function `apply`, return the sum of the elements of the matrices contained in the first two dimensions (you should obtain a 2×2 matrix).

1.3. Manipulating and indexing matrices
Fill a 6×6 square matrix with a regular sequence of 36 numbers. Using logical indexing, extract values of the upper half triangle (diagonal excluded). Transpose these values and paste them in the lower half triangle to obtain a symmetric matrix.

1.4. Using R functions to find critical and p-values
Using the online help and random normal function, find the p-values for a F-value of 2.01 corresponding to a bilateral test, with $df1 = 3$ and $df2 = 12$. Find the code for retrieving the F-value from degrees of freedom and p-value.

1.5. Loops
Write a function that uses loops for extracting column and row indices for the values of a matrix that are above a critical value. Store the results in a two-row matrix.

2

Acquiring and Manipulating Morphometric Data

The first step of any statistical or morphometric analysis is to gather and organize raw data. R offers a graphical interface that allows diverse datasets to be directly captured from digital images, as we will see. Some basic image manipulation and analysis is introduced as well. This chapter explains how to gather morphometric data in an appropriate way and how to assign them to R objects. The quality of data acquisition determines part of the quality of the results: measurement error results both from the user and from the accuracy of the different tools used for measuring data. It may happen that data are incomplete for some objects (e.g., some objects can be broken, and all the landmarks or distances cannot be captured). The last part of the chapter explains how to handle missing data and to estimate measurement error.

2.1 Collecting and Organizing Morphometric Data

2.1.1 Collecting Data

Traditionally, morphometric data are sets of distance, angle, perimeter, surface, or volume measurements. One can obtain them in theory from coordinates of landmarks or pseudolandmarks (see Chapter 1). Manual tools (rulers, calipers), hardware (digitizers, tracing tables), or position of a pointer on digital image allows these coordinates to be recorded in the x, y, and eventually z-dimensions.

Collecting Distances

Although one can directly record distances from digitized pictures, one usually collects these measurements directly from objects using calipers or rulers, or any other manual device. Later, these data are stored in handsheets or directly in a computer file. For microscopic and bigger objects (like geographic data), distances are obligatorily obtained through images acquired through different kinds of lenses and mirrors (microscope, telescope, magnifying lens, etc.) and captured to computer files using a digital camera, or to photographic films using a conventional camera. It is necessary here to photograph a size standard (a ruler or any other known distance) together with the pictured object for retrieving the size of structures. Nowadays, images can be numerized and their properties can be analyzed using elementary computer image analysis. Distances can be calculated using the coordinates of endpoints for a given measurement.

Collecting Coordinates of Points

Any image-analyzing system can collect 2D cartesian coordinates of points on a picture file. It is important to estimate the size of the image by using a scale such that

$$True\ coordinates = \frac{Image\ coordinates \times Size\ of\ the\ scale\ on\ the\ image}{True\ size\ of\ the\ scale}.$$

This relationship holds for matrices of coordinates as well (configuration matrices), meaning that the scalar multiplication is applied to the matrix of coordinates.

The relative positions of coordinates for an object can be estimated from coordinates of points digitized using a cartesian reference. Users are thus invited to define the origin (which can be any landmark of the object or outside the object), and the x-axis, y-axis, and possibly the z-axis directions to fix the orientation of the coordinate system. Defining the orientation of the coordinate system all at once avoids any further problems with reflection between configurations.

For some purposes, we need to define the cartesian system directly from the object. If every landmark you want to digitize on your object is not accessible to your digitizing device without repositioning the object, a protocol is necessary (you may wish to record coordinates on both dorsal and ventral surfaces of an object). For example, you may need to reverse the object to localize landmarks on the ventral side. For keeping the relative position between points digitized on the ventral side and dorsal side invariant to object reposition, you can record coordinates on the whole object by defining your system coordinates with three landmarks shared by the ventral and dorsal side.

One can directly record coordinates on objects using hardware (2D or 3D digitizers) connected to the computer. Some 3D hardware (confocal microscope, Computed Tomography system, etc.) records a series of images that are basically equally spaced slices of the objects. Coordinates of pixels are recorded by the way of 3D image analysis.

Collecting Surfaces and Perimeters

For 2D objects, surfaces and perimeters can be appraised by letting the computer count the pixels of the structure of interest. The surface corresponds to the number of pixels of the object on the image multiplied by a scaling factor. The perimeter corresponds to the number of pixels involved in the outline multiplied by a scaling factor. The pixel is a unit of surface measurement, thus the scaling factor should take into account the width and length of the pixel. For polygonal and known geometrical shapes, classical geometric addition or multiplication of distance measurements obtained from landmark coordinate data allows the calculation of surfaces or perimeters.

Collecting 3D Surfaces or Volumes

A 3D surface corresponds to the sum of the pixels involved in outlines of each slice belonging to the surface of the object scaled by the inter-image space and the pixel length and width ratio. For volume, the traditional way is to submerge the object in a liquid and measure the volume of liquid that has been displaced, following the Archimedes principle. In addition, volumes can be estimated as the sum of the pixels of the object for each image multiplied by an appropriate scale factor. The surface unit is the pixel so the measure of the volume should take into account the width and length of the pixel and the inter-image space. One can use voxel size to estimate the volume, if one works on voxel formatted files.

Collecting Images

One can obtain most morphometric properties of objects based on pixel (automated) counting or on pixel coordinates with elementary computer image analysis. A large number of inexpensive digital cameras are now available on the market. Hardware used for collecting 3D properties of objects is more expensive and consists of different machines: scanners, stereographic devices, 3D digitizers .... Three dimensional scanners provide a series of images.

Given an appropriate image format, location and color of the pixels of the image can be stored and analyzed. Binary or black and white images are defined by a series of pixels that take two values (0 and 1); gray-scale or monotonic images have pixel values ranging from 1 to 2^n. Here an image file can be organized as a matrix object (column and row indices corresponding to indices of pixel coordinates, and cells to the pixel value). For color images, each pixel location is related to three values, each ranging from 1 to 2^n. In this latter case, the data can be organized in a three-dimensional array where cell values correspond to the level of one of the three basic color channels (red, green, blue). Similarly, each color channel can be stored in a matrix.

Images for morphometric analysis can be stored in files and reworked using your favorite application software for image analysis, but this latter task can be achieved with the help of R as we will see in Section 2.2.6.

2.1.2 Organizing Data

R can navigate in your repertories to read files. R has a working directory returned through the command getwd(). This repository is set for a session using the following command for Windows:

```
> setwd("C:/data")
```

or for Linux:

```
> setwd("/home/juju/data")
```

For files that are not in the working directory of R, you have to specify the path as a string for the function to find their location (e.g., "/home/juju/myfile.R").

It is very important to organize your data files to optimally work through R. If your data are gathered on several files, it is important to keep the same organization throughout these files, to allow repetitive operations to be easily run by your computer. Additionally, it is good to name your file so it can clearly be recognizable. If the names of all files in a directory follow a given logic, R can open a series of related files using loops or logical indexing in the correct order. For this purpose, we can use R to generate sequences and concatenate strings. It is worth using loops if you need to perform the same operation on several files. It will be easier if part of the name of your files follow some regular sequence, without which you will have to create and write a probably long vector containing the name of all your files, or eventually an extra file.

We will first learn how to organize data generated with R before we learn how to read and organize data that are outside of the R environment. As an example, I present the possible ways to store the essential information about a configuration (its names and the coordinates of landmarks) in various R objects. Configurations of landmarks usually correspond to matrices M of p rows for landmarks and of k columns for dimensions. As for distances, the input of landmarks must follow the same order for each configuration to allow comparisons between configurations.

```
>juju<-scan()
   1:   0.92 100.00 0.99 100.25 1.07 99.99 1.26 99.99 1.11
   10: 99.87 1.16 99.70 1.00 99.86 0.87 99.72 0.88 99.89
   19: 0.74 99.98
 Read 20 items
>JUJU<-matrix(juju, 10, 2, byrow=T)
>colnames(JUJU)<-c("x", "y")
>rownames(JUJU)<-paste("Lan", 1:10, sep="")

>JUJU
          x       y
Lan1   0.92 100.00
Lan2   0.99 100.25
Lan3   1.07  99.99
Lan4   1.26  99.99
Lan5   1.11  99.87
Lan6   1.16  99.70
Lan7   1.00  99.86
Lan8   0.87  99.72
Lan9   0.88  99.89
Lan10  0.74  99.98
```

Alternatively, the configuration can be defined to a 1 by $k \times p$ matrix that will store a succession of coordinates x, y, and z, for the p landmarks. The $k \times p$ configuration matrix (M) can be coerced in the corresponding m vector. m is the vectorized form of the M matrix.

```
>JOJO<-matrix(c(0.72,100.32,0.75,100.36,0.77,100.32,0.81,
+   100.32,0.77,100.29,0.77,100.24,0.73,100.28,0.7,100.26,
+   0.7,100.3,0.67,100.33), 10, 2, byrow=T)
>colnames(JOJO)<-c("x", "y")
>rownames(JOJO)<-paste("Lan", 1:10, sep="")
>t(JOJO)
      Lan1    Lan2    Lan3    Lan4    Lan5    Lan6    Lan7    Lan8
x     0.72    0.75    0.77    0.81    0.77    0.77    0.73    0.70
y   100.32  100.36  100.32  100.32  100.29  100.24  100.28  100.26
  Lan9   Lan10
x    0.7    0.67
y  100.3  100.33

>as.vector(t(JOJO))
 [1]    0.72 100.32    0.75 100.36    0.77 100.32    0.81 100.32
 [9]    0.77 100.29    0.77 100.24    0.73 100.28    0.70 100.26
[17]    0.70 100.30    0.67 100.33
>clname<-expand.grid(colnames(JOJO),rownames(JOJO))
```

The function expand.grid *creates a* data.frame *object using all combinations of a group of supplied vectors. It has been used here for creating the new row names of the second matrix.*

```
>JOJO1<-matrix(t(JOJO), 1, 20)
>rownames(JOJO1)<-"JOJO"
>colnames(JOJO1)<-paste(clname[,1], clname[,2], sep="-")
>JOJO1

        x-Lan1 y-Lan1 x-Lan2 y-Lan2 x-Lan3 y-Lan3 x-Lan4
JOJO     0.72 100.32    0.75 100.36    0.77 100.32    0.81
        y-Lan4 x-Lan5 y-Lan5 x-Lan6 y-Lan6 x-Lan7 y-Lan7
JOJO   100.32    0.77 100.24    0.73 100.28    0.7 100.26
        x-Lan8 y-Lan8 x-Lan9 y-Lan9 x-Lan10 y-Lan10
JOJO     0.7  100.3   0.67 100.33    0.77  100.29
```

A collection of n configurations can be stored in an array object of p, k, n dimensions if all configurations $M_{1 \to n}$ contain the same numbers of landmarks and dimensions. The full array can be easily transformed in a data.frame object with rows corresponding to objects and columns to the succession of x, y (and z for 3D) coordinates for each landmark. Alternatively, the configuration set can be stored as a list if objects contain different numbers of landmarks and dimensions. Here we organize the collection of the configurations through R in three different ways.

Example of configuration set assigned to an `array` *object:*

```
>array(cbind(JUJU, 2, JOJO, 2), dim=c(10, 2, 2))
, , 1

        [,1]    [,2]
 [1,]  0.92 100.00
 [2,]  0.99 100.25
 [3,]  1.07  99.99
 [4,]  1.26  99.99
 [5,]  1.11  99.87
 [6,]  1.16  99.70
 [7,]  1.00  99.86
 [8,]  0.87  99.72
 [9,]  0.88  99.89
[10,]  0.74  99.98


, , 2

        [,1]    [,2]
 [1,]  0.72 100.32
 [2,]  0.75 100.36
 [3,]  0.77 100.32
 [4,]  0.81 100.32
 [5,]  0.77 100.29
 [6,]  0.77 100.24
 [7,]  0.73 100.28
 [8,]  0.70 100.26
 [9,]  0.70 100.30
[10,]  0.67 100.33
```

Example of configuration set assigned to a `data.frame` *object:*

```
>JJ<-data.frame(rbind(as.vector(t(JUJU)),
+               as.vector(t(JOJO)))))
>rownames(JJ)<-c("JUJU", "JOJO")
>colnames(JJ)<-paste(clname[,1], clname[,2], sep="-")
>JJ

      x-Lan1 y-Lan1 x-Lan2 y-Lan2 x-Lan3 y-Lan3 x-Lan4
JUJU    0.92 100.00   0.99 100.25   1.07  99.99   1.26
JOJO    0.72 100.32   0.75 100.36   0.77 100.32   0.81
      y-Lan4 x-Lan5 y-Lan5 x-Lan6 y-Lan6 x-Lan7 y-Lan7
JUJU   99.99   1.11  99.87   1.16  99.70   1.00  99.86
JOJO  100.32   0.77 100.29   0.77 100.24   0.73 100.28
      x-Lan8 y-Lan8 x-Lan9 y-Lan9 x-Lan10 y-Lan10
JUJU    0.87  99.72   0.88  99.89    0.74   99.98
JOJO    0.70 100.26   0.70 100.30    0.67  100.33
```

Example of a configuration set assigned to a `list` *object:*

```
>list(JUJU=JUJU, JOJO=JOJO)
$JUJU
         x       y
Lan1  0.92 100.00
Lan2  0.99 100.25
Lan3  1.07  99.99
Lan4  1.26  99.99
Lan5  1.11  99.87
Lan6  1.16  99.70
Lan7  1.00  99.86
Lan8  0.87  99.72
Lan9  0.88  99.89
Lan10 0.74  99.98

$JOJO
         x       y
Lan1  0.72 100.32
Lan2  0.75 100.36
Lan3  0.77 100.32
Lan4  0.81 100.32
Lan5  0.77 100.29
Lan6  0.77 100.24
Lan7  0.73 100.28
Lan8  0.70 100.26
Lan9  0.70 100.30
Lan10 0.67 100.33
```

Other ways to organize data are possible. Later in the text, I will usually distinguish sets of configurations (arrays) from single configuration (matrix) with the respective letters A and M. In the following section, we will see how to import data files in the environment of R.

2.2 Data Acquisition with R

2.2.1 Loading and Reading R Datafiles

The function `data` loads data files followed by certain extensions (.R, .r, .rda, .rdata, .txt, .csv ...) and is searching by default for sets in every currently loaded package. If the dataset is not assigned to a new object, the name of the data object corresponds to the name of the dataset without any extensions.

```
>data(iris)
>iris[1,]
  Sepal.Length Sepal.Width Petal.Length Petal.Width Species
1          5.1         3.5          1.4         0.2  setosa
```

Several morphometric sets have been stored in R (notably in the packages shapes and ade4) and can be used as tutorials. It is also possible to use the function `data` for opening your own datasets.

2.2.2 Entering Data by Hand

We can enter data by hand using the CLI and/or by filling arguments of some functions. However, it is not very convivial, and the number of entries can become very large and boring. R provides a data editor on some platforms using the function `de` and typing the code `de(NA)`. Once data are typed, the function assigns the value to a new object of the `list` class. Using the data editor depends on the operating system, and users are invited to read the online help. Note that like any other function, arguments can be passed through the prompt. Similarly, an empty data frame can be edited by the command `edit(as.data.frame(NULL))`, and assigned to an object. It is then filled by hand by the user. The `transform` function allows you to transform certain objects in a `data.frame`, and/or to append new columns calculated by manipulation on the original columns.

```
>transform(matrix(rnorm(4),2,2), new=X1*X2)
          X1        X2         new
1  0.54885879 0.3121743  0.171339622
2 -0.01887397 0.4368677 -0.008245428
```

The package **Rcmdr** ("R commander") provides a graphical interface with several menus that allows direct operations using the menus and mouse buttons. It can do the job of a data editor as well.

```
> library(rcmdr)
> Commander()
```

Working with "R commander" is really straightforward and there is no need for long explanations. My opinion is that you should avoid using it if you are a newcomer to R, because it does not ask you to think too much when producing any kind of beautiful graph. You must be able to think to become a star in shape statistical analysis.

2.2.3 Reading Text Files

Usually raw data are recorded in a data frame where columns represent variables while rows represent individuals. For R to read the file easily, each element of the dataset stored on your computer files must be separated by the same separator character. Blank spaces are among the most commonly used, but R handles any other kinds, like tabulations. It is important that no elements or strings of your data contain the separator character used for separating elements.

The functions `read.table` and `scan` can read ascii files. The first argument of these functions contains the name of the file (with its extension) and the path (if the file is not in the working directory). The following series of arguments can be completed to indicate among others: number of lines to skip, data types, field separator character, and decimal separator.

`read.table` reads tabular data (to avoid error messages, be careful that each line contains the same number of observations and check your field separator; in the case of missing data, fill empty cells with "NA" (NA for nonavailable). A number of related functions can be used (`read.csv`, `read.delim`, and `read.fwf`) that may have some advantages regarding the way that the initial file is stored.

`scan` is much more flexible and can interpret the file as a vector or a list.

Functions of some packages have been developed for importing and reading files in other formats (Excel, SAS, SPSS, matlab, newick, html...), and access SQL-type databases. If you are a Matlab lover, the R.matlab package allows communications between R and a Matlab servers. A similar package exists for using R language and packages on Matlab (RLink).

2.2.4 Reading and Converting Image Files

R does not read just text files; it can also read and display an image file on a graphical device, thanks to the development of several packages and functions (Table 2.1). Accessing pixel values and their coordinates consists of extracting either the slots or the appropriate components of objects returned by these functions.

Table 2.1. Functions for importing image files and related packages

Function	Package	Image Format	Returned Object
read.pnm	pixmap	pbm, pgm, ppm	objects of diverse pixmap classes
read.jpeg	rimage	jpeg	image.matrix object
readTiff	rtiff	tiff	pixmap object

Some code is provided below to demonstrate how R handles various image files with different packages.

Working with the package rimage.

```
>library(rimage)
>x <- read.jpeg(system.file("data", "cat.jpg",
+    package="rimage"))
>dim(x[,,1])
>is.array(x)
[1] TRUE
```

Returning dimensions of the Red level matrix.

```
>dim(x[,,1])
[1] 420 418
```

Working with the Package pixmap.

```
>library(pixmap)
>x <- read.pnm(system.file("pictures/logo.ppm",
+    package="pixmap")[1])
>str(x)
Formal class 'pixmapRGB' [package "pixmap"] with 8 slots
```

```
..@ red     : num [1:77, 1:101] 1 1 1 1 1 1 1 1 1 1 ...
..@ green   : num [1:77, 1:101] 1 1 1 1 1 1 1 1 1 1 ...
..@ blue    : num [1:77, 1:101] 1.000 1.000 0.992 ...
..@ channels: chr [1:3] "red" "green" "blue"
..@ size    : int [1:2] 77 101
..@ cellres : num [1:2] 1 1
..@ bbox    : num [1:4] 0 0 101 77
..@ bbcent  : logi FALSE
>dim(x@red)
[1]  77 101
>is.matrix(x@red)
[1] TRUE
>x@red[35,8]; x@green[35,8]; x@blue[35,8]
[1] 0.4392157
[1] 0.4392157
[1] 0.4078431
>x@red[35,8]*255
[1] 112
```

Pixels of a gray-scale image take values comprised between zero and one with a step of 1/255, while RGB images have pixels with three values ranging in the same way. Functions in the above packages easily convert color images to gray-scale images.

```
>library(rimage)
>x <- read.jpeg(system.file("data", "cat.jpg",
+     package="rimage"))
>y <- (rgb2grey(x))
>rm(x)
>dim(y)
[1] 420 418
```

The functions that work with image files usually return very long objects that consume the memory. These objects can be removed from the environment using the function rm, if they are no longer useful. As a frequent user, I recommend not using JPEGs exceeding 100 Ko.

Some image file formats are not readable for R, thus it is necessary to convert the format of images. This operation can waste a lot of time if there is a need to convert a large series of images. R offers the possibility to invoke a system command from the CLI with the function shell. One can therefore call the command of an image converter program directly from R. Among software for manipulating image files, Imagemagick[1] is free and available online and can be installed on many operating systems to convert and manipulate image files. Once you have installed Imagemagick, you can directly work from your R environment as follows:

```
>setwd("/usr/lib/R/library/rimage/data")
>shell("convert cat.jpg cat.bmp")
```

---

[1] www.imagemagick.org

2.2.5 Graphical Visualization

Visualizing data is often necessary to check whether you have correctly collected your data. The generic function `plot` plots diverse R objects. Its first arguments can be two vectors that contain the x and y coordinates of points of an object. Alternatively, a two-column matrix can be passed as argument to produce the same result. Many arguments (see the online help) can be entered directly through the function, including arguments concerning parameters of the graphical device. Alternatively, many parameters (font labels, margin widths, frame size, x and y-axis ranges, title positions, background color, and many other) can be set before to open the graphical device through the function `par`. Low-level plotting commands can add objects on the graph using a series of functions (`points`, `abline`, `text`, `segments`, `locator`, `arrows`, `polygon`, `legend`, `rectangle`, `axis`, `lines`, etc.).

Here is an example of script for plotting the configurations called JUJU and JOJO (see Fig. 2.1).

Draw the two configurations, with two different landmark symbols.
Add the title "Sea stars" to the graph.

```
>plot(rbind(JUJU, JOJO), pch=c(rep(21, 10), rep(20, 10)),
+       asp=1,main="Sea stars")
```

Draw a polygon with vertices corresponding to the configuration coordinates. Use different line types for each configuration.

```
>polygon(JUJU, lty=3)
>polygon(JOJO, lty=1)
```

Add landmark labels for the JUJU configuration.

```
>text(JUJU, labels=1:10, pos=c(3,2,3,3,3,4,3,2,3,3))
```

Alternatively, we can specify the position of labels relative to the coordinates they design in the plot with the function identify, *by left clicking (to the left, right, top, or bottom) near the landmark of interest.*

```
>identify(JUJU, labels=1:10, pos=TRUE)
```

Add eyes to JUJU.

```
>points(c(0.95,1.05,0.955,1.055),rep(99.95,4),pch=21,
+     cex=c(1,1,2,2),bg=c(1,1,NA,NA))
```

Add a mouth to JOJO.

```
>lines(c(0.76, 0.74,0.72), c(100.3,100.29,100.3))
```

Note that the function lines *can draw outlines as well.*

The function `persp` can display 3D plots. The first useful step is often to define the space that is necessary for the display, then points and lines are drawn using low-level plot commands, with the function `trans3d`.

Assign a 3D configuration matrix to the tetra *object.*

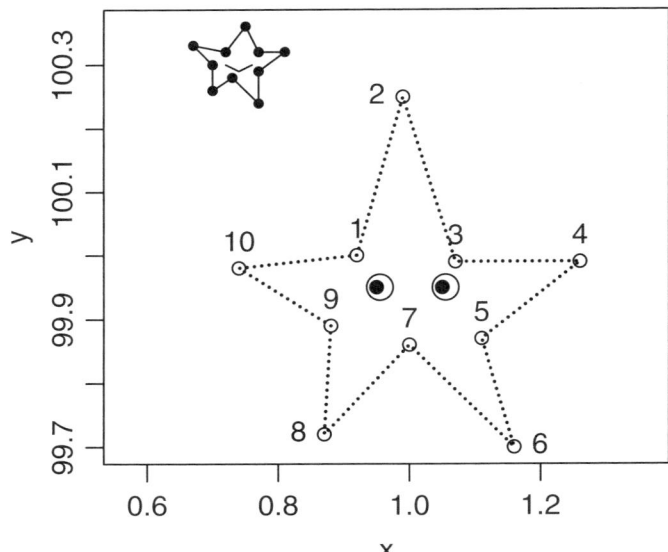

Fig. 2.1. Plotting 2D configurations with the function `plot`, and low-level commands

```
>tetra<-matrix(c(0,2,1,1,0,0,1,0.5,0,0,0,1),4,3)
```

Define the range of the space for the visualization and indicate the orientation of the cartesian system to be projected on the screen.

```
>x<-seq(-0.5,2.5, length=10)
>y<-x
>z<-matrix(-0.5, 10, 10)
>res<-persp(x,y,z,zlim=c(-0.5, 1.5),theta=30,phi=15,r=10,
+           scale=F)
```

Plot the landmarks of the configuration.

```
>points(trans3d(tetra[,1],tetra[,2],tetra[,3],pm=res),
+       col=1,pch=16)
```

Add segments ("links") between the landmarks of the configuration.

```
>lines(trans3d(tetra[-3,1],tetra[-3,2],tetra[-3,3],pm=res),
+       col=1,lw=2)
>lines(trans3d(tetra[-c(2,3),1],tetra[-c(2,3),2],
+       tetra[-c(2,3),3],pm =res),lw=2)
>lines(trans3d(tetra[-1,1],tetra[-1,2],tetra[-1,3],pm=res),
+       lty=3,lw=2)
>lines(trans3d(tetra[-c(2,4),1],tetra[-c(2,4),2],
+       tetra[-c(1,4),3],pm=res),lty=3,lw=2)
```

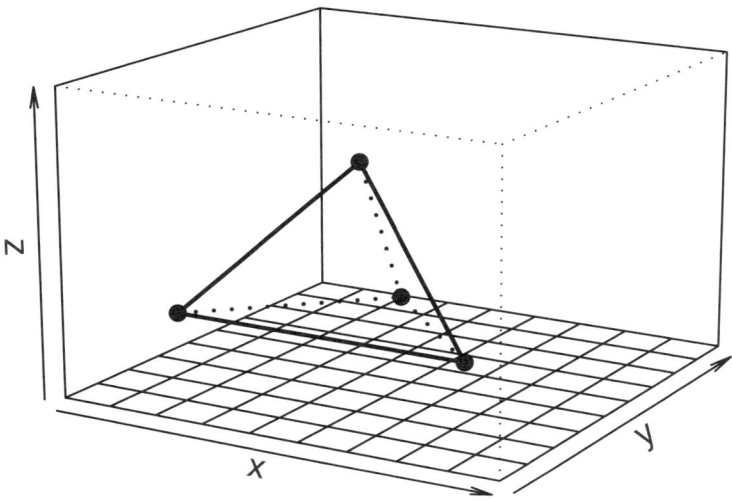

Fig. 2.2. Plotting 3D configurations with the function `persp`

The `Xll()` command opens a supplementary graphical device, and several graphs can be presented on a single device with the `layout` function. The `dev.set (devicename)` and `dev.off(devicename)` commands, respectively, close and activate the device named "devicename." One returns the list of devices by typing the command `dev.list()`. The script below opens three graphs on a single devices and displays x, y, and z-projections of a 3D configuration, the resulting plots are displayed in Fig. 2.3.

```
>layout(matrix(1:4, 2,2))
>res<-persp(x, y, z, zlim=c(-0.5, 2.5),theta=30,phi=30)
>points(trans3d(tetra[,1],tetra[,2],tetra[,3],pm = res),
+        col=1,pch=16)
>lines(trans3d(tetra[-3,1],tetra[-3,2],tetra[-3,3],pm=res),
+      col=1,lw=2)
>lines(trans3d(tetra[-c(2,3),1],tetra[-c(2,3),2],
+      tetra[-c(2,3),3],pm =res),lw=2)
>lines(trans3d(tetra[-1,1],tetra[-1,2],tetra[-1,3],pm=res),
+      lty=3,lw=2)
>lines(trans3d(tetra[-c(2,4),1],tetra[-c(2,4),2],
+      tetra[-c(1,4),3],pm=res),lty=3,lw=2)
>plot(tetra[,2:3],asp=1,xlab="y",ylab="z",
+      main="xprojection")
>polygon(tetra[,2:3])
>plot(tetra[,-2],asp=1,xlab="x",ylab="z",
+      main="yprojection")
>polygon(tetra[,-2])
>lines(tetra[c(2,4),-2])
```

```
>plot(tetra[,1:2],asp=1,xlab="x",ylab="y",
+      main="zprojection")
>polygon(tetra[,1:2])
>lines(tetra[c(2,4),-3])
>lines(tetra[c(1,3),-3])
```

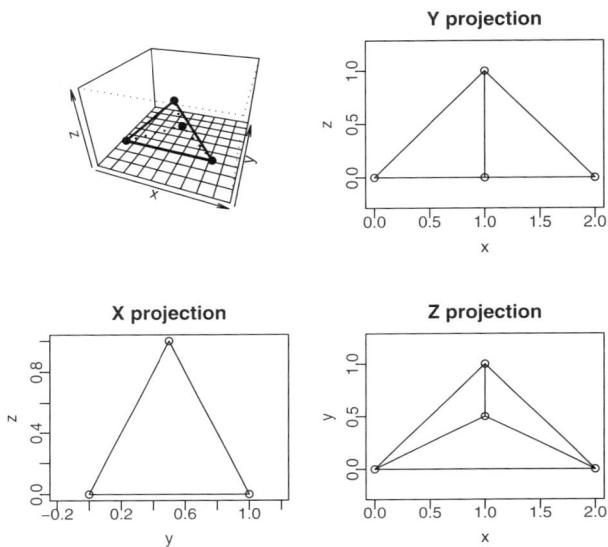

Fig. 2.3. Partitioning the graphical device using the function `layout`

The `scatterplot3d` function of the **scatterplot3d** package produces 3D plots with the coordinates of points. Arguments and scripts are somehow simpler than the `persp` function if the aim of users is to produce a cloud of points (there is no need for specifying the volume of the space on which points will be plotted; coordinates of points can be passed through a vector or a three-column matrix).

The `locator` and `identify` functions directly interact with the graphs. Other possibilities are offered by the **dynamicGraph** and **rgl** packages which build interactive graphs and where users can modify some parameters using the mouse.

The **rgl** package is obviously a useful tool for visualizing and manipulating 3D configurations or surfaces (Fig. 2.4). Let's see some script for visualizing the configuration of the previously defined tetrahedron. In addition, animation of the scene is easily produced.

Load the package rgl, *and clear the scene of the* rgl *graphical device.*

```
>library(rgl)
>rgl.clear()
```

```
>coll<-palette(gray(seq(0.4,.95,length=4)))
```

Set the color of the background.

```
>bg3d("white")
```

Draw triangle surfaces using their coordinates.

```
>for(i in 1:4)
+     {rgl.triangles(tetra[-i,1],tetra[-i,2],tetra[-i,3],
+                    colors=coll[i])}
```

Add spheres at the location of landmarks.

```
>rgl.spheres(tetra[,1],tetra[,2],tetra[,3]
+             ,radius=0.1,col=1)
```

Create an animation.

```
>for (i in 1:360) {rgl.viewpoint(i,45)}
```

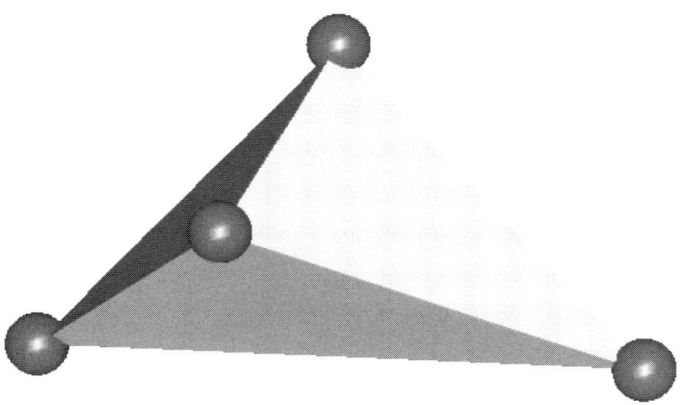

Fig. 2.4. The rgl graphical device

Of course, more complex configurations can be displayed. Fig. 2.5 corresponds to the location of landmarks recorded on the bony shell of the turtle *Dermatemys*. To obtain the graph[2], I adapted the same commands, and I underlined contacts between bones with the rgl.lines function.

In addition to reading image files, the **pixmap**, **rtiff**, and **rimage** packages offer a way to display the image classes as graphs using the generic plot function. The image function of the **base** package, with care about pixel size, can provide the same results. Indeed, image creates a grid of colored rectangles with colors corresponding to the values of a matrix. Note that x and y-axes have to be inverted and that the plot must be further reflected to display a very similar result (Fig. 2.6). Let's see some applications with the picture of sea shell mytilus.jpg[3] that we have

---

[2] The matrix of landmark coordinates and the code are available in the online supplement.

[3] Available in the online supplement

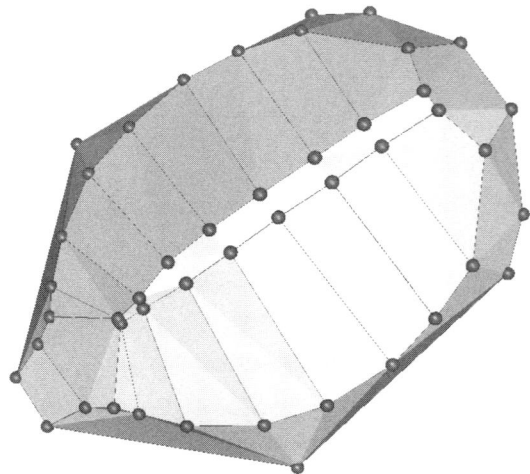

Fig. 2.5. Three dimensional display of a configuration digitized on a turtle shell with the `rgl` graphical device

formerly converted into a *.ppm format. This file is sufficiently small (less that 50 Ko) for being read and rapidly interpreted by R.

```
>setwd("/home/juju/morph")
>shell("convert mytilus.jpg mytilus.ppm")
>library(pixmap)
>M<- read.pnm("mytilus.ppm")
>plot(M)
```

Convert the RGB image to a gray-scale image.

```
>M<- as(M, "pixmapGrey")
>plot(M)
>layout(matrix(1:4, 2,2))
```

Similar operations using the `image` *function. Note the way to scale the color in the second argument, and the* `asp` *argument set to 2/3, or 3/4, or 9/16 for avoiding distortion due to pixel width on length ratio (see below). It is possible to play with the depth of the color scale for displaying various images.*

```
>image(t(M@grey[dim(M@grey)[1]:1,]),col=gray(0:255/255),
+    asp=9/16,axes=F,main="Gray-scale: 8-bits")
```

Plot the same image with a 2-bit gray-scale depth, and with a binary depth.

```
>image(t(M@grey[dim(M@grey)[1]:1,]),col=gray(0:3/3),
+    asp=9/16,axes=F,main="Gray-scale: 2 bits")
>image(t(M@grey[dim(M@grey)[1]:1,]),col=gray(0:1/1),
+    asp=9/16,axes=F,main="Monochrome: 1 bit")
```

Exploring possibilities offered by the `contour` *function .*

```
>contour(t(M@grey[dim(M@grey)[1]:1,]),asp=9/16,axes=F,
+    levels=0:10/10, main="Contour plot",drawlabels=F)
```

The `image` and `contour` functions displays interesting graphs when their argument is a gray-scale image matrix, or any tone channel matrix.

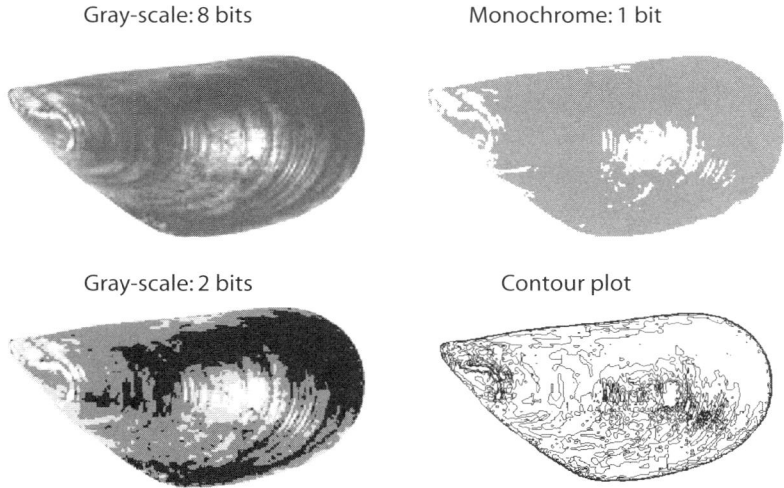

Gray-scale: 8 bits   Monochrome: 1 bit

Gray-scale: 2 bits   Contour plot

Fig. 2.6. Displaying images with various gray-scale depths, or by using contour plot

2.2.6 Image Analysis and Morphometric Data Acquisition with R

The `locator` function reads the position of the graphic cursor when the left mouse button is pressed. It can digitize coordinates of points on any displayed image file by the graphical interface. Specifying arguments in the `locator` function offers further possibilities: printing digitized points, linking them with a line, etc. (see Fig. 2.7). In addition, one can combine `locator` with low-level graphical functions such as `points` or `lines`, or `polygon` to directly interact and draw on the graphical device.

```
>par(mar=c(1,1,1,1))
```

The `mar` *graphical parameter sets the margin of the plot; one can set many other graphical parameters with the* `par` *function. You can find the image file in the online supplement.*

```
>library(rimage)
>x<-read.jpeg("/home/juju/morph/wing.jpg")
>plot(x)
>ji<-locator(5,type="p",pch=3)
```

`locator` *returns an object of the* `list` *class with a vector of* x *and a vector of* y*-coordinates.*

```
>ji
$x
[1]  327.0152 443.4394 662.5909 618.0758 814.9697

$y
[1]  269.3939 207.7576 240.2879 322.4697 320.7576
>text(ji, pos=2, labels=1:5)
>ju<-locator(5,type="l")
>polygon(ju, density=12)
```

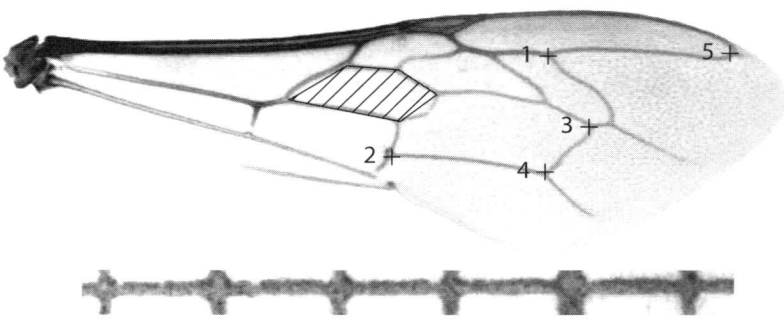

Fig. 2.7. Digitizing landmark locations with R

Calibrating the actual size is possible if one knows any interlandmark distance on the image . I invite users to photograph the object together with a scale (ruler, micrometer . . .). One can then obtain actual interlandmark distances using the cartesian coordinates of the scale (see the script and Fig. 2.8).

We first open an image,[4] and we invert it for an easier digitization process. We produce the inverted image by taking the absolute value of the difference of pixel values minus one.

```
>library(rimage)
>x<-read.jpeg("/home/juju/morph/jawd.jpg")
>x<-rgb2grey(x)
>x<-1-abs(x)
>plot(x)
```

Use locator *for localizing landmarks separated by 1 cm on the ruler.*

```
>a<-locator(2,type="o",pch=8,lwd=2,col="grey60",lty="11")
```

Determine the size of a known distance (1 cm) on the graph.

```
>scale1<- sqrt(sum(diff(a$x)^2+diff(a$y)^2))
```

Return the vector of scaled coordinates.

---

[4] The image file of the example is available in the online supplement

```
>b<-unlist(locator(10,type="p",pch=21,bg="white"))/scale1
```

Return the scaled configuration matrix.

```
>matrix(b, 5, 2)
```

An alternative manipulation.

```
>d<- locator (10, type="p"))
>d<- rbind(d$x, d$y)/scale1
```

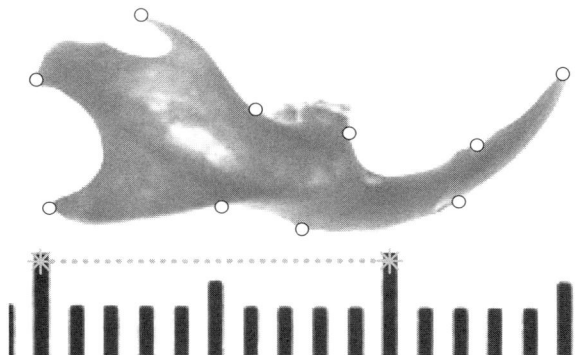

Fig. 2.8. Digitizing landmarks, and measuring the virtual size of the scale with the `locator` function. The digitized scale is indicated by gray stars and dotted segment, while circles indicate digitized landmarks

Depending on the camera or on the screen display, users must be careful with pixel size and non-square pixels that may produce a 3:2, 4:3, or 16:9 aspect ratio (see [68]). Correcting for pixel distortion requires multiplying or dividing matrix indices by corresponding ratios, or resizing one of the two dimensions.

Most of the time, one wants to simplify the image to find pixels of interest more easily. These simplifications involve defining a threshold pixel values and binarizing the images (Fig. 2.9). The following example illustrates these simple manipulations of pixel values. Using thresholds can help in calculating the surface of a specified area, especially when this area is well contrasted from the surrounding background. Logical indexing and the threshold value can be used to specify a desired surface. We can check whether the threshold is reliable for our task by assigning a new value or new color to the pixels of interest, and by plotting the binarized image. I tried three different thresholds on the *Mytilus* shell image.

```
>setwd("/home/juju/morph")
>x<- read.pnm("mytilus.ppm")
>y<- as(x, "pixmapGrey")
>rm(x)
>par(mar=c(1,3,2,1))
```

```
>Y<-y@grey
>layout(matrix(1:4, 2,2))
>plot(y, main="Gray-scale image")
>y@grey[which(Y>=0.1)]<-1
>y@grey[which(Y<0.1)]<-0
>plot(y, main="Bin image, threshold=0.1")
>y@grey[which(Y>=0.3)]<-1
>y@grey[which(Y<0.3)]<-0
>plot(y, main="Bin image, threshold=0.3")
>y@grey[which(Y>=0.9)]<-1
>y@grey[which(Y<0.9)]<-0
>plot(y, main="Bin image, threshold=0.9")
```

Estimate the number of pixels included in the surface of interest (which have values below the threshold in this case).

```
>length(y@grey[which(Y<0.9)])
[1] 29102
```

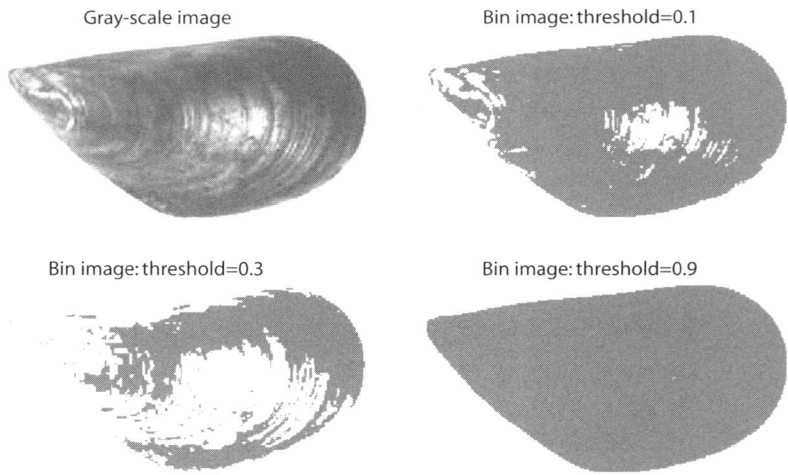

Fig. 2.9. Image binarization using a homemade threshold filter

Applying a threshold to pixel values of an image confers some other advantages for manipulating image files with R. Color image files, indeed, usually require an important memory size; therefore binarizing offers possibilities for compressing the size of the image object. One can extract the coordinates or the indices of the pixel of interest in a two-column matrix to release useless information contained in the background. The plot function can easily restitute the part of the image that is interesting for us. These manipulations are especially useful for processing with outline or surface extraction. The image file used for the example is available online.

```
>x <- read.jpeg("/home/juju/morph/jaw3.jpg")
>plot(x)
>dim(x)
[1] 378 891
```

Calculate the matrix indices of pixels of interest, with the integer division and the modulo operators.

```
>xx<-which(x<=0.5)%%378
>yy<-which(x<=0.5)%/%378
>plot(yy, -xx, type="p",pch=24, cex=0.1)
```

Following the same methodology, it is possible to export the slice number (if there is a series of images at different depths for a 3D object) together with the xx and yy in an object of the array or list classes for defining 3D surfaces or 3D volumes.

Inverting image can reveal features that could have been less visible on the normal image. For this, we have simply to invert the value of pixels for each channel: the biggest becomes the smallest and vice versa. Since the minimal and maximal pixel values are between zero and one, respectively, the inversion is straightforward. We may wish to accentuate the value of a single color channel, or to remove a channel as in the following example.

```
>x<-read.pnm(system.file("pictures/logo.ppm",
+    package="pixmap")[1])
```

Invert the red channel.

```
>x@red<-abs(1-x@red)
```

Invert the green channel.

```
>x@green<-abs(1-x@green)
```

Invert the blue channel.

```
>x@blue<-abs(1-x@blue)
>plot(x)
```

Accentuate the red channel.

```
>x@red<-0.5+x@red/2
```

Remove the green channel.

```
>x@green<-0
```

During image acquisition, it is possible to accentuate the contrast between the object and the background. This makes some manipulation easier, such as contour extraction. If the range of pixel values is less than 1, enlarging the full range to 1 will make details of the object more contrasted to identify specific features (although it does not increase the number of possible pixel values).

```
>x<- read.pnm(system.file("pictures/logo.ppm",
+    package="pixmap")[1])
>x<- as(x, "pixmapGrey")
>x@grey<- x@grey /diff(range (x@grey))
>x@grey <-x@grey - min (xgrey)
>plot(x)
```

We easily modify the brightness of the picture using the power operator.

```
>x<- read.pnm(system.file("pictures/logo.ppm",
+    package="pixmap")[1])
>x<- as(x, "pixmapGrey")
>y<-x
>y@grey<- y@grey^2
>plot(y)
```

Other possibilities are offered for modifying a picture. Displaying the histogram of pixel values (with the `hist` function) can help to understand how to transform pixel values appropriately for a given channel. Histograms can reveal undesirable values of pixels (for example, marginal values) that can be easily eliminated using a few functions, similarly to the way that we have processed for thresholding pixel values). One can also change the distribution of pixels so it could become uniform, or one can modify the symmetry of the distribution using one of the link functions to rescale pixel values between 0 and 1.

Some morphometric techniques are dedicated to the analysis of outline. I present here a small function called `Conte`. It extracts the coordinates of pixels defining an outline from a picture file. The function is the transcription of an outline extraction algorithm. The function starts with a point and looks for the nearest neighbor pixels, rotating and extracting coordinates on the outline in a clockwise way. Notice that we here use a threshold of 0.3 for finding the first point of the outline, and 0.1 for finding the nearest neighbor. One can modify these subjective values to adapt for the outline one wants to extract. You can store the function on an ASCII file somewhere in your computer, and paste it on the computer when you need it.

Function 2.1. `Conte`

Arguments:

      `x`: *Vector of the x and y-coordinates for the starting point. This starting point must be chosen on the left part and inside the object.*

      `imagematrix`: `imagematrix` *object (the picture for which one wants to extract the outline).*

Values:

      `X`: *x-coordinates of the outline.*

      `Y`: *y-coordinates of the outline.*

```
1  Conte<-function(x, imagematrix)
2  {I<-imagematrix
3  x<-rev(x)
4  x[1]<-dim(I)[1]-x[1]
```

The first step consists of moving a cursor from the selected pixel to the left until you find two pixels that significantly differ in their values for setting the "true starting point" of the outline. The function is not finding the closest pixel to the selected starting points but the pixel of the outline located on the left side of the selected location. This pixel will be the first point of the outline.

```
5  while (abs(I[x[1],x[2]]-I[x[1],(x[2]-1)])<0.1){x[2]<-x[2]-1}
6  a<-1
```

The M *matrix contains the indices (coordinates) of pixels that are located around the current pixel; the current pixel values are momently set to (0,0).*

```
7  M<-matrix(c(0,-1,-1,-1,0,1,1,1,1,1,0,-1,-1,-1,0,1),
8            2,8,byrow=T)
9  M<-cbind(M[,8],M,M[,1])
```

The X, Y, x1, x2, SS, *and* S *values are initialized before the contour extraction to start.*

```
10  X<-0; Y<-0;
11  x1<-x[1]; x2<-x[2]
12  SS<-NA; S<-6
```

The index of the pixel corresponds to a. *It is incremented by 1 every time the next pixel of the contour is found in a clockwise way.* X *and* Y *record the successive coordinates; the algorithm evaluates the following pixel value turning a block of three pixels clockwise around the current pixel, and progresses pixel by pixel until the location of the next pixel belonging to the outline is found.*

```
13  while ((any(c(X[a],Y[a])!=c(x1,x2) ) | length(X)<3))
14   {if (abs(I[x[1]+M[1,S+1],x[2]+M[2,S+1]]-I[x[1],x[2]])<0.1)
15    {a<-a+1;X[a]<-x[1];Y[a]<-x[2];x<-x+M[,S+1]
16    SS[a]<-S+1; S<-(S+7)%%8}
17  else if (abs(I[x[1]+M[1,S+2],x[2]+M[2,S+2]]
18              -I[x[1],x[2]])<0.1)
19    {a<-a+1;X[a]<-x[1];Y[a]<-x[2];x<-x+M[,S+2]
20    SS[a]<-S+2; S<-(S+7)%%8}
21  else if (abs(I[x[1]+M[1,(S+3)],x[2]+M[2,(S+3)]]
22              -I[x[1],x[2]])<0.1)
23    {a<-a+1;X[a]<-x[1];Y[a]<-x[2];x<-x+M[,(S+3)]
24    SS[a]<-S+3; S<-(S+7)%%8}
25  else S<-(S+1)%%8}
```

Return the resulting objects of the function under the form of a list *containing the* X *and* Y *vectors of* x *and* y*-coordinates for pixels of the outline.*

```
26  list(X=(Y[-1]), Y=((dim(I)[1]-X))[-1])}
```

We can use this new function to extract the outline coordinates of the shell on the *Mytilus* image.

Binarize the image.

```
>y<- read.pnm("mytilus.ppm")
>y<- as(y, "pixmapGrey")
>y@grey[which(y@grey>=0.9)]<-1
>y@grey[which(y@grey<0.9)]<-0.7
>par(mar=c(1,1,1,1))
>plot(y)
```

Use `locator` *for defining the starting point.*

```
>start<-locator(1)
>Rc<-Conte(c(round(start$x),round(start$y)),y@grey)
>lines(Rc$X, Rc$Y, lwd=4)
```

Draw an arrow at the starting point.

```
>arrows(0,Rc$Y[1],Rc$X[1],Rc$Y[1],length=0.1)
```

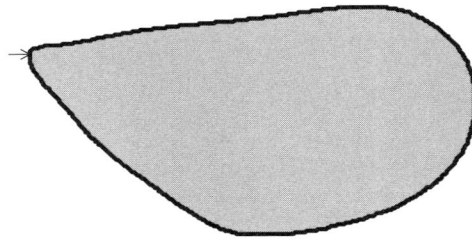

Fig. 2.10. Automated outline extraction. The arrow indicates the starting point

Defining a function for digitizing open curves is rather easy and uses most parts of this algorithm. There is simply a need for inputting the coordinates of the ending point, or rather their approximate position. Notice that you must first have prior knowledge about how coordinates of pixels are handled and how they are plotted on the graphic device; indeed, typing `plot(y)` here displays something quite different than the `image(y@grey)` command.

The `Conte` function can be adapted for a series of scanner images for extracting a 3D surfaces. It presents the advantage that the indices of pixel coordinates correspond to the elementary curvilinear abscissa.

2.3 Manipulating and Creating Data with R

After the acquisition of raw data, it is often necessary to reorganize or slightly transform them to start operationally analyzing the morphometric data. Here are some of the very usual data manipulations: scaling an image, obtaining angles or distances from coordinates of points. We will also program some more specific tools such as one function for obtaining regularly spaced landmarks on a given outline.

2.3.1 Obtaining Distance from Coordinates of Points

The distance d_{EF} between two landmarks E and F in two or three dimensions is the square root of the sum of the squared differences between each coordinate. It is given by the relationship

$$d_{EF} = \sqrt{\sum_{i=1}^{k}(E_i - F_i)^2} \; .$$

The transcription in R language is simple:

Acquisition of the coordinates of the landmarks E *and* F.

```
>E<-c(1,  4)
>F<-c(6,  8)
```

Computation of the interlandmark distance between E *and* F.

```
>sqrt(sum((E-F)^2))
[1] 6.403124
```

We implement this relationship in the ild function. This function computes the distance between any pair of landmarks:

Function 2.2. ild

Arguments:
    E: *x and y-coordinates of the first point as a* vector *object.*
    F: *x and y-coordinates of the first point as a* vector *object.*
Value:
    Distance between the two points.

```
ild<-function(E, F){ sqrt(sum((E-F)^2))}
```

```
>ild(E, F)
[1] 6.403124
```

When the function is loaded on the computer, it correctly computes the distances between two landmarks, provided that landmarks have the same number of dimensions and that their coordinates are written in the same order.

2.3.2 Calculating an Angle from Two Interlandmark Vectors

The angle θ between two vectors $\overrightarrow{AB}$ and $\overrightarrow{CD}$ is defined as the difference of their arguments using $\mathbb{C}$ vectors.

```
>CD<-c(2, 4)
>AB<-B-A
>ABc<-complex(real=AB[1], imaginary=AB[2])
>CDc<-complex(real=CD[1], imaginary=CD[2])
>ABc; CDc
[1] 5+4i
[1] 2+4i
```

The `Arg` *function returns the argument of a complex number. The angle between vectors corresponds to the difference of their arguments. The function uses properties of complex numbers for calculating the angle between the* `CD` *and* `AB` *vectors.*

```
>Arg(ABc)
[1] 0.674741
>Arg(CDc)
[1] 1.107149
>Arg(ABc)-Arg(CDc)
[1] -0.4324078
```

Calculate the result in degrees.

```
>Arg(ABc)-Arg(CDc) / pi * 180
[1] -62.76021
```

The `angle2d` function calculates the angle between two 2D `v1` and `v2` vectors.

Function 2.3. `angle2d`

Arguments:
      `v1`: *2D vector of* `numeric` *mode.*
      `v2`: *2D vector of* `numeric` *mode.*
Value:
      Angle between the two vectors in radians.

```
1  angle2d <- function(v1,v2)
2  {v1<-complex(1,v1[1],v1[2])
3  v2<-complex(1,v2[1],v2[2])
4  (pi+Arg(v1)-Arg(v2))%%(2*pi)-pi}
```

To calculate the angle θ between two vectors of higher dimensions, one must use the relationship between their norm and their dot product such that

$$|\theta| = \frac{\overrightarrow{CD} \cdot \overrightarrow{AB}}{\|\overrightarrow{AB}\|\|\overrightarrow{CD}\|} \; .$$

The `angle` function uses the norm and dot product relationship to calculate the angle between two vectors. However, the orientation of this angle will not be signed.

Function 2.4. `angle`

Arguments:
 `v1`: *Vector of* `numeric` *mode.*
 `v2`: *Vector of* `numeric` *mode and of length equal to the length of* `v1`.
Value:
 Angle between the two vectors in radians.

```
1  angle<-function(v1, v2)
2  {temp <- sum(v1*v2)/( sqrt(sum(v1^2))*sqrt(sum(v2^2)) )
3    acos (temp)}
```

For 3D vectors, one must check the sign of the determinant of a 3×3 matrix with the first row being a triple unit vector ($\mathbf{1}$), and the next two rows corresponding to the vector coordinates. This operation corresponds to a triple scalar product such that

$$det \begin{vmatrix} \mathbf{1} \\ \overrightarrow{AB} \\ \overrightarrow{CD} \end{vmatrix} = \mathbf{1} \cdot (\overrightarrow{AB} \times \overrightarrow{CD}) \,,$$

where "$\times$" denotes the vector cross-product. The `angle3` function calculates the signed angle between two 3D vectors, it depends on the `angle` function:

Function 2.5. `angle3`

Arguments:
 `v1`: *3D vector of* `numeric` *mode.*
 `v2`: *3D vector of* `numeric` *mode.*
Value:
 Signed angle between the two vectors in radians.

```
1  angle3<-function(v1, v2)
2  {a<-angle(v1, v2)
3    b<-sign( det(rbind(1, v1, v2)) )
4    if (a == 0 & b == 1){jo<-pi/2}
5      else if (a == 0 & b == -1){jo<- - pi/2}
6      else {jo<- a * b}
7  (pi+jo)%%(2*pi)-pi}
```

2.3.3 Regularly Spaced Pseudolandmarks

In morphometrics, in particular with Fourier analysis of outlines (see Chapter 4), prior operations are usually performed on the collection of coordinates of pixels

defining the outline. One of these operations is to obtain equally spaced pseudolandmarks on the digitized outline (see Fig. 2.12). Our `Conte` function extracts coordinates of points on an outline with a one pixel in length curvilinear abscissa. If there
are enough pixels, one can extract a given number of equally spaced pixels using the
regular sequence-generating function of R.

Obtaining 32 equally spaced pseudolandmarks on the outline of the Mytilus shell (Rc$X and
Rc$Y *are coordinates of pixels of the outline).*

```
>layout(matrix(c(1,2), 1,2))
>Rc32x<-(Rc$X[seq(1,length(Rc$X),length=33)])[-1]
>Rc32y<-(Rc$Y[seq(1,length(Rc$Y),length=33)])[-1]
>plot(Rc$X, Rc$Y, type="l", lwd=1.5, asp=1, axes=F
+    , main = "curvilinear")
>points(Rc32x, Rc32y)
```

One can also digitize equally spaced pseudolandmarks on any kind of curve that
one has approximated by digitizing several points by hand. For acquiring points on
the curve, one uses the `locator` function . Indeed, `locator` allows one to collect
coordinates that can define lines or segments. Then, one can obtain regularly spaced
landmarks on lines or surfaces with the `spsample` function of the **sp** package.
We will apply this exercise to the curve depicting the lower part of the rodent jaw.[5]
We will sample pseudolandmarks between two well-known landmarks (incisor-bone
contact and extremity of the angular apophysis) (see Fig. 2.11).

```
>library(rimage)
>layout(matrix(c(1,2), 1,2))
>par(mar=c(0,1,0,0))
>x<-read.jpeg("/home/juju/morph/jaw2.jpg")
>plot(x)
>dig<-locator(type="o",col="white",lwd=1.5)
```

Draw successive segments with the mouse for digitizing the curve of interest.

```
>DIG<-matrix(unlist(dig),length(dig$x),2)
>library(sp)
>Ldig<-Line(DIG)
```

Transform the object of the Line *class for sampling pseudolandmarks on successive segments.*
The matrix of sampled coordinates are inside the @coords *slot returned by* spsample.

```
>pseudo<-spsample(Ldig,16,type="regular",
+                 offset=c(0,1))@coords
```

Do not forget to remove the last landmark digitized in the sample (because of the offset
argument), and plot landmarks and pseudolandmarks on a new graph.

```
>plot(x)
>points(DIG[c(1,dim(DIG)[1]),],cex=1.5,pch=20,
+              frame=F, axes=F,asp=1)
>points(pseudo[-nrow(pseudo),],pch=21,bg="white")
```

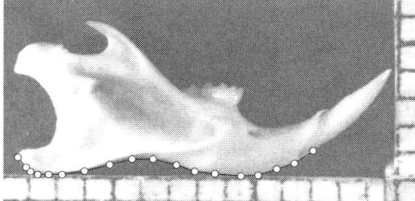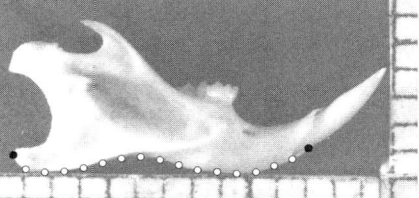

Fig. 2.11. Obtaining equally spaced pseudolandmarks using curve digitizing; pseudoland-marks are white circles, and ending landmarks are black rounds on the right side graph

One can eventually adjust the sampling process by smoothing the original curve (see later in this section).

Rather than selecting equally spaced points according to the curvilinear abscissa, one may prefer to select landmarks that are spaced with a regular sequence of angles taken between the outline coordinates and the centroid. Let say that the origin O is the first digitized point. We must therefore transform every cartesian coordinate into polar coordinates using complex numbers and operations for our task. The `regularradius` function returns n points on equally spaced radii.

Function 2.6. `regularradius`

Arguments:

    `Rx`: *Vector containing the x-coordinates of the outline.*

    `Ry`: *Vector containing the y-coordinates of the outline.*

    `n`: *Number of points to be sampled.*

Values:

    `pixindices`: *Vector of radius indices.*

    `radii`: *Vector of sampled radii lengths.*

    `coord`: *Coordinates of sampled points arranged in a two-column matrix.*

```
1  regularradius<-function(Rx, Ry, n)
2  {le<-length(Rx)
3  M<-matrix(c(Rx, Ry), le, 2)
4  M1<-matrix(c(Rx-mean(Rx), Ry-mean(Ry)), le, 2)
5  V1<-complex(real=M1[,1], imaginary=M1[,2])
6  M2<-matrix(c(Arg(V1), Mod(V1)), le, 2)
7  V2<-NA
```

The following code finds the indices of the nearest pixel on the outline using the angular increment.

```
8   for (i in 0:(n-1))
9       {V2[i+1]<-which.max((cos(M2[,1]-2*i*pi/n)))}
10  V2<-sort(V2)
11  list("pixindices"=V2,"radii"=M2[V2,2],"coord"=M1[V2,])}
```

---

[5] The image file is available in the online supplement

To visualize the outline and the equally spaced radii, we first need to calculate the coordinates of the centroid of the outline. Here we work on the *mytilus* shell image.

The centroid coordinates Xc *and* Yc *of the outline are defined as the mean of* x *and* y*-coordinates sampled on the outline and are computed straightforwardly:*

```
>Xc <- mean(Rc$X)
>Yc <- mean(Rc$Y)
>plot(Rc$X,Rc$Y,type="l",lwd=1.5,asp=1,axes=F,main="polar")
>points(Xc, Yc, pch=4)
```

Using a loop, we draw the successive segments linking the centroid to the points sampled on the outline.

```
>ju<-regularradius(Rc$X, Rc$Y, 32)
>points(ju$coord[,1]+Xc, ju$coord[,2]+Yc)
>for (i in 1:32){
+   {segments(0+Xc,0+Yc,ju$coord[,1]+Xc,ju$coord[,2]+Yc)}
```

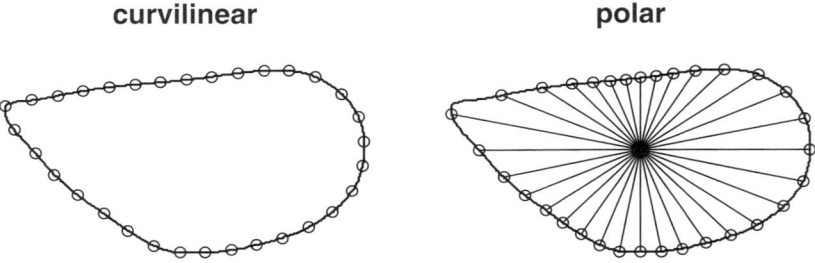

Fig. 2.12. Diverse types of pseudolandmarks automated digitizations with the outline of the *Mytilus* shell. On the left, pseudolandmarks are equally spaced following an equal curvilinear abscissa, while on the right, pseudolandmarks are spaced according to equally spaced angles between segments departing from the centroid to the outline

2.3.4 Outline Smoothing

Depending on the resolution and the sensitivity of an automated outline extraction, it is often necessary to smooth the outline for further analyses. When digitized outlines with high resolution produce undesirable irregularities, Haines and Crampton [43] recommend smoothing the outline based on the following formula:

$$(x,y)_i^{new} = \frac{1}{4}(x,y)_{i-1}^{old} + \frac{1}{2}(x,y)_i^{old} + \frac{1}{4}(x,y)_{i+1}^{old} .$$

A function that performs this operation requires two arguments; the first is the outline coordinates to be smoothed, and the second is the number of iterations. We want to apply n times this smoothing function to the raw configuration matrix. We program this under the `smoothout` function:

Function 2.7. `smouthout`

Arguments:
    M: *x and y-coordinates of the outline arranged in a two-column matrix.*
    n: *Number of iterations.*
Value:
    Matrix of smoothed coordinates.

```
1  smoothout<-function(M, n)
2  {p<-dim(M)[1]
3  a<-0
4  while (a<=n)
5  {a<-a+1
6  Ms<-rbind(M[p,],M[-p,])
7  Mi<-rbind(M[-1,],M[1,])
8  M<-M/2+Ms/4+Mi/4}
9  M}
```

Low resolution of images can be a source of error during automatic image digitization. Artificially inflating the number of landmarks can provide some approximation of the reality. It can be achieved by interpolating supplementary landmarks that correspond to the mean coordinates of two adjacent landmarks and writing the `landmark.addition` function. One can later smooth this outline using the `smoothout` function, if needed.

Function 2.8. `landmarkaddition`

Arguments:
    M: *x and y-coordinates of the outline arranged in a two-column matrix.*
    n: *Number of iterations.*
Value:
    Matrix of original and interpolated coordinates.

```
1   landmark.addition<-function(M, n)
2   {a<-0
3   while(a<=n)
4   {p<-dim(M)[1]
5   k<-dim(M)[2]
6   N<-matrix (NA,2*p,k)
7   N[((1:p)*2)-1,]<-M
8   N[(1:p)*2,]<-(M+(rbind(M[-1,],M[1,])))/2
9   M<-N}
10  M}
```

2.4 Saving and Converting Data

The easiest way to save objects with R is probably the `save` function which writes a binary R file in a specified folder. The extension can be of any type. For reloading the saved file, the `load` function loads the object on the working environment or follow a specified path entered as argument. The original name of the object is loaded with its value.

The `write.table` function is convenient for storing data frames. Finally, the `cat` function can convert and concatenate objects to character strings, separating them by a specified separator. In specifying `"\n"`, one can write data on different lines.

R can read many formats, most of them being primarily ascii files. Here my goal is to show how one opens, converts, and interprets them through R, and how to perform the reverse operation (for example, digitizing landmarks on R, and then exporting them in the appropriate format).

The *.NTS format is one of the more commonly used in geometric morphometrics. Software performing morphometric analyses (Morpheus,[6] the TPS family[7] etc.) can save or convert data in this format. The data are stored as a matrix with the rows corresponding to the configurations and the columns to the coordinates of each landmark. The first line of the file contains four or five arguments. The first and fourth are fixed, while the second and third respectively correspond to the number of specimens and the number of landmarks multiplied by the number of dimensions $(k \times p)$. The fifth and optional argument specifies the number of dimensions, and follows the string "DIM=". By correctly filling the arguments of the `scan` function, we open this type of file quite easily. If the string or the character for comments is the double quoting mark along the file, the conversion requires not more than two lines as illustrated below.

The RATS.NTS dataset can be found in the data sets of the freely available software tpsRegr.[8]

```
>jo<-scan("/home/juju/morph/RATS.NTS",what="char",
+        quote="",sep="\n", comment.char="\"")
```

Obtain the number of dimensions k.

```
>jo1<-jo[1]
>l1<-unlist(strsplit(jo1, "="))
>l2<-unlist(strsplit(jo1, " "))
> if(length(l1)==1) {k<-2}
+ else {k<-as.numeric(l1[2])}
>k
[1] 2
```

Extract coordinates.

---

[6] http://life.bio.sunysb.edu/morph/morpheus/
[7] http://life.bio.sunysb.edu/morph/
[8] http://life.bio.sunysb.edu/morph/

```
>jo2<-jo[-1]
>cat(jo2, file="jo2.txt")
>data<-matrix(scan("jo2.txt"), as.numeric(l2[2]),
+    as.numeric(l2[3]), byrow=T)
>unlink ("jo2.txt")
```

Delete the temporary file jo2.txt.

Notice the sep="\n" argument specified for the scan function; it indicates that the separator corresponds to a new line.

A more complex format is the format *.tps that has been developed for a series of programs by James Rohlf.[9] Here is the code for importing configurations, and later for importing names and coordinates that define curves. In the code below, we finally plot configurations, and curves. Similarly one can extract and assign other attributes to a list or an array.

The file sneathd.tps *can be obtained from the datasets of the freely distributed software tpsRelw.*

```
>jo<-scan("/home/juju/morph/sneathd.tps", what="char",
+          quote="", sep="\n", strip.white=T)
>jo<-casefold(jo, upper=F)
```

Find the indices where each configuration starts.

```
>sp<-grep("lm=", jo)
```

Find the n *number of configurations.*

```
>n<-length(sp)
```

Find the p *number of landmarks for each configuration, knowing that it is indicated after each* "=".

```
> p <-as.numeric(unlist(strsplit(unlist(strsplit
+    (jo[sp[1]], "=")) [2], " ")) [1])
```

Find the k *number of dimensions.*

```
>k<-length(unlist(strsplit(jo[sp[1]+1],split=" +")))
```

Prepare an empty matrix that you assign to the new config *object for storing coordinates of configurations.*

```
>config<-matrix(NA, n, p*k)
>for (i in 1:n)
+    {config[i,]<-as.numeric(unlist(strsplit(
+    jo[(sp[i]+1):(sp[i]+p)], split=" +")))}
```

Read and store the coordinates of the first curve and of the first object.

```
>curve1<-grep("curves", jo)
```

Find the q *number of outlines in the object.*

---

[9] http://life.bio.sunysb.edu/morph/

```
>q <-as.numeric(unlist(strsplit(unlist(strsplit(
+    jo[curve1[1]],"="))[2]," "))[1])
>point1<-grep("points", jo[sp[1]:sp[2]])
```

The nb *object is a vector that will contain the number of landmarks stored for each curve.*

```
>nb<-NA
>for (i in 1:length(point1))
+    {nb[i]<- as.numeric(unlist(strsplit(unlist
+    (strsplit(jo[point1[i]], "="))[2], " "))[1])}
```

Store the coordinates of the curve points in a list *of* q *vectors.*

```
>out1<-list()
>for (i in 1:q)
+    {out1[[i]]<- as.numeric(unlist(strsplit(jo[(
+    point1[i]+1):(point1[i]+nb[i])], split=" +")))}
>l<-length(unlist(out1))
```

Prepare the space required for the x *and* y*-axes in the graph.*

```
>m1<- min(unlist(out1)[(1:(l/2))*2])
[1] 90.21
>m2<- min(unlist(out1)[(1:(l/2))*2-1])
[1] 222.15
>M1<- max(unlist(out1)[(1:(l/2))*2])
[1] 1492.27
>M2<- max(unlist(out1)[(1:(l/2))*2-1])
[1] 1860.8
```

Plot the landmarks of the first configuration.

```
>par(mar=c(4,1,1,0))
>plot(matrix(config[1,],p,k,byrow=T),pch=20,
+    cex=1.5, xlim=c(m2-10,M2+10), ylim=c(m1-10,
+    M1+10), asp=1, xlab="Homo", ylab="",axes=F)
```

Plot the curves of the first configuration.

```
>for (i in 1:q)
+    {points(matrix(out1[[i]], nb[i],k, byrow=T),
+    type="l", lw=2, lty="11")}
```

Remark the casefold *function in the script. It translates a character vector in upper or lower* case *(useful for standardizing a dataset). The result is presented in Fig. 2.13. The* grep *and* sub *functions are used for the conversion of parts of the ascii file because they return indices and replace strings in a vector of characters respectively.*

Exporting R data in one other format corresponds to the reverse operation and is straightforward. The cat function with the separator argument specified as "\n" allows manipulation of data for exporting an ascii file in an appropriate way. Given two configuration matrices in R, M and N, each containing three 2D landmarks, one converts these objects in *.NTS format as follows:

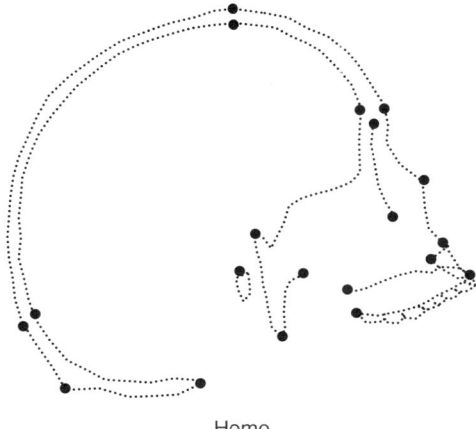

Homo

Fig. 2.13. Opening and printing a *.tps file

```
>M<-matrix(round(rnorm(6)),3, 2)
>M
     [,1] [,2]
[1,]   -2    1
[2,]    0   -1
[3,]    1   -3
>N<-matrix(round(rnorm(6)),3, 2)
>N
     [,1] [,2]
[1,]    2    0
[2,]    0    0
[3,]   -1    0
>config<-c(paste(t(M),collapse=" "),
+                      paste(t(N), collapse=" "))
>commentl<-paste("\"","configurations ","M ","and N")
>firstl<-paste(1, 2, 6, 0, "dim=2")
```

Save configurations in a *.NTS file.

```
>cat(firstl,commentl,config,sep="\n",
+          file="/home/juju/morph/juju.NTS")
```

Print the file as it is.

```
>cat(firstl, commentl, config, sep="\n")
1 2 6 0 dim=2
" configurations  M  and N
1 0 -3 0 2 -1
1 2 0 0 1 0
```

Writing a function for exporting data in a desired format involves looping around this theme. Note the backslash for invoking quotation marks or new line in the code.

2.5 Missing Data

Many statistical functions in R contain an argument specifying how to handle missing data. This argument is on the `na.rm`, `na.action`, or `use` forms, and can be specified to change its default value. For example, `na.rm` expects a logical for indicating whether nonavailable data are dropped. Look at the help file of functions such as `cor`, `mean`, and `lm` to understand the different available options. Although one can choose to exclude one measurement or one landmark, one can sometimes estimate missing values from original data.

2.5.1 Estimating Missing Measurements by Multiple Regression

When the studied sample shows some homogeneity in shape variation, measurements are often related. This is often the case with a collection of biological distances. Distances have often a high degree of intercorrelation because they correspond to different body parts that often grow harmoniously during the development. Other causes of the high degree of correlation can be biomechanic, architectural, or genetic. This high correlation can help us infer missing values with regression techniques. It is often better to use all the data to perform these inferences. Multiple regression can be applied as in the following example:

Simulate four distance measurements a, b, c, *and* d *for 30 individuals.*

```
>a<-abs(rnorm(30))
>b<-2*a+rnorm(1, 5)+rnorm(30, 0, 0.5)
>c<-7*a+rnorm(1, 6)+rnorm(30, 0, 4)
>d<-12*a+rnorm(1, 3)+rnorm(30, 0, 3)
>data<-cbind(a, b, c,d)
```

Generate 10 missing values.

```
>data[sample(1:120)[1:10]]<-NA
```

Using the scale *function scales the data with a variance=1 and a mean=0.*

```
>datas<-scale(data)
```

Find the missing values and return the indices of their row with integer division operators and the which *function.*

```
>indic<-which(is.na(data))
>ro<-indic%%30
```

The unique *function removes duplicated values in* ro.

```
>ro<-unique(ro)
```

Change the 0 indices to be 30.

```
>if (any(ro==0)){ro[which(ro==0)]<-30}
>for (i in 1:length(ro))
+    {ind<-which(is.na(data[ro[i],]})
```

The generic `predict` *function is used for appraising fitted values according to a given model (here a multiple linear model). The missing values are estimated with the prediction from the other data. For removing a raw or a column in the matrix, notice the "−" for negative indexing.*

```
>for (j in 1:length(ind))
+    {MOD<-lm(data[, ind[j]] ~ datas[,-ind])
+    data[ro[i],ind[j]]<-predict(MOD,data.frame(datas
+    [,-ind])[ro[i],])[ro[i]]} }
```

2.5.2 Estimating Missing Landmarks on Symmetrical Structures

Several objects exhibit bilateral symmetry. It is often useful to estimate the location of landmarks missing on one side prior to morphometric or statistical analyses. This is not too difficult because symmetric landmarks have a mirror copy. The location of these landmarks is obtained by an appropriate geometric reflection.

The first step for obtaining a mirror image requires estimating the axis or plane of symmetry of the object. Let missing values correspond to NA, and let the M configuration be a matrix of p rows and k columns. We will define the indices of landmarks theoretically lying on the symmetry plane or axis in the `imp` vector, those of left landmarks in the `pa1` vector and those of the right landmarks in the `pa2` vector. The axis (or plane) that passes through midline landmarks and at the midline of the segments defined by paired landmarks is the axis (or plane) of symmetry. We define the N matrix as containing the coordinates of the landmarks expected to be on the symmetry axis, plus the coordinates of points defined at the midline between paired and symmetric landmarks. Because all landmarks of the configuration contribute to the definition of the symmetry axis, coordinates of points defined by left and right sides are weighted twice. To estimate N, we have to keep the landmarks of the axis and calculate the midline coordinates of remaining paired landmarks. We obtain N using indexing and a few computations.

Obtain separate matrices for the midline, left-, and right-side coordinates.

```
>N1<-M[imp,]
>NL<-M[pa1,]
>NR<-M[pa2,]
```

Obtain the indices of the missing landmarks. The `unique` *function is used for removing duplicated elements, and* `sort` *allows a vector to be sorted in ascending order.*

```
>no<-sort(unique(c(which(is.na(M[pa1,1])),
+    which(is.na(M[pa2,1]))))))
```

Calculate coordinates of the points at the midline of segments defined by corresponding left and right landmarks.

```
>N2<-(NL[-no,]+NR[-no,])/2
>N<-rbind(N1,N2, N2)
```

In practice, coordinates of landmarks of N are rarely aligned along a straight line (or on a plane, for 3D data). Indeed, there is usually some variation in the position of landmarks on the axis of symmetry in comparison with their expected position (biological asymmetries due to development, measurement error). For defining our transformation, we could estimate the plane or axis equation using the coordinates contained in N, and use this equation to perform our transformation. However, the shortest way is to rotate our configuration and axis so that its coordinates will fit to the line of coordinates $y = 0$ (for 2D data) or for the plane defined by $z = 0$ (for 3D data). This axis or plane is minimizing distances between N (the raw rotated axis coordinates) and the symmetry axis we will use for the reflection. The rotation is easily appraised by decomposing the variance and covariance of the 2D or 3D coordinates of N. We write the corresponding code in the function eigenrotation:

Function 2.9. eigenrotation

Arguments:
> M: *k-column matrix of landmark coordinates (missing landmarks excluded) to be rotated.*
> N: *k-column matrix of coordinates that belongs to the symmetry axis or plane (median plane or midline landmarks, plus midline points estimated from paired symmetrical points).*

Value:
> *k-column matrix of rotated coordinates.*

```
1  eigenrotation<-function(as.matrix(N), as.matrix(M))
```

The eigen *function computes the eigenvectors and corresponding eigenvectors for a rectangular matrix.* eigen *returns a* list *object with the first element being the normalized eigenvector loadings.*

```
2  sN<-eigen(var(N))$vectors
```

Data are projected on the eigenvectors of their own variance-covariance matrix. This performs a rotation of the original data. These are then translated so that axis or the plane of reflection includes the origin (remark that this may introduce undesired reflections).

```
3  k<-dim(N)[2]
4  p<-dim(M)[1]
5  Nn<-N%*%sN
6  Mn<-M%*%sN
```

Compute the vector that translates the landmark of the configuration so their centroid becomes $(0, 0)$.

```
7  uNn<-apply(Nn, 2, mean)
```

Translate the rotated data.

```
8  Mnf<-Mn-rep(1,p)%*%t(uNn)
9  Mnf}
```

When the configuration is rotated, we have first to check whether the orientation between coordinates of landmarks in our data has been preserved. For 2D data, we can check the angle between the first three landmark coordinates. If the sign of the vector has changed, we have to multiply the y- (or z-) coordinates by -1. The sign of the angle can be checked using the `angle2d` function for 2D data, or the `angle3` function for 3D data (see Section 2.3). Once reflection is checked, we appraise coordinates of missing landmarks multiplying the corresponding landmark y (for 2D) or z (for 3D) coordinates by -1, and we duplicate the x. To obtain the reflected 3D missing landmarks, the sign of the third coordinate of the corresponding paired landmark is multiplied by -1.

Other solutions for estimating missing landmarks can involve functions like thin-plate spline (see Section 4.3).

2.6 Measurement Error

Measurement error is defined as "*the variability of repeated measurements of a particular character taken on the same individual, relative to its variability among individuals ...*" for quoting Bailey and Byrnes [5]. The source of error is multiple.

2.6.1 Sources of Measurement Error

Measurement Error Due to the Measurement Device (Precision)

Measurement error and precision are close concepts. Most digitizing devices and digtizers like observers produce an error that corresponds to their imprecision (precision is the level of similarity among the same, repeated measurement) and to their inaccuracy (inaccuracy is measured as the difference between the measured value and the true value). The cause of this error depends mostly on the reliability and the sensitivity of the measuring device. If you digitize landmarks and measurements on an image, the error depends on the size (number of pixels) and on the resolution of the image (the resolution corresponds to the smaller details that can be seen on a picture).

Measurement Error Due to the Definition of the Measure

Although the definition of the landmark position can be unequivocal (e.g., position of a foramen, intersection between nervations), there can still remain a small variation around the landmark position. If the position of the landmark is as precise as 0.1 mm, this will generate a variation among measurements of the same object and will contribute to the total measurement error. This may have some incidence when the observer wants to compare objects that differ in size, and if recognizable landmarks have the same imprecision between objects. For instance, we could consider a landmark that corresponds to a foramen, and that this landmark would be digitized among small and large skeletons; let the size of the foramen be similar in size for large and small objects, consequently the percent of error variation of the smaller object will be inflated. Actually, this depends on the range of size variation among objects you are exploring, and you can expect that error to decrease with the size of objects.

Measurement Error Due to the Quality of the Measured Material

Some part of the error is inherent to the data themselves. For example, fossils can show different levels of preservation that may affect the way we measure the object. The imprecision in digitizing a given landmark will depend on the preservation of objects. In this case, one expects a positive relationship between error and level of preservation.

Measurement Error Due to the Measurer

Give a measuring tape to three different people and ask them to measure your own hip circumference with a precision of one mm and you will probably get three different measurements. People necessarily differ in the way that they measure distances, and this depends on their individual condition: their degree of concentration, eye health, stress, or knowledge and interest in the objects they measure may affect the outcome of multiple measurements differently. Among users, the error terms vary not only in intensity, but also in the geometrical way that the error is produced (for example, some people will produce more error in positioning a landmark more laterally than vertically, which will affect the covariation pattern in the error component of the variation).

Measurement Error Due to the Environment of the Measurer

Another source of error comes from the direct environment of the observer. One can expect that a noisy or peaceful environment differently influences the observer. Change in luminosity and lighting may compete with the quality of the measurement made by the observer. This is not only true with direct measurements (using a caliper or a ruler) but also with indirect measurements obtained from pictures. In the latter case, the source and intensity of light should be similar for each capture of image (limiting errors arising from differences in shadows or contrasts). More generally, reducing error requires keeping the same conditions for measurement acquisition throughout the full session.

Measurement Error Due to the Measurement Protocol

The better the measurement protocol is established, and the lower the error is. Consider a photography of a 3D object on which landmarks are later digitized; an important part of error may appear depending on the position of the object under the camera. It is useful to set a reference plane from the object for photographing all similar objects according to the same orientation. Errors are inherent to most optical devices because lenses usually slightly deform the shape of objects, especially when focal distances are very short. Generating variation by focusing differently when capturing an image inherently inflates error variation. To limit this source of error, the user should set focal distance once and for all before digitizing.

2.6.2 Protocols for Estimating Measurement Error

Estimating measurement error involves taking into account most of its origins. It is often interesting to explore the different possible sources of error variation in categorizing the variation. For some analyses of variation, it is strongly recommended to estimate measurement error (especially when the investigated signal of variation is expected to be low). Yezerinac et al. [124] suggest an ANOVA design to compute a percent measurement error that allows further comparisons between different studies. Using a percent of measurement error allows results to be independent of the range or the units of measured objects.

Repeating measurements on the same objects is always necessary to correctly estimate measurement error. It is best repeat ALL measurements at least twice (measurement error is influenced by objects themselves; we can think that some will be more difficult to accurately measure than others, and the economy of time passed through the digitization of a subsample may change our way of estimating the measurement error). However, estimating measurement error using a smaller representative sample is a possible alternative.

The percentage of measurement error is defined as the ratio of the within-measurement component of variance on the sum of the within- and among-measurement component. Percent of measurement error $\%ME$ can be obtained as follows:

$$\%ME = \frac{s^2_{\text{within}}}{(s^2_{\text{within}} + s^2_{\text{among}})} \times 100 \ .$$

Components of variance [76] are themselves derived from the mean squares of the one-way ANOVA considering the factor individual as a source of variation. The among and within variation are estimated from the mean sum of squares:

$$s^2_{\text{among}} = \frac{\text{MSS}_{\text{among}} - \text{MSS}_{\text{within}}}{m} \ ,$$

and

$$s^2_{\text{within}} = \text{MSS}_{\text{within}} \ ,$$

where m corresponds to the number of repeated measurements.

Doing different sessions under different conditions, with different observers, or using different measurement devices is a way to inspect origins and contributions of the putative candidates that may inflate error. Unfortunately, we can regret that this boring stage is still not systematically present in scientific contributions performing morphometric analyses. One can check whether there are differences in mean measurements between both sessions with the significance of the session effect. I supply a short simulated example below invoking R commands in the case of a univariate measurement.

Simulation of a set of 20 real distances following a normal distribution, with two measurement sessions and with an error term normally distributed.

```
>truemeasures<-rnorm(20, 20, 3)
>measure1<-truemeasures + rnorm(20, 0, 1)
>measure2<-truemeasures + rnorm(20, 0, 1)
>sessionfactor<-gl(2, 20)
>individualfactor<-as.factor(rep(1:20, 2))
>totalobservation<-c(measure1, measure2)
>summary(aov(totalobservation~sessionfactor))
              Df Sum Sq Mean Sq F value Pr(>F)
sessionfactor  1   1.77    1.77  0.1188 0.7323
Residuals     38 566.47   14.91
```

The one-way ANOVA reveals that there is not a strong influence of session on the measurement and that variation between sessions is lower than within session in this example.

```
> mod<-summary(aov(totalobservation~individualfactor))
> mod
                 Df Sum Sq Mean Sq F value    Pr(>F)
individualfactor 19 544.87   28.68  24.533 6.438e-10 ***
Residuals        20  23.38    1.17
---
Signif. codes:  0 *** 0.001 ** 0.01 * 0.05 . 0.1   1
```

This second one-way ANOVA shows that interindividual variation is much larger than within individual variation, we can consider that measurement error is low enough to interpret interindividual variation.

```
>s2within<-mswithin<-mod[[1]][2,3]
>mod[[1]][2, 3]
[1] 1.168932
>MSamong<-mod[[1]][1, 3]
>s2among<-(MSamong-MSwithin)/2
>s2within/(s2within+s2among)*100
[1] 7.833057
```

The percent measurement error is 7.8% in this example.

If you reiterate this example a large number of times, the averaged measurement should be close to $1/9$. This corresponds to the square of the division of the interindividual variance by the error variance that we introduce in our simulated data ($1/3$).

Problems

2.1. Extracting cartesian coordinates and initializing the cartesian system

Write a function for extracting cartesian coordinates on an image using the `locator` function. This function must set the cartesian system to a 0 origin with x and y-axes that are initialized by clicking on the picture. The three first clicks correspond to the position of the origin, the direction of the x-axis, and the orientation of the y-axis.

2.2. Calculating the surface of a digitized polygon
Use the splancs package and the `area.pl` function to calculate the surface of a polygon drawn with the `locator` function.

2.3. Changing the class of R objects containing a collection of configurations
Write a function to change the organization of a collection of configuration from an `array` to a `data.frame` object. Write the inverse function (transforming the `data.frame` object into an `array` object).

2.4. Manually acquiring outline coordinates
Write a function that draws and acquires the coordinates of equally spaced points on an outline that is digitized as the succession of small segments digitized by clicking on the image. In this respect, you have to gather not only their coordinates, but also compute the distance of each segments; read the help file of the `locator` and `seq` functions for achieving this aim.

2.5. A magic tool for selecting an area on a picture
Write a function that finds indices of pixels that have values close to the pixel selected by a left click (with the `locator` function).

2.6. Digitizing open curves
Adapt the `Conte` function (defined in Section 2.2.6) for digitizing open curves. The coordinates of the starting and of ending points are entered as arguments with the image file, or by clicking on the picture.

2.7. Writing the `read.nts` function
Use the code of Section 2.4 to write a `read.nts` function that opens morphometric data stored with the *.nts extension.

2.8. Writing the `export.nts` function
Use the code of Section 2.4 to write a `export.nts` function that saves morphometric data acquired with R.

2.9. Writing a comprehensive function for estimating coordinates of missing landmarks of symmetric structures
Use the code of Section 2.5.2 to write a function that estimates the coordinates of missing landmarks on either one or the other side for both 2D or 3D configurations. The number of dimensions must be first estimated to apply adapted functions.

3

Traditional Statistics for Morphometrics

Chapter coauthored by: **Julien Claude, Michel Baylac, Tristan Stayton**

Traditional morphometrics deals with linear measurements of objects. Although morphometrics and statistics saw considerable theoretical and technical developments at the end of the 19th Century and in the early 20th Century, they only became united more recently. In the second half of the last century, multivariate statistics became standard for analyzing large sets of measurements from different samples. The development of statistics, especially in the field of biology, for integrating growth or developmental features aided the development of the science of shape and size. Statistical developments concerning the concepts of shape, size, and growth emerged in a series of seminal papers aimed at providing a rigorous methodology for the study of size and shape (e.g., [50, 49, 16, 75]); these were initiated earlier by statisticians like Galton, Teissier, and Fisher. The analysis of complex shapes has benefited from the development of computers in the 1960s when publications of applied morphometrics increased exponentially in number. Although more modern tools are available today, research is still necessary in the theory and methods of understanding the relationships between the raw set of measurements and both concepts of size and shape variations.

3.1 Univariate Analyses

Univariate statistics and univariate plots can be very easily produced with little knowledge to analyze a simple distance measurement (i.e., the size of an object). Some of the most useful functions are summarized in Table 1.1. More specific functions are presented in Table 3.1. These functions usually apply to objects of the `vector` class containing numeric elements but most of them are generic and accept other specific arguments. Here we supply only the basics, and the user can refer to references [25] or [120] for more details.

Table 3.1. Useful functions

Function	Package	Short Description
var	stats	computes the unbiased estimator of variance
sd	stats	unbiased estimator of the standard deviation
median	stats	computes the median
quantile	stats	produces sample quantile
summary	base	returns summarized univariate statistics
scale	base	normalizes a vector or a matrix
range	base	returns the minimum and maximum values
sum	base	returns the sum of elements
diff	base	returns a vector of iterated differences
cumsum	base	returns a vector of cumulative sums
cumprod	base	returns a vector of cumulative products

3.1.1 Visualizing and Testing the Distribution

The distribution of a variable can be visualized with a histogram using the `hist` function. The `freq` argument of this function expects a logical. If `freq=T`, the resulting graph is a representation of frequencies; if `freq=F`, the graph represents a probability density (see Fig. 3.1). This function allows a basic exploration of the distribution. For example, it can help for determining the number of modal values in the distribution. For this task, using the `breaks` argument allows the number of bins of the histogram to be specified. One can calculate a density line from the data using the `density` function. Here is an example:

```
>layout(matrix(c(1,2),1,2))
```

Compute a random normally distributed variable.

```
>variable<-rnorm(1000, 2, 1)
```

Draw a histogram of frequencies.

```
>ju<-hist(variable, freq=T, main="Histogram of frequencies",
+    xlab="value")
```

Obtain the width of the histogram cells.

```
>wid<-ju$breaks[2]-ju$breaks[1]
>lines(density(variable, weights = rep(wid, 1000)), lw=2)
```

Draw a histogram of probability densities.

```
>hist(variable, freq=F, main="Histogram of
+    density probabilities", xlab="value")
>lines(density(variable), lw=2, lty=3)
```

There are several possible conformity tests on univariate distributions. For instance, the Shapiro-Wilk test for normality is run with the `shapiro.test` function on the numeric vector of interest. The `ks.test` function performs one or two-sample Kolmogorov-Smirnov tests. The Kolmogorv-Smirnov test is recommended

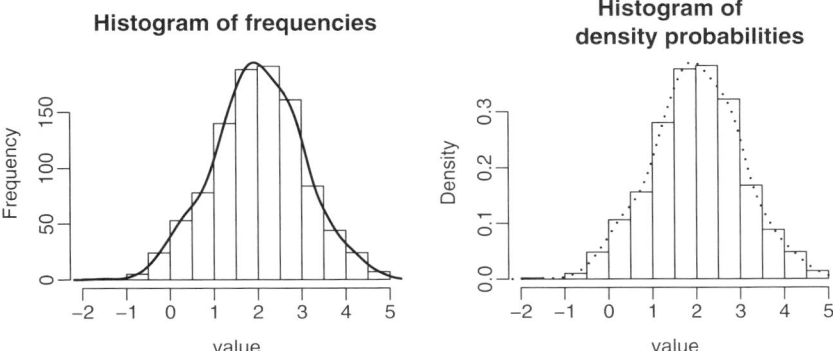

Fig. 3.1. Producing frequency and density histograms with the `hist` and `density` functions

when you want to determine whether two samples are drawn from the same distribution, or whether a sample is drawn from a given distribution. You can run the examples of these functions to further understand how to pass their arguments.

The observations (measured values) can belong to different categories. We will simulate a dataset that could correspond to mouse tail lengths of three different species. The data are summarized in a table of two columns: one containing values of the measured tails, the other being a category specifying a species (in R, it is a `factor` object). To obtain the number of observations within each category, we can use the `table` function ; in addition, we can obtain summarized statistics of the `data.frame` object with the `summary` function as exemplified below.

```
>mussp1<-rnorm(24, 7, 0.5)
>mussp2<-rnorm(24, 7.7, 0.5)
>mussp3<-rnorm(36, 8, 1)
>sp<-c(rep("sp1",24),rep("sp2",24),rep("sp3",36))
>tailsize<-c(mussp1, mussp2, mussp3)
>mus<-data.frame(tailsize, sp)
>table(mus$sp)
 sp1 sp2 sp3
 24  24  36
> summary(mus)
    tailsize        sp
 Min.   :4.986   sp1:24
 1st Qu.:7.154   sp2:24
 Median :7.557   sp3:36
 Mean   :7.657
 3rd Qu.:8.165
 Max.   :9.920
```

3.1.2 When Data are Organized in Several Groups

Graphics

The `barplot` and `piechart` functions produce barplots and pie charts that represent graphically the numbers of observations or the relative frequency for each group. There are many possibilities for filling cells or slice portions with colors or with equally spaced lines as in the code below (Fig. 3.2).

```
>layout(matrix(c(1,2),1,2))
>pie(table(mus$sp),density=c(10,20,50),
+    angle=c(0,90,45),col="black",main="pie chart")
>barplot(table(mus$sp),density=c(10,20,50),
+    angle=c(0,90,45),col="black",main="bar plot")
```

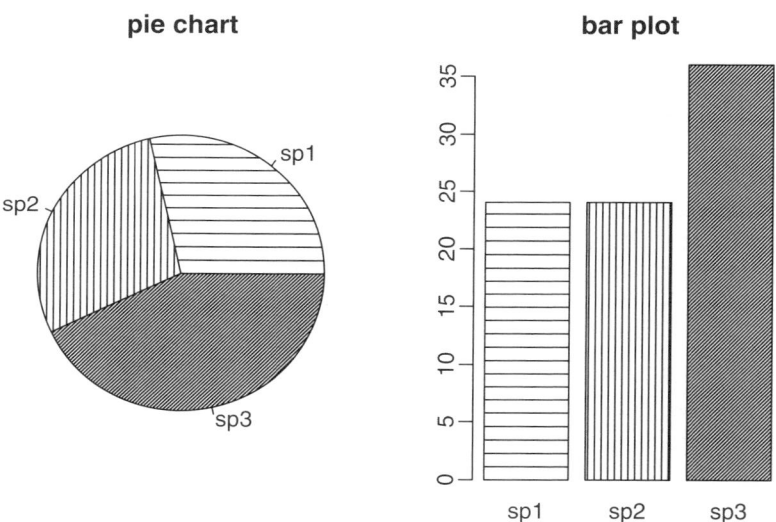

Fig. 3.2. Pie chart and bar plot for categorized observations

One can extract values for each category using either logical indexing, or the `subset` function. The `apply`, `tapply`, and `lapply` functions iterate the same function according to the data frame organization or to a factor.

Summarized statistics for each group can be represented using box plots. You can produce them easily with the `boxplot` function; it will draw the median and the interquartile range and return an object of the `boxplot` class. Producing plots with mean and standard deviation and/or standard error is elementary with R as well (Fig. 3.3). An example is provided below. The notch option, activated with the default arguments, allows rapid checking of whether there are significant differences between medians [19].

```
>layout(matrix(c(1,2),1,2))
```

Box-plot with median, interquartile, and full range.

```
>ju<-boxplot(tailsize~sp, range=0,data=mus, notch=T,
+    main="median, interquartile, range")
```

Box-plot with mean, standard deviation, and 95% and 99% confidence intervals. For producing the graph, we changed the values of some elements in the object returned by the boxplot *function.*

```
>musmn<-tapply(mus$tailsize, mus$sp, mean)
>mussd<-tapply(mus$tailsize, mus$sp, sd)
>musse<-mussd/sqrt(as.numeric(table(mus$sp)))
>ju$stats[3,]<-musmn
>ju$stats[c(2,4),]<-rbind(musmn-qnorm(0.995)*musse,
+    musmn+qnorm(0.995)*musse)
>ju$stats[c(1,5),]<-rbind(musmn-mussd, musmn+mussd)
>ju$conf<-rbind(musmn-qnorm(0.975)*musse, musmn+
+      qnorm(0.975)*musse)
>bxp(ju, notch=T, main="mean, 95%, 99% intervals, sd")
```

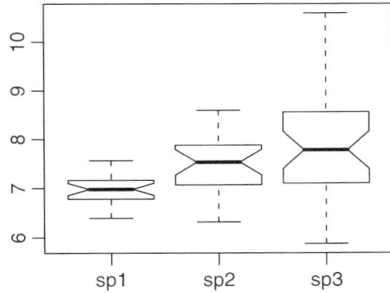 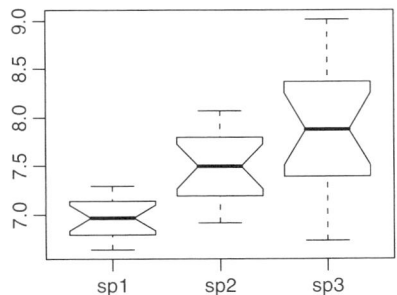

Fig. 3.3. Box plots for categorized observations

Looking at the distribution of the individual measurements belonging to several categories usually raises the following questions: are sample variances and sample means equal, greater, or smaller relative to each other? These questions are usually addressed with analysis of variance and related F-tests. For samples with more than 30 observations, the standard error estimate of the mean se_M is calculated as

$$se_M = \sqrt{\frac{\sigma}{n-1}} \, ,$$

with σ being the standard deviation, and n the number of observations.

The confidence interval for the mean corresponds to $[\mu - t \times se_M, \mu + t \times se_M]$, with μ being the sample mean, σ being the standard deviation of the sample, and t corresponds to the upper critical value of the student-t distribution with $n-1$ degrees of freedom. For sample size above 30, t can be estimated by the upper critical value of the normal distribution. Here is an example in which we compute the confidence interval for petal length of one *Iris* species that is stored in the R `iris` dataset.

```
>data(iris)
>setosa<-iris[iris$Species=="setosa",]$Petal.Length
>mset<-mean(setosa); sset<-sd(setosa); lset<-length(setosa)
> cat(paste("95% Conf. limits: [",
+     round(mset-qt(1-(0.05/2),lset-1)*sset/(lset-1),
+     3),",",round(mset+qt(1-(0.05/2),lset-1)* sset/(lset
+     -1),3),"]\n",sep=""))
95% Conf. limits: [1.455,1.469]
```

Note the use of the `cat` *function and of* `\n` *to return the result and to skip a new line.*

The confidence interval for the sample variance is based on the Chi-square distribution. Indeed, $\chi^2 = (\frac{}{n} - 1)s^2/\sigma^2$ follows a χ^2 distribution with $n-1$ degrees of freedom. Therefore, if α is the critical value, the confidence interval is calculated as $P(\chi^2 \leq \chi_1) = \alpha/2; P(\chi^2 \geq \chi_2) = \alpha/2$. The variance is contained in the interval

$$I_\alpha = [(n-1)s^2/\chi_2, (n-1)s^2/\chi_1]$$

where χ_1 and χ_2 corresponds to the lower and upper tail critical values of the χ^2 distribution with $n-1$ degrees of freedom. This interval of confidence is computed below for the petal length of the species *Iris setosa*.

Lower bound at 95% for the variance of the petal length in I. setosa.

```
>var(setosa)*49/qchisq(0.975, 49)
[1] 0.02104456
```

Upper bound at 95% for the variance of the petal length in I. setosa.

```
>var(setosa)*49/qchisq(0.025, 49)
[1] 0.04683264
```

For other parameters like the median, there is no known distribution for estimating confidence intervals. Instead, it is possible to obtain a bootstrap uncertainty estimate. To estimate the null distribution of the statistics (here the variance), data are resampled several times (usually 500 or 1,000 times), and the statistic is recalculated. The borders of lower and upper tails specify the endpoints of the confidence interval. The `sample` function, allowing for resampling of the data, permits us this estimation. Here is an example where we obtain a 95% median confidence interval by bootstraping the data 1,000 times.

```
>bootstrap<-numeric(1000)
>for (i in 1:1000)
+    {bootstrap[i]<-median(sample(setosa, replace=T))}
>sort(bootstrap)[c(25, 975)]
```

Note the `numeric(1000)` code for specifying an initial zero vector of length 1,000.

From the computation of simple confidence intervals, we can now develop tests to discover whether two samples differ in their statistical parameters. The means M_1 and M_2 of two different samples with variances s_1^2 and s_2^2 and of size n_1 and n_2 are different if the value

$$t = \frac{M_1 - M_2}{\sqrt{\frac{s_1^2}{n_1} + \frac{s_2^2}{n_2}}}$$

is below or above the critical value of the Student distribution (sample size ≤ 30) with $n_1 + n_2 - 2$ degrees of freedom. This statistic follows the normal distribution of parameters (0,1) for sample size > 30. The mean of samples can be compared with the t-test using the `t.test` function of the **stats** package. In the arguments, the variances have to be assumed to be equal; if not, R will correct degrees of freedom using the so called "Welch" modification.

For comparing the variance of two samples, we use their ratio – the largest variance (or the expected one) being the numerator and the lowest being the denominator. This gives an F-value that can be compared to the F-distribution with $n_1 - 1$ and $n_2 - 1$ degrees of freedom.

Analysis of Variance

The basic idea behind analysis of variance (ANOVA) is to partition the observed variance according to an explanatory variable that is primarily a factor. There is a large number of statistical models behind ANOVA. The simplest one is testing whether there is a difference between the means of different samples. One uses ANOVA to determine whether two or more groups differ in their mean. When performing an ANOVA, the total sum of squares is decomposed into components related to the effects. The remaining sum of squares is usually called the error term, and is not explained by the model. In simple models, the sum of squares of effects add up with the error term equaling the total sum of squares, while variances (sum of squares divided by appropriated degrees of freedom) do not. The variances of the different effects are compared in the analysis of variance to estimate the significance of an effect. The variance of the effect (variation that is explained by the model) is usually compared with the error term to appraise the significance of the between-group differences. In the simplest model, we have j populations, and we want to test whether the means of these populations are equal. Let Y_{ij} be the ith observation (or individual measurement) within the jth population. Y_{ij} can be decomposed as

$$Y_{ij} = \mu + \alpha_j + e_{ij} \; ,$$

where μ is the grand mean, α_j is the deviation from the grand mean for the jth treatment (α is sometimes called the effect term), and e_{ij} is the residual variation, assumed to follow a normal distribution with mean zero.

Alternatively, Y_{ij} can be expressed as

$$Y_{ij} = \mu_j + e_{ij} \; ,$$

where μ_j corresponds to the mean of the jth population. From these simple models, we can produce an ANOVA table for analyzing variation due to effects. Assume we have a treatments and a total of N observations, then the table can be summarized as in Table 3.2.

Table 3.2. One-way ANOVA table

Var. Source	df	SS		MSS	F
effect	$a-1$	$SSA = \sum_{j=1}^{j=a} n_j(\mu_j - \mu)^2$		$\frac{SSA}{df A}$	$\frac{MSSA}{MSSE}$
error	$N-a$	$SSE = \sum_{j=1}^{j=a}\sum_{i=1}^{i=n_j} Y_{ij} - \mu_j$		$\frac{SSE}{df E}$	
total	$N-1$	$SST = \sum_{i=1}^{i=n_j}\sum_{j=1}^{j=a}(Y_i j - \mu)^2$		$\frac{SST}{df T}$	

The sum of squares of effect and the sum of squares of error correspond respectively to the between and within-population sum of squares. The F-test compares between and within variations. If variation between groups is greater than variation within groups, then the F-value is greater than 1, and differences in mean values of groups will be significantly greater than within groups. The following example illustrates how to perform an ANOVA-test with R:

```
>summary(aov(mus$tailsize~as.factor(mus$sp)))
                  Df Sum Sq Mean Sq F value    Pr(>F)
as.factor(mus$sp)  2 12.433   6.217  10.296 0.0001037 ***
Residuals         81 48.907   0.604
---
Signif. codes:  0 *** 0.001 ** 0.01 * 0.05 . 0.1   1
```

The ANOVA table summarizes the analysis: the first column corresponds to degrees of freedom of effects and the residual error term, the second gives the sum of squares of the effect, the third corresponds to the mean squares (sum of squares divided by the respective df, analogs to the variance), and then the forth gives the F-value (note that R divides mean squares of the effect with that of residuals when `summary.aov` is used), and the last the p-value. The same table would have been returned using `anova(lm(mus$tailsize~as.factor(mus$sp)))`.

One can further investigate the differences between categories with post-hoc tests such as the "Honestly Significantly Different" Tukey test. The `TukeyHSD` function performs the test and creates a set of confidence intervals for the differences between the means of the levels of a factor with the specified family-wise probability of coverage. The resulting object returned by this function can be plotted using the function `plot`.

One speaks about multi-way ANOVA when more than one effect is tested. For two-way (or more effects) ANOVA, one must distinguish between fixed, mixed, or random models that consider fixed, fixed and random, or random effects. This terminology is that of hierarchical linear modeling. For factors, we speak about fixed effects, when observations of the sample take all the possible modalities of the factor we investigate (sex or ecotype are likely to be fixed factors). There is no need of extra- or interpolation to determine the significance of fixed-effect models. We speak about random effect for a factor when just part of the modalities are available, or drawn from a sample to extrapolate for the whole population. A random effect is often the factor "individual," on which we apply several treatments, or that we measure at several intervals during his existence; indeed we rarely have all the individuals of one population. But, actually, if you want to test the differences between these individuals, "individual" becomes a fixed effect, because here we will not infer any hypothesis concerning the whole population of individuals. When at least one fixed and one random effect is included in the model, we speak about mixed effect models.

Table 3.3. Two-way ANOVA table: fixed effects (γ and ϵ denotes the interaction and the error term)

Source of Variation	df	SS	MSS	F
α	$a - 1$	$SSA = nb \sum_{i=1}^{i=a} (\mu_i - \mu)^2$	$\frac{SSA}{df\,A}$	$\frac{MSSA}{MSSE}$
β	$b - 1$	$SSB = na \sum_{j=1}^{j=b} (\mu_j - \mu)^2$	$\frac{SSB}{df\,B}$	$\frac{MSSB}{MSSE}$
γ	$(a-1)(b-1)$	$SSAB =$	$\frac{SSAB}{df\,AB}$	$\frac{MSSAB}{MSSE}$
		$n \sum_{i=1}^{i=a} \sum_{j=1}^{j=b} (\mu_{ij} - \mu_i - \mu_j + \mu)^2$		
ϵ	$ab(n-1)$	$SSE = \sum_{i=1}^{i=a} \sum_{j=1}^{j=b} \sum_{k=1}^{k=n} Y_{ijk} - \mu_{ij}$	$\frac{SSE}{df\,E}$	
total	$N - 1 = nab - 1$	$SST = \sum_{i=1}^{i=a} \sum_{j=1}^{j=b} \sum_{k=1}^{k=n} (Y_{ijk} - \mu)^2$	$\frac{SST}{df\,T}$	

The strategy for testing models with several factors is the same as for the one-way ANOVA except that we have to consider more effects and their type. The distinction between random-, fixed-, and mixed-effect models is not just a question of vocabulary because they differ in what is included in the error variance. We will illustrate the two-way case. Rather than determining the significance of a single effect on the distribution of data, several effects and their interaction are tested. In practice, pure random effect models are rare, and morphometric statistical analysis is restricted to mixed and fixed models.

We start with fixed models and call α the first effect with i modalities ranging from 1 to a and β, the second effect with j modalities ranging from 1 to b; γ is the interaction between both effects. The total number of observations is N, and there is n observations for each combination of factors. Models become more complicated when group sizes differ between groups are defined by combination of effects. An observation Y_{ijk} can be expressed as

$$Y_{ijk} = \mu + \alpha_i + \beta_j + \gamma_{ij} + e_{ijk} \,.$$

We summarized the two-way ANOVA table for fixed factors in Table 3.3.

When entering several effects, we use an argument of the `formula` class using the $+$, $*$, or/operators between effects. The second operator takes into account the interaction between terms, and the third ensures that one factor (left of the slash) is nested within a second factor (on the right).

Table 3.4. Operators in formulae

Operator	Meaning	Usage	Equivalent
+	add effect B	A + B	
:	interaction of A and B	A:B	
*	crossed effects	A * B	A + B + A:B
−	remove effect B	A*B-B	A + A:B
/	B nested effect in A	A/B	A + A:B

Here we investigate whether the distribution of tooth size is influenced by both a genetic parameter (population) and sex in the dataset for the species *Mus domesticus*. Populations differ in their chromosomal formula and the modalities of the factor correspond to chromosome number. One population has only individuals with 40 chromosomes while the other population is composed by individuals with 22 chromosomes.[1]

```
> musdom1<-read.table("musdom1.txt", header=T)
> LM1<-musdom[,4]
> Pop<-as.factor(musdom[,3])
> Sex<-as.factor(musdom[,2])
> summary(aov(LM1~Pop*Sex))
             Df    Sum Sq   Mean Sq F value      Pr(>F)
Pop           1 0.049562 0.049562 22.8378 1.180e-05 ***
Sex           1 0.007722 0.007722  3.5583   0.06409 .
Pop:Sex       1 0.000129 0.000129  0.0596   0.80793
Residuals    60 0.130210 0.002170
---
Signif. codes:  0 *** 0.001 ** 0.01 * 0.05 . 0.1   1
```

The model.tables *function computes a summary table with mean value for each category.*

```
> model.tables(aov(LM1~Pop*Sex), type="means")
Tables of means
Grand mean

1.445609

 Pop
Pop
     22      40
1.4734 1.4178
```

---

[1] The dataset is available in the online supplement.

```
 Sex
Sex
     f       m
1.4346 1.4566

 Pop:Sex
    Sex
Pop  f       m
  22 1.4639 1.4830
  40 1.4054 1.4302
```

Females have smaller anterior molars than males in both populations; however, this difference is not significant. The population with 40 chromosomes has smaller anterior molars than does the population with 22 chromosomes.

In the case of mixed models, the situation becomes different especially if the interaction term becomes significant. The simple additive model is used loosely in this situation, since the interaction contains a random factor term. Therefore, the fixed effect is tested against the interaction, while the random effect is tested against the error. For purely random models, the effects are usually tested against the interaction term, especially if this is important. However, whether testing the factors against the interaction or the residual error should not be considered as a rule: this depends on the hypothesis that the observer wants to test. It is especially true for studies dealing with fluctuating asymmetry (random difference between left and right sides of genotypes), when this is considered as the interaction between individual (genotype) and side. We rarely have all the individuals of a given population; the effect individual is then regarded as the random factor, while side is regarded as the fixed factor. Palmer and Strobeck [78, 79] explain how to appraise the significance of directional asymmetry and fluctuating asymmetry, and interindividual differences using a mixed model. However, in this model (when the interaction is found significant: i.e., fluctuating asymmetry), not only the side effect (directional asymmetry), but also the genotype effect (symmetric individual variation) should be tested against the interaction. Indeed, when we test for individual differences, we want to know whether they are higher than intra-individual differences. It is likely that intra-individual variation will be greater in terms of fluctuating asymmetry rather than in terms of measurement error introduced by the user.

The example below uses the length of the first upper molar of *Mus musculus domesticus*. The table consists of five columns, the first indicating the individual code number, the second the sex, the third the side, the fourth the replication session number, and the last the measurement. We have measured the teeth twice to allow for the estimating of measurement error.[2]

```
>musdom<-read.table("musdom.txt",header =T)
>LM1<-musdom[,5]
>side<-as.factor(musdom[,3])
>ind<-as.factor(musdom[,1])
```

---

[2] The dataset is available in the online supplement.

```
>musaov<-summary(aov(LM1~ind*side))
            Df   Sum Sq  Mean Sq  F value     Pr(>F)
ind         23 0.294082 0.012786 243.5460 < 2.2e-16 ***
side         1 0.000301 0.000301   5.7341   0.02059 *
ind:side    23 0.004868 0.000212   4.0318 2.296e-05 ***
Residuals   48 0.002520 0.000053
```

Note that the test does not consider interaction as the error term.
F-test for individual variation:

```
>FIND<-musaov[[1]][1,3]/musaov[[1]][3,3]
>FIND
[1] 60.40554
>pf(FIND,musaov[[1]][1,1],musaov[[1]][3,1],lower.tail=F)
[1] 1.606294e-15
```

F-test for directional asymmetry:

```
>FSIDE
[1] 1.422208
>pf(FSIDE,musaov[[1]][2,1],musaov[[1]][3,1],lower.tail=F)
[1] 0.2451990
```

In this example, it is evident that the choice of the error term becomes the determinant for conclusions concerning the significance of directional asymmetry. Actually, not considering interaction as the error term in this design has consequences, because we could conclude from the test that there is a significant directional asymmetry in the studied sample although there is not.

If the combination of factors defines groups of equal size, there is no problem with interpreting the table of results. Things becomes more difficult if effects are unbalanced. In this latter case, results will depend on the order of entry of factors. In other words, orthogonality between effects is missing and the principle of marginality precludes results to be easily interpreted in one single analysis. The car package offers some possibilities for exploring these kinds of design. Remember that increasing the complexity of the ANOVA design usually yields results more difficult to interpret.

3.2 Bivariate Analyses

3.2.1 Graphics

One examines the relationships between two distance measurements with simple bivariate plots. The plot function accepts either the vectors x and y or the first two columns of a matrix as arguments for abscissa and ordinate coordinates of observations in the scatterplot. One can assign different symbols, colors, and size to the points of the plot by entering specific arguments. This is particularly useful for highlighting distributions according to one or more categories (for example, sex and species). Plot is a generic function, it depends on the class of the argument. For example, with an object of the `groupedData` class returned by functions of the nlme package, you will not produce a bivariate but a treilli plot.

The `pairs` function is useful for displaying all possible bivariate relationships between variables of a data frame.

```
>library(MASS)
>data(crabs)
>plot(crabs$FL,crabs$RW,bg=c("grey50", "white")[crabs$sex]
+        ,pch=c(21,22)[crabs$sp],cex=c(1.8,1)[crabs$sp])
```

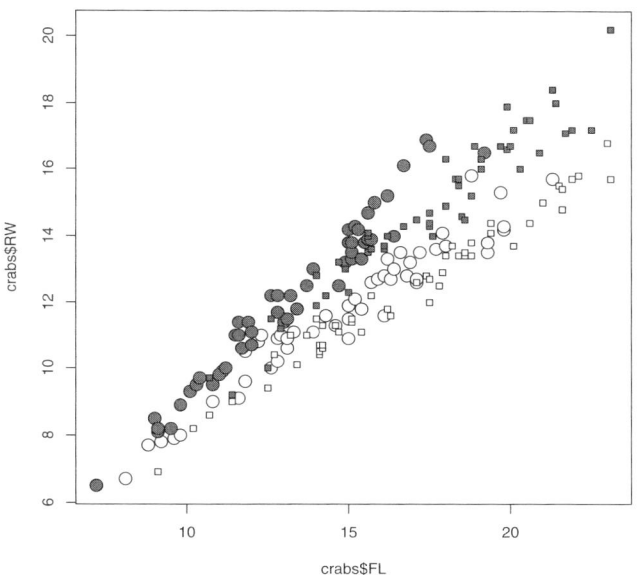

Fig. 3.4. Scatterplot of crab measurements of the genus *Leptograpsus* according to the species and sex: 'FL' frontal lobe size (mm) and 'RW' rear width (mm); males are gray, females are white symbols, squares and circles correspond respectively to each species; data from Campbell [17]

3.2.2 Analyzing the Relationship Between two Distance Measurements

Regression

The relationship between one dependent and one independent variable is usually studied by regression analysis using a simple linear model. Regression of one normally distributed variable on one other is achieved using a linear model with the `lm` function, which accepts a formula as its first argument. This function returns an object of the `lm` class and prints some statistics. The result summary (produced with the `summary` function) provides more properties concerning the parameters and for the fit (slope, intercept, correlation, and goodness-of-fit).

```
>summary(lm(crabs$RW~crabs$FL))

Call:
lm(formula = crabs$RW ~ crabs$FL)

Residuals:
     Min        1Q     Median        3Q       Max
-2.05794 -0.88826 -0.08583   0.85724   2.94821

Coefficients:
              Estimate Std. Error t value Pr(>|t|)
(Intercept)    2.33302    0.35187    6.63 3.11e-10 ***
crabs$FL       0.66775    0.02204   30.30  < 2e-16 ***
---
Signif. codes:  0 *** 0.001 ** 0.01 * 0.05 . 0.1   1

Residual standard error: 1.087 on 198 degrees of freedom
Multiple R-Squared: 0.8226,      Adjusted R-squared: 0.8217
F-statistic: 918.3 on 1 and 198 DF,  p-value: < 2.2e-16
```

The formula entered as a unique argument determines effects that are estimated by the model. If an operation, a command, or a function is applied on a predictor of the model, it has to be specified with the `I` function. One can use the `anova` function to perform an ANOVA of predictors in the model. When `anova` receives two or more formulae, it tests the models against one another in the specified order. The interaction between terms, numeric variables, and the intercept can be implicitly removed in the formula using $0+$ or -1 in the formula.

Plotting an object of the `lm` class with the function `plot` provides different displays to examine outliers or residuals. This strategy provides a diagnostic tool to examine whether the conditions are filled for using the model. The `abline` function can plot the regression line if one specifies the slope and intercepts in its arguments. A variety of diagnostic tests for simple linear models exists in the lmtest package. Examining fitted values and residuals is indeed important for determining the validity of the model that explains the data. The residuals of a given model are returned using the `residuals` function applied to the `lm` object; and the fitted values using the `predict` function. Furthermore, it is also possible to fit nonlinear models in R with the `nls` function, and to include mixed random and non-random effects using various linear and nonlinear modeling functions (`lme`, `nlme` of the nlme package). The latter could be used in morphometrics in the case of longitudinal data (e.g., measurements that are taken during the growth of a given animal). Instead of linear models with fixed effects, you can alternatively use mixed-effect linear models with the `lme` or `lmer` functions of the nlme and lme4 packages. The models implemented in these functions are becoming common and very useful. One can also apply nonlinear mixed models on the data with `nlme` (nlme: nonlinear mixed modeling). These models can be appropriate for analyzing grouped data and have not yet received enough attention, although they permit interesting applications in traditional and geometric morphometrics (especially for the analysis of growth patterns and longitudinal data). Relative theory and practice are beyond the scope of this book, but you can find additional information in references [119, 85, 121].

In our simple linear model, the R^2 value describes the percent of variation in one variable explained by the other variable. The syntax `anova(lm(my_model))` returns the estimate of variance components. One can obtain the R^2 from the variance of the fitted and residuals values returned by the linear model.

```
>model<-(lm(crabs$RW~crabs$FL))
```

Compute the R^2 with fitted and residual values.

```
>fv<-model$fitted.values; resv<-model$residuals
>sum((fv-mean(fv))^2)/(sum((fv-mean(fv))^2)+sum(resv^2))
[1] 0.8226265
```

Note that `predict` *can return fitted values as well.*
Compute the R^2 coefficient by performing an analysis of variance on the model.

```
>anova(model)
Analysis of Variance Table

Response: crabs$RW
            Df   Sum Sq Mean Sq F value      Pr(>F)
crabs$FL     1 1084.05 1084.05  918.29 < 2.2e-16 ***
Residuals  198  233.74    1.18
---
Signif. codes:  0 *** 0.001 ** 0.01 * 0.05 . 0.1   1

>anova(model)[1,2]/(anova(model)[1,2]+anova(model)[2,2])
[1] 0.8226265
```

Correlation

One can calculate the R^2 coefficient using the covariance and the variances. In the case of regression, the R coefficient corresponds to a measure of association between both variables.

The covariance measures how two random variables vary together. It is calculated as follows

$$\sigma_{xy} = \mathrm{cov}(x, y) = \frac{1}{n} \sum_{i=1}^{n} (x_i - \bar{x})(y_i - \bar{y}) \,,$$

where $\bar{x}$ is the mean of x. The unit of covariance is the product of the units of x and y measurements.

The variance and covariance of two or more variables are computed with the `var` function of R. If the variables vary in the same way, the covariance is positive. It is negative when one variable increases, and the other decreases; and it is 0 when variables vary independently.

The $R(x, y)$ Pearson correlation coefficient is a linear, dimensionless measure of dependence between variables. One computes this coefficient in "normalizing" the covariance by the product of variances of each variable such that

$$R(x, y) = \frac{\sum (x_i - \overline{x})(y_i - \overline{y})}{\sqrt{\sum (x_i - \overline{x})^2}\sqrt{\sum (y_i - \overline{y})^2}} \ .$$

By default, the `cor` function returns this correlation coefficient, which indicates the strength and direction of the linear relationship between two variables. Unlike the covariance, this measure is independent of the variance of each variable. In specifying the method in the arguments, R alternatively computes the Spearman or Kendall correlation coefficients. One can estimate the significance of the correlation coefficient using a test of association between paired samples, this is returned by the `cor.test` function.

One can interpret the square of the sample correlation (R^2) as the fraction of the variance in one variable explained by a linear fit of the variable with the other. The R^2 coefficient is a measure of association between the variables but is independent on the direction of the association. Here we check this using the `var` and `cor` functions.

```
>ju<-lm(crabs$RW~crabs$FL)
> (var(crabs$RW)-var(ju$residuals)) / var(crabs$RW)
[1] 0.8226265
>cor(crabs$RW, crabs$FL)^2
[1] 0.8226265
```

3.2.3 Analyzing the Relationship Between Two Distance Measurements in Different Groups

We are interested in determining whether or not groups significantly overlap in their distance measurements. We can draw a 95% interval with the function `ellipse` of the car package. The Rmorph package contains an `ellipse` function that calculates and draws confidence ellipses too. We can even program our own function: `ELLI`". It has to consider how many percents of the distribution we want to include in the interval, the data (typically x and y-coordinates), and the number of points used to plot the ellipse. We will fix some default arguments of `ELLI`: the default confidence interval is 95% and the default number of points sampled on the plotted ellipse is 50.

Function 3.1. ELLI

Arguments:

　　x: *First variable (a numeric vector).*
　　y: *Second variable (a numeric vector).*
　　conf: *Confidence level in %.*
　　np: *Number of sampled points on the ellipse.*
Value:
　　Coordinates of points sampled on the ellipse.

```
ELLI<-function(x,y,conf=0.95,np=50)
  {centroid<-apply(cbind(x,y),2,mean)
  ang <- seq(0,2*pi,length=np)
  z<-cbind(cos(ang),sin(ang))
  radiuscoef<-qnorm((1-conf)/2, lower.tail=F)
  vcvxy<-var(cbind(x,y))
  r<-cor(x,y)
  M1<-matrix(c(1,1,-1,1),2,2)
  M2<-matrix(c(var(x), var(y)),2,2)
  M3<-matrix(c(1+r, 1-r),2,2, byrow=T)
  ellpar<-M1*sqrt(M2*M3/2)
  t(centroid + radiuscoef * ellpar %*% t(z))}
```

We plot confidence ellipses considering the factors sex and species in the crabs morphological dataset.

```
>coul<-rep(c("grey40","black"),2)
>lwe<-c(2,2,1,1)
>plot(crabs$FL,crabs$RW,bg=c("grey50","white")[crabs$sex],
+          pch=c(21,22)[crabs$sp],cex=c(1.8,1)[crabs$sp])
>for (i in 1:4){a<-levels(crabs$sp:crabs$sex)[i]
+ lines(ELLI(crabs$FL[crabs$sp:crabs$sex==a],
+ crabs$RW[crabs$sp:crabs$sex==a]),col=coul[i],lwd=lwe[i])}
```

In Fig. 3.5, the species factor seems to be related to the overall size of the crabs since both measurements seem slightly smaller in one species.

　　We see several features that are usually associated with morphometric measurements in Fig. 3.5.

- Measurements are strongly related, and this is likely to be because of size variation; we have small and large individuals for each sample.
- Small individuals are more similar than large individuals. Differentiation according to sex and species seems to occur because of growth.
- Some differences seem to be more related to the relationships between variables rather than in the dispersion of the observations. Relationships between variables are the expression of shape difference since they correspond to differences in proportion or in proportional change during growth.

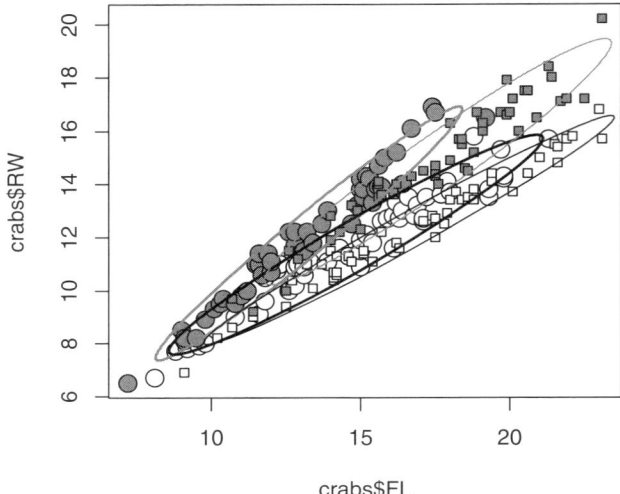

Fig. 3.5. 95% confidence ellipses for the crab measurements of the genus *Leptograpsus* according to species and sex, with the same abbreviations as for Fig. 3.4. The different colors of ellipses are for sexes, while the different line widths are for species

- If one estimates regression parameters using a linear model, one usually finds that the intercept is significantly different from zero; this is because the relationships between measurement variables are not constant or not the simple expression of a linear expression during growth.

Continuing with the `crabs` dataset, we could be tempted in this case to use a more complex model and mix categories (sex and/or species) and the relationship between measurements to examine shape and growth difference between species. This could be resolved by analysis of covariance (ANCOVA), and could be easily computed with the functions `anova` and and `lm`. Although categories are balanced in this sample, the relationship between variables seems to be different between groups, and in this latter case, we may not have equality between slopes of the different populations. The covariance model in this case is not appropriate since it violates marginality principles [77]. Instead of this, it is possible to compare equality between regression lines.

The `summary.lm` function returns slope values and standard errors. We can therefore apply a Student's *t*-test for slope difference. This *t*-value is computed as the difference between the two slopes divided by the square root of the sum of the squared standard errors and can be expressed as

$$t = \frac{\beta_1 - \beta_2}{\sqrt{se_{\beta_1}^2 + se_{\beta_2}^2}} \; .$$

The test follows a t-distribution and has $n_1 + n_2 - 4$ degrees of freedom.

We will compare, for example, the regression slopes between sexes of the first species of the `crabs` dataset.

Regression parameters and statistics for the first group.

```
>a<-summary(lm(crabs$RW[1:50]~crabs$FL[1:50]))$coefficients
>a
                 Estimate  Std. Error    t value        Pr(>|t|)
(Intercept)     2.2490836  0.36496360   6.162488  1.415187e-07
crabs$FL[1:50]  0.6379812  0.02404744  26.530109  2.485878e-30
```

Regression parameters and statistics for the second group.

```
>b<-summary(lm(crabs$RW[51:100]~
+    crabs$FL[51:100]))$coefficients
>b
                  Estimate  Std. Error  t value     Pr(>|t|)
(Intercept)       0.071914  0.3623808   0.1984  8.4353e-01
crabs$FL[51:100]  0.909275  0.0267982  33.9305  3.3365e-35
```

Computation of the t-value.

```
>tt<-(a[2,1]-b[2,1])/sqrt(a[2,2]^2+b[2,2]^2); tt
[1] -7.534725
```

Computation of the p-value in the case of a bilateral test.

```
>(1-pt(abs(-tt),(length(crabs$FL[1:100])-4)))*2
[1] 2.706035e-11
```

There is, however, a more sophisticated and elegant way to test differences between slopes and regression functions. Note that one can compare models giving different linear models as arguments of the `anova` function. When several models are entered as arguments in the `anova` function, the latter performs the general linear test. If, instead of variances, models return likelihood or deviance, `anova` performs an analysis of deviance between models. The `anova` function is thus interesting to compare full and reduced models and to estimate whether the parameters added to the reduced model significantly improve the explained part of variance. The general linear test corresponds to an F-ratio between sums of squares of error-variance of the full and the reduced model:

$$F = \frac{\text{SS}_{\text{error}} - \text{SS}_{\text{effect}}}{df_{\text{error}} - df_{\text{effect}}} \Big/ \frac{\text{SS}_{\text{effect}}}{df_{\text{effect}}} \; .$$

This value follows an F-distribution with $(df_{\text{error}} - df_{\text{effect}})$, df_{effect} degrees of freedom. In our example, sex and species are categorical factors; we can add them in the model, and compare with simpler models to test their influence on morphology. When several predictor variables are introduced in the model, we say that we perform a multiple regression. We can remove predictor one by one from multiple to simple regression models, and compare successive models to test the effect of predictor variable on the variance. We can use a similar strategy to estimate whether sex or species interact with regression parameters. If the interaction between the covariate and the factor is negligible, it means that the slopes are similar between groups. In other words, testing whether this interaction is negligible is similar to comparing models with and without interaction. We will consider only the first species of `crabs`, and determine whether sex influences morphology and interacts with growth.

```
>mod1<-lm(crabs$RW[1:100]~crabs$FL[1:100])
>mod2<-lm(crabs$RW[1:100]~crabs$FL[1:100]+
+    crabs$sex[1:100])
>fullmod<-lm(crabs$RW[1:100]~crabs$FL[1:100]*
+    crabs$sex[1:100])
>anova(fullmod,mod1)
Analysis of Variance Table

Model 1: crabs$RW[1:100]~crabs$FL[1:100] * crabs$sex[1:100]
Model 2: crabs$RW[1:100]~crabs$FL[1:100]
  Res.Df      RSS Df Sum of Sq      F    Pr(>F)
1     96   25.613
2     98   99.708 -2   -74.095 138.86 < 2.2e-16 ***
---
Signif. codes:  0 *** 0.001 ** 0.01 * 0.05 . 0.1   1
```

The regression functions differ between sexes.

```
>anova(fullmod,mod2)
Analysis of Variance Table

Model 1: crabs$RW[1:100]~crabs$FL[1:100] * crabs$sex[1:100]
Model 2: crabs$RW[1:100]~crabs$FL[1:100] + crabs$sex[1:100]
  Res.Df      RSS Df Sum of Sq      F    Pr(>F)
1     96   25.613
2     97   40.496 -1   -14.883 55.783 3.715e-11 ***
---
Signif. codes:  0 *** 0.001 ** 0.01 * 0.05 . 0.1   1
```

The slopes are different between both groups. Note that both tests yield very similar p-values.

Instead of confidence ellipses on scatterplots, one can estimate confidence interval bands depending on the model fitted to the data. The `ci.plot` function of the HH package plots them directly from the `lm` object. The `predict` function is a generic function for predictions based on the results of various model-fitting functions; it can estimate band interval as well. The `matplot` function is useful for quickly plotting all the lines of the model since it plots the columns of one matrix against those of another. Here we plot confidence intervals on both the prediction and regression slopes.

```
>x <-crabs$FL[1:50]
>new<-data.frame(x=seq(max(crabs$FL[1:50]),
+    min(crabs$FL[1:50]), length=50))
>pred.int<- predict(lm(crabs$RW[1:50] ~ x), new,
```

```
+      interval="confidence")
>pred.fit<- predict(lm(crabs$RW[1:50] ~ x), new,
+      interval="prediction")
>plot(crabs$FL[1:50], crabs$RW[1:50],
+      pch=21,cex=1.8, bg="grey50",asp=1)
>matplot(new$x,cbind(pred.int),lty=c(1,2,2),type="l",
+      col=1,add=T)
>matplot(new$x,cbind(pred.p),lwd=c(1,2,2),lty=c(1,2,2),
+      col=c("black",rep("grey50",2)),type="l",add=T)
```

The default value for confidence interval level in `predict.lm` is 95%. We use the argument "level" to modify this default value.

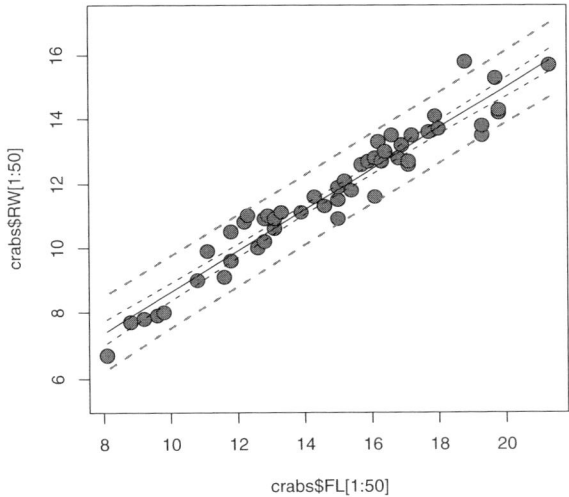

Fig. 3.6. Regression of the measurement 'RW' on 'FL' for the males of the blue species of the crab *Leptograpsus*. 95% confidence band interval on the slope of the regression is printed with black dashed lines, and confidence band interval on the prediction of the model is printed with gray dashed lines, with the same abbreviations as for Fig. 3.4

3.2.4 A Short Excursion to Generalized Linear Models

R fits linear and nonlinear models, but can also fit generalized linear models with the `glm` function and generalized estimating equations with functions of the gee package. GEEs are an extension of GLMs that are appropriate when the data do not consist of independent observations and/or when they consist of pseudoreplications. GLMs are useful for some special morphometric applications and for logistic regression, especially when one wants to explain a property by the morphology. This will not be developed in detail here, but GLMs and related models can sometimes deal with problems that cannot be solved with simple linear models. This is the case when the response or the residuals are not normally distributed or lack homoscedasticity. With GLMs, one can assume nonconstant variance for the response. For this, the variance must be a function of the predictor variables through the mean response. One important application of GLMs is in fitting logistic regression. In logistic regression, the response takes two values, while the predictor variables can be of different types.

    We can simulate an example where two possible diets are predicted from the mouth size of an hypothetical animal. The larger the mouth size, the higher the probability is that the animal is carnivorous, and the smaller the mouth size, the higher is the probability that the animal is herbivorous. We assume that this probability increases with mouth size. The problem is that this probability cannot exceed one, and cannot be below zero. For a very large size, the probability of being carnivorous will thus be nearly equal to one, and for a very small mouth size it will be nearly equal to zero. Assume that for a mouth size smaller than five cm, the probability of eating animal food is low and reaches 0.05, while the probability of predating only on animals for a mouth larger than 10 cm in size is high and reaches 0.95. Between both values the probability of being carnivorous varies linearly with mouth size. Code for simulating the data is provided below.

We first simulate mouth size of a sample with 40 individuals having mouths ranging from 2 to 25 cm.

```
>msize<-runif(40, 2, 25)
```

We associate the probability of their diet according with their mouth size.

```
>prob<-numeric(40)
>for (i in 1:40){
+    if (msize[i] < 5) {prob[i]<-0.05}
+    else{if (msize[i] > 10) {prob[i]<-0.95}
+    else {prob[i]<-((msize[i]-5)/5)*0.95}}}
```

We simulate diet according to their mouth size with the rbinom *random generator following a binomial law of parameters (events, and probability).*

```
>diet<-rbinom(40,1,prob)
>plot(msize, diet, xlab="Mouth size", ylab="Diet", axes=F)
>box(); axis(1)
>axis(2, at=c(0,1), labels=c("Herbivorous","Carnivorous"))
```

    Once data are simulated, we have to fit a model to the observed data. For this, we use the glm and predict.glm functions. The summary.glm function returns some statistics on the parameters.

```
> summary(glm(diet~msize, family="binomial"))

Call:
glm(formula = diet ~ msize, family = "binomial")

Deviance Residuals:
```

```
      Min          1Q      Median         3Q          Max
-1.52322    -0.14915    0.01390    0.11695    1.88194


Coefficients:
             Estimate Std. Error z value Pr(>|z|)
(Intercept)   -7.3910     2.9872  -2.474   0.0134 *
msize          0.8335     0.3313   2.516   0.0119 *
---
Signif. codes:  0 *** 0.001 ** 0.01 * 0.05 . 0.1   1

(Dispersion parameter for binomial family taken to be 1)

    Null deviance: 50.446  on 39  degrees of freedom
Residual deviance: 13.261  on 38  degrees of freedom
AIC: 17.261

Number of Fisher Scoring iterations: 8
```

Draw the logistic regression line.

```
>z<-seq(min(msize),max(msize),length=50)
>pred<-predict(glm(diet~msize, family="binomial"),
     data.frame(msize=z),type="response")
>lines(z,pred, type="l")
```

Here the `family` *argument specifies a binomial response. It also states that the probability* p
is linked linearly to the predictor variable by the link function $\log(p/(1-p))$.

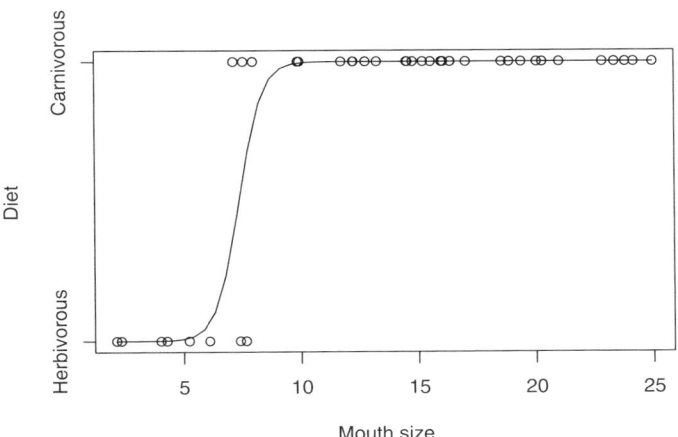

Fig. 3.7. Fitting a GLM to perform logistic regression

The link function is essential in GLMs and is the way to adapt the data so that they can be fitted by a linear and simpler model. Families other than the *logit* one exist for fitting GLMs (see the help file of the `glm` function). In addition to providing basic statistics about model parameters (slope and intercept values are significant in our example), `summary.glm` returns also two deviance values (null deviance and residual deviance) and an AIC-value (Akaike Information Criterion), which is not only a measure of goodness-of-fit of the statistical models, but also takes into account the number of parameters included in the model. The deviance is equal to -2 multiplied by the log of the likelihood of the model. The deviance is a variation of the log-likelihood from the perfect model (a perfect model would assign a 0 or 100% probability for each observation). The deviance of the perfect model is zero. The variation in log-likelihood between the perfect model and the null model is the null deviance. The variation in log-likelihood between the model and the perfect model is the residual variance. The smaller it is, the better the fit is. By contrast with the R^2 value, the AIC is penalized by the number of parameters, therefore it increases as we include more predictors. The AIC equals the residual deviance plus twice the number of parameters of the model. The `anova.glm` function allows the effect of the factors to be tested. The difference between the null deviance and the residual deviance follows a Chi-square distribution.

```
>glm1<-predict(glm(diet~msize, family="binomial")
>anova(glm1, test="Chisq")
Analysis of Deviance Table

Model: binomial, link: logit

Response: diet

Terms added sequentially (first to last)


        Df Deviance Resid. Df Resid. Dev P(>|Chi|)
NULL                      39      40.032
msize    1   32.835       38       7.197 1.003e-08
```

We conclude from the analysis of the deviance table that the size effect is significantly related to the diet.

Multinomial logistic regression can be performed with the `multinomial` function of the **VGAM** package, in which the user can find other general linear and general additive models. We have applied GLMs in the case of logistic regression; but you must keep in mind that GLMs can fit types of response other than binomial ones.

3.2.5 Interspecific Measurements and Phylogenetic Data

It is common in biology to study differences between species; this is the purpose of comparative anatomy. When analyzing morphological data at the population level, there is usually no major statistical problem because we consider that observations are independent. Statistical problems occur when we want to explain interspecific variation in relation to one or more explanatory variables. Indeed, one can not consider data recorded on species as independent because of the evolutionary relationships between species. Somehow, interspecific data contain a longitudinal component. Felsenstein illustrated the problem well in 1985 [30]. Simply speaking, species that share recent common ancestors will look more alike than species sharing a more distant ancestor. An elephant will look more similar to a mammoth and a human more similar to a chimpanzee, while difference between members of each group will be more spectacular. Consider the relationship between hindlimb and nose (or proboscis) size in a sample mostly constituted of elephants and mammoths to which you have added data from apes. It is likely that the covariation will be high, although in fact there are no functional relationships between limb size and nose size. The problem could be solved by taking into account the evolutionary relationship between the species.

Several methods have been developed in the last 20 years to overcome these problems: autocorrelation, contrast analysis, GLS, GEE, etc. Most of them have been programmed for R in the ape package [83]; one can be find some other in the PHYLOGR and ade4 packages. The relative functions are summarized in Table 3.5.

Table 3.5. Comparative methods in R

Method	Authors	Function	Package
Phylogenetic independent contrasts	[30]	`pic`	ape
Cheverud's method	[20]	`compar.cheverud`	ape
Lynch's method	[67]	`compar.lynch`	ape
Generalized estimating equations	[82]	`compar.gee`	ape
Phylogenetic ANOVA	[40, 26]	`variance.phylog`	ade4
GLS linear model	[34]	`phylog.gls.fit`	PHYLOGR

A complete book has been devoted to the analysis of phylogenetic data with R [81]; application of comparative methods to biological measurements (morphological or not) is provided as well.

Here I illustrate how to perform a contrast analysis between interspecific measurements, and how to obtain ancestral character states. The method of phylogenetically independent contrast was developed by Felsenstein in 1985. The idea was to examine paired differences between direct descendants of every node of the phylogeny. Values at ancestral nodes are approximated using weighted means of descendant taxa. When comparing several continuous characters, the contrasts are used instead of the original variables, which avoids the inflation of degrees of freedom introduced by the dependencies of data in the phylogeny. The data used here are taken from Garland and Janis [35]. Data are log-transformed for improving normality and homoscadisticity. Note that this log transformation has a direct impact on the estimation of ancestor morphologies or other phenotypic properties. We follow what was done in the initial paper.

```
>library(ade4)
>library(ape)
>hl<-log(carniherbi49$tab2$hindlength)
>speed<-log(carniherbi49$tab2$runningspeed)
>cor.test(hl,speed)

        Pearson's product-moment correlation

data:  hl and speed
t = 2.9593, df = 47, p-value = 0.004817
alternative hypothesis: true correlation is not equal to 0
95 percent confidence interval:
 0.1295547 0.6095775
sample estimates:
      cor
0.3963135
```

There is a significant correlation between both variables when phylogeny is not taken into account.

```
>tre1<-read.tree(text=carniherbi49$tre1)
>pic.hl<-pic(hl, tre1)
>pic.speed<-pic(speed, tre1)
>summary(lm(pic.hl~pic.speed-1))

Call:
lm(formula = pic.hl ~ pic.speed - 1)

Residuals:
     Min       1Q   Median       3Q      Max
-0.16534 -0.02411  0.00868  0.07365  0.14900

Coefficients:
          Estimate Std. Error t value Pr(>|t|)
pic.speed   0.1644     0.1602   1.027     0.31

Residual standard error: 0.0784 on 47 degrees of freedom
Multiple R-Squared: 0.02193,  Adjusted R-squared: 0.001119
F-statistic: 1.054 on 1 and 47 DF,  p-value: 0.3099
```

Note that the regression line passes through the origin for contrast data. This is reasonable and to be expected since contrasts for the same species or very closely related species are expected to be null.

This example shows that taking phylogeny into account yields different results from analyses ignoring this kind of dependencies.

Since morphological characters are often only available from fossils, it is often necessary to appraise ancestral morphologies. The ace function of **ape** estimates ancestral character states. Here we investigate whether estimating ancestral character states differs whether raw or log-transformed data are used.

```
> exp(ace(log(hl), tre1, method="pic")$ace[1:5])
      50       51       52       53       54
63.31105 41.36165 34.37632 33.51877 70.78191
>ace(hl, tre1, method="pic")$ace[1:5]
      50       51       52       53       54
71.35941 46.40598 39.39102 40.52667 71.21333
```

Whether the log-transformation or the measurement itself evolves according to a given model has two different meanings in morphological evolution. Thus, one must be careful when one uses the log-transformation before applying phylogenetic correction.

3.2.6 Allometry and Isometry

The relationship between two measured variables is not always linear, and this is especially true for biological measurements during growth. Most organisms exhibit growth. Growth can be isomorphic (shape is constant during growth – variation between forms only concerns differences in size) or allometric (shape changes during growth). If growth is allometric, growth gradients (or relative growth) are not equal between organs or within a given organ. An organ that becomes longer relative to its width during growth is said to exhibit an allometric pattern. One should not confuse the concept of allometry with the study of growth curves and rates during ontogeny, since here we are not determining the relationship between one measurement and age.

Considering that the ratios of relative growth rates in organs to the relative growth of the body remain constant, one can express the relationship between two distances in an organism using the formula published by Huxley [47]

$$y = \beta x^\alpha \, ,$$

where x and y are two measurements, α is the constant differential growth ratio, and β is a constant. This equation is also known as the Huxley allometry equation (although Huxley was speaking about "heterogeny" in his original publications [46, 47]). Huxley and Teissier coined the term "allometry" later in 1936 in a joint publication [36, 48]. α is also the differential rate of change in y according to x as shown by the following equation:

$$\alpha = \frac{dy}{y} \bigg/ \frac{dx}{x} \, .$$

In taking logs of both sides of the equation, we obtain

$$\log y = \alpha \log x + \log \beta \, .$$

Note the recurrent use of logarithms in classic morphometrics. Note, however, that it is important to understand whether the transformation is necessary. There is no reason to transform the data if it is not justified. One can use logarithms for two purposes:

1. Linearizing relationships between variables.
2. Altering the distribution of variables to obtain normally distributed observations and residuals respecting homoscedascicity.

Before performing this transformation, one should examine the linear model constructed with raw data, and examine whether logarithms have resolved problems with homoscedascicity and normality with residuals; if not, one should find one solution in nonlinear models with the `nls` function for example. Finally, one should remember that one analyzes the relationships between logarithms and not those between raw measurements.

It is possible by linear regression to determine whether both measurements are linked by some allometric relationship. If the slope is 1 between both logs of measurements, we can conclude that isometry is present. However, the concept of allometry is obviously when one variable (size) is a predictor of one other (shape). The statement of allometry or isometry implicitly requires a size standard to be defined if one estimates these patterns by regression. The standard is usually one measured distance that serves as size. When two measurements are available, the one that the morphometrician considers as size or shape is arbitrary, and performing regression is not necessarily straightforward. Residual variation in the regression model is only variation in response, and it may be reasonable to think that this residual variation is shared by both measurements. In this case, the axis passing through the bivariate scatterplot should reduce the net distances of observations to the fitted regression. One can estimate this axis and rotate data accordingly using major axis or reduced major axis methods. These methods are equivalent to principal component analysis for bivariate data, and the main axis parameters are calculated so that it passes through the centroid of the bivariate scatters. While the major axis method works on raw data and on the variance-covariance matrix, the reduced major axis method works on centered and reduced variables and on the correlation matrix. The second method is recommended when one scales measurements that have noncomparable units (e.g., weight and metric measurements). Warton et al. [123] have recently reviewed bivariate line fitting in the context of allometry. Table 3.6 summarizes estimates for the slope and intercept in the ordinary regression, major axis, and reduced major axis methods.

Table 3.6. Estimation of the slope a and intercept b parameters in ordinary regression, major axis, and reduced major axis methods

Parameter	Regression	Major Axis	Reduced Major Axis
a	$\frac{s_{xy}}{s_x^2}$	$\frac{s_y^2 - s_x^2 + \sqrt{(s_y^2 + s_x^2)^2 + 4s_{xy}}}{2s_{xy}}$	$\text{sign}(s_{xy})\frac{s_y}{s_x}$
b	$\bar{y} - a\bar{x}$	$\bar{y} - a\bar{x}$	$\bar{y} - a\bar{x}$

Warton et al. have also developed a full package (smatr) for estimating slopes and intercepts according to the major axis and reduced major axis methods. The major axis Method is preferred over regression because one can interpret the major axis as a size axis. The line.cis function estimates the slopes and the intercepts according to the reduced major axis, major axis, or ordinary least-squares methods. One can apply different tests to each fit according to the method: testing whether the slope is different from one (testing for isometry in the case of log-transformed data), whether slopes are different between samples (differences in allometric patterns between populations), whether there is a lateral shift between the major axes, or whether there are different intercepts. Some of the functions are summarized below in Table 3.7. Other functions of the package estimate common axes and provide

Table 3.7. Test for bivariate line-fitting methods

Test	Function
Slope equals a given value	slope.test
Are axes sharing a common slope	slope.com
Shift along the major axis	shift.com
Intercept equals a given value	elev.test
Equal intercept among several lines	elev.com

confidence intervals for fit parameters.

We reuse our crabs dataset with both sexes of the first species to make inferences in their growth and shape changes.

```
>library(MASS); library(smatr)
>data(crabs)
>MRW<-crabs$RW[1:50]; FRW<-crabs$RW[51:100]
>MFL<-crabs$FL[1:50]; FFL<-crabs$FL[51:100]
```

Computation of major axis slopes and intercepts for males and females of the blue species.

```
>line.cis(log(FFL), log(FRW), method="MA")
            coef(MA) lower limit upper limit
elevation 0.1309508 0.004097290   0.2578042
slope     0.9833790 0.933546783   1.0358243
>line.cis(log(MFL), log(MRW), method="MA")
            coef(MA) lower limit upper limit
elevation -0.2550968  -0.4541924 -0.05600117
slope      1.1980736   1.1198780  1.28278668
```

Test for isometry

```
>unlist(slope.test(log(FFL), log(FRW), method="MA"))
          r           p  test.value           b
-0.09325563  0.51948183  1.00000000  0.98337900
        ci1         ci2
```

```
 0.93354678  1.03582432
>unlist(slope.test(log(MFL), log(MRW), method="MA"))
          r                p  test.value                b
6.139065e-01 2.124640e-06 1.000000e+00 1.198074e+00
        ci1             ci2
1.119878e+00 1.282787e+00
```

Test whether there is a difference in allometric growth between sexes.

```
>unlist(slope.com(crabs$FL[1:100],crabs$RW[1:100],
+        groups=crabs$sex[1:100],method="MA"))
           LR                p                b             ci1
4.306581e+01 5.292922e-11 1.205568e+00 1.114607e+00
          ci2             varb           lambda             bs1
1.346128e+00 1.155037e-03 1.000000e+00 1.079202e+00
          bs2              bs3              bs4             bs5
1.017174e+00 1.145318e+00 1.536700e+00 1.425999e+00
          bs6
1.659867e+00
```

The tests for isometry are based on the sample correlation between residuals and fitted values. Here, they show that while females have a growth pattern not different from isometry, males show a significant allometric growth pattern. The test for common slopes shows that growth gradients for both measurements are different between males and females.

3.3 Size: A Problem of Definition

Until now, we have not tried to isolate the shape and size components from a collection of measurements. Morphometric analyses attempt to decompose the form of an object into one size and several shape components when several measurements have been collected. The form variation can correspond to the total amount of variation in a collection of measurements. The choice of size is arbitrary, and the definition of shape depends on that of size, as we will see. Since shape and size are more arbitrary concepts, one must define them prior to any analysis aiming to quantify shape and size variation.

Size is a linear measure of some kind that has a unit of length. When a single measurement is available, one measures size variation. When more measurements are available, the user should define a scalar for size. Indeed, things are not intuitive when defining the size of a triangle by comparison to the size of a segment: Which triangle side should be chosen for scaling the others? Should we use the square root of the area? In the case of more complex shapes, the size reference becomes even more difficult to define.

When more than two measurements are present, a combination all distance measurement $x_{i \to p}$ seems to be more objective for defining size than the selection of a specific standard measurement for the size variable. Several options exist: among them, the arithmetic and geometric means of measurements, and the square root of the sum of squared measurements are the more common. The respective formulae are supplied below:

$$1/p \sum_{i=1}^{p} x_i \; ; \qquad \prod_{i=1}^{p} x_i^{1/p} \; ; \qquad \sqrt{\sum_{i=1}^{p} x_i^2} \; .$$

The geometric mean has several advantages when one deals with growth features and allometries (see [75]).

Defining the shape component of forms is not obvious. In the bivariate case, the size and shape relationships are estimated using the proportion of one variable on the other considered to be the size variable. Actually, this concept of size is close to that of scale, and the concept of shape to a proportion or a ratio in this case. The shape variation of a rectangle will correspond to the variation of the ratio between width and length. Length becomes a scaling factor, and can be used for measuring size, while shape variation depends on variation of the width to length ratio. Here, since the definition of size is dependent on what you use as a scaling function (you could have chosen some other scaling option, using for example the square root of the product of length times width), the definition of shape becomes dependent on your size definition. In scaling data with one of these size functions, we have what Mosimann [75] calls a shape vector. The ratios or proportions have no physical dimensions and are shape descriptors.

$$z = \frac{1}{g(x)} x \; ,$$

with X being the vector of distance measurements, Z the shape vector, and $g(X)$ the size function. Two individuals are said to have the same shape if they have the same shape vector. Mosimann [75] introduced the important following lemma: "Given any set of measurement vectors x, then any shape vector $Z_j(x) = z_j$ is a function of any other shape vector $Z_i(x) = z_i$, and in particular $z_j = Z_j(z_i) = z_i/g_j(z_i)$ for all values z_i, z_j, and for any choice i, j of shape vector." Calculating Mosimann shape vectors is easy with the `apply` function that iterates the same function to one margin of a `matrix`, `data.frame` or `array` object. We will use the `turtles` dataset of Jolicoeur and Mosimann as an example [50]. We can find this dataset in the Flury package. As an example, we compute the size and shape ratios as in the following code.

```
>library(Flury)
>data(turtles)
>geosize<-(apply(turtles[,2:4],1,prod))^(1/length
+        (turtles[1,2:4]))
>shaperatio<-as.matrix(turtles[,2:4]/geosize)
>pairs(log(shaperatio),pch=21,cex=1.5,
+        bg=c("black","white")[turtles[,1]])
```

For illustrating all kinds of bivariate relationships between shape ratios, we use the `pairs` function (Fig. 3.8). In our example, we see that males differ from females in being relatively higher, narrower and shorter.

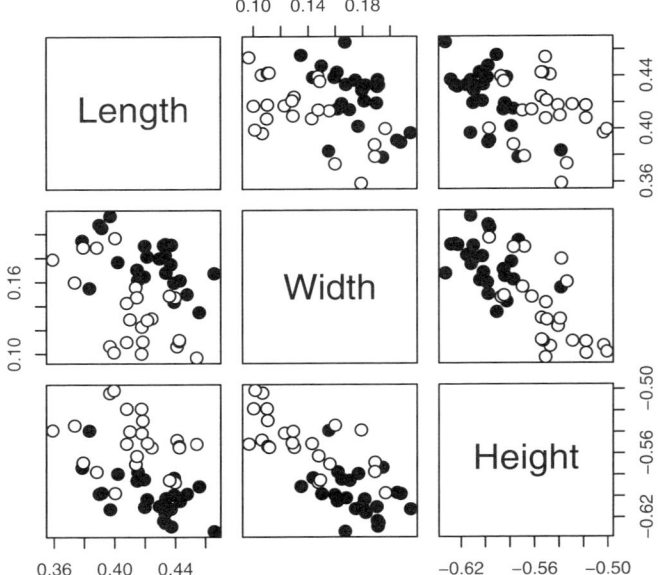

Fig. 3.8. Log-shape ratios applied to the `turtles` dataset. Males correspond to black rounds, while females correspond to circles

One can log-transform the shape ratios, and compare them with log-sizes to check for allometric or isometric variation. The analysis of variance of size variation explained by log-shape variables in a multivariate regression provides a test for isometry.

```
>anova(lm(I(log(geosize[1:24,]))~I(log(shaperatio[1:24,]))))
Analysis of Variance Table

Response: I(log(geosize[1:24]))
                               Df    Sum Sq   Mean Sq
I(log(shaperatio[1:24, ]))      2  0.093277  0.046639
Residuals                      21  0.082110  0.003910

                             F value     Pr(>F)
I(log(shaperatio[1:24, ]))    11.928  0.0003461 ***
Residuals
---
```

The I *function allows a variable to be directly transformed in a formula. The relationship between variables is not purely isometric, as demonstrated by the significant F-value.*

The `lmosi` and `iso.lsr` functions of the **Rmorph** package compute logshape ratios and test for isometry respectively. One can determine the significance of allometric relationships by fitting shape ratios on size. Actually most work that has been done in multivariate statistics for "dividing" the form into size and shape is a matter of describing proportional relationships since size is expected to be scalar. The approaches presented in this chapter are those that are "thought of as working on static mode" while modern morphometrics uses "a dynamic situation by means of oriented procedure," quoting Reyment [91]. Actually, Reyment [91] notices that both approaches are not exclusive.

Methods exist for extracting shape components other than scaling data by size. In addition, ratios are not exempt from problems. The advantages of ratios are that their computation is simple and that one can easily interpret them in geometric terms of shape variation. However, several papers have pointed out that working with ratios introduces spurious correlations between variables (see [2, 3] and related papers in the same issue of Systematic Zoology). The increase of correlation comes because data become dependent after being standardized. Scaling affects the geometry of the shape space, so that it becomes non-Euclidean. Although it removes the size parameter, using ratios increases the correlation between data.

A second way to conceptualize shape and size is to consider shape as the remaining variation once variation explained by size has been filtered. In this second case, size becomes the predictor and shape is contained in the residual variation. One can filter size out of the variation with regression (look at Atchley et al. [2] and Rohlf [103]). Considering size as a linear function of the measurements, one can theoretically remove its effect by multivariate regression. Shape will correspond to the residual variance. The `summary.aov` and `anova.lm` functions applied on the regression model return multiple and multivariate tests of variance for estimating the effect of size on form variation respectively. This approach has the disadvantage of being more difficult than the former one for understanding variation in geometric terms. Integrating several populations in the regression model is more complex and has to take into account a group factor (see Section 3.2.3).

```
>regmod<-lm(as.matrix(turtles[1:24,2:4])~geosize[1:24])
>pairs(regmod$residuals,labels=(c("res.length","res.width",
+     "res.height")), pch=21, bg="black", cex=1.5)
```

In filtering size by regression, one obtains a new combination of variables on which one can appraise the variation explained by the scalar size. Actually, this variation is isometric and also allometric variation because we authorized the intercept to be estimated by the model. A solution is to force the model to pass through the intercept.

```
>regmod1<-lm(as.matrix(turtles[1:24,2:4])~geosize[1:24]-1)
>pairs(regmod$residuals,labels=(c("res.length","res.width",
+     "res.height")))
```

We compute the percentage of variance explained by size.

```
>sum(diag(var(regmod$fitted.values)))/sum(diag(
+     var(as.matrix(turtles[1:24,2:4]))))
[1] 0.9683624
>sum(diag(var(regmod1$fitted.values)))/sum(diag(
```

```
+       var(as.matrix(turtles[1:24,2:4])))))
[1] 0.8453677
```

More than 96% of the morphological variation is explained by size alone in the first model, while only 84.5% is explained in the second. The first `regmod` model reallocates some allometric variation into the variation explained by size.

When analyzing to measurements and one size scalar, one can expect that allometric relationships exist somewhere when a covariance persists between the residuals of each measurement regressed on size. This allometric part is not caused by the general growth trend contained in the data, and will remain "invisible" since the size variation has been removed. Fig. 3.9 illustrates the residual variation of the first model. In this figure, we can observe that residual variation in height and width covary in opposite directions. Although this relationship occurs not only because of overall size variation, it means that there is a trend in turtle carapace variation (flatter and higher shell) that may be related to some allometric relationship: in becoming wider, turtles become flatter.

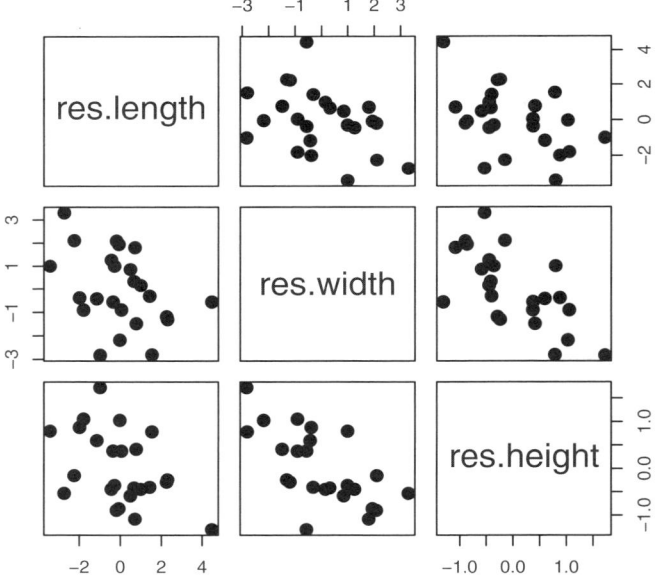

Fig. 3.9. Filtering size by regression with the `turtles` dataset (males only)

We expect that one dimension is lost because of regression. Examining eigenvalues of the variance-covariance matrix of residual variation of the first model tells us what happens to the dimensionality of the new dataset. The function decomposes `svd` a variance-covariance matrix and yields singular values.

```
>svd(var(regmod$residuals))$d
[1] 5.985543e-04 3.663968e-04 3.366490e-21
```

The last singular value is actually equal to zero (at tolerance approximation of the function `svd`). This is not the case in the second model. This is because of the degeneracy of regression.

```
svd(var(regmod1$residuals))$d
[1] 7.579346e-01 2.420254e-01 3.999751e-05
```

Working on log-transformed data allows possible allometric relationships to be identified. In this case, the expectation for an isometric dataset will be equal slope for every variable. A possible statistic for isometry would correspond to the angle formed between the vector of slopes with a vector of similar size of all elements equal to a given constant. The can be passed to a `coli` function that checks for collinearity between two vectors. The function evaluates the angle between two vectors (`ev1` and `ev2`) and compares it with the distribution of angles obtained from random vectors of similar size. We can use the property of the scalar product in this function to compute the angle between vectors. The user can interact with this function by changing the number of permutations, and by telling whether the distribution graph should be displayed.

Function 3.2. `coli`

Arguments:
   `ev1`: *Numeric vector.*
   `ev2`: *Numeric vector of same length as* `ev1`.
   `nperm`: *Number of permutations.*
   `graph`: *Logical indicating whether a graph should be returned.*
Values:
   `z.stat`: *Cosine of the angle between the two vectors.*
   `p`: *p-value.*
   `angle`: *Angle between the two vectors in radians*

```
1  coli<-function(ev1, ev2, nperm=1000, graph=T)
2  {dist<-numeric(nperm)
3    n<-length(ev1)
4  Angle<-function(v1, v2)
5  {sum(v1*v2)/(sqrt(sum(v1^2))*sqrt(sum(v2^2)))}}
```

The internal `Angle` *function computes the cosine between the* `ev1` *and* `ev2` *vectors. Then the function store the random cosine obtained for* `nperm` *permutations.*

```
6  for (i in 1:nperm)
7    {X1<-runif(n, -1, 1); X2<-runif(n, -1, 1)
8    dist[i]<-angle(X1, X2)}
```

Compare the observed value with the null distribution.

```
9   zobs<-Angle(ev1, ev2)
10  pv<-length(dist[dist>zobs])/nperm
```

Produce a graph if the user asks for one.

```
11  if (graph)
12  {hist(dist,breaks=50,
13   main="Distribution␣of␣the␣cosine␣of␣the␣angle␣between␣2
14  ␣␣random␣vectors", xlab="Z␣statistic",ylab="#␣of␣vect.
15  ␣␣éalaloire", sub=paste("Actual␣z-obs␣=",round(zobs,5),":
16  ␣␣p<",round((1-abs(0.5-pv)),5)))
17     abline(v=zobs)}
18  list(z.stat=zobs,p=1-(abs(0.5-pv))*2, angle=acos(zobs))}
```

Here we check for isometry in one sex of our turtle example. For this task, we test whether the angle between the vector containing the slope parameters of one sex of turtles and a vector of one is sufficiently small compared angled obtained from a random distribution of vectors.

```
>regmod<-lm(as.matrix(log(turtles[1:24,2:4]))~
+        log(geosize[1:24]))
>regmod$coefficients[2,]
   Length      Width     Height
1.1912115  0.8935417  0.9152468
>unlist(coli(regmod$coefficients[2,],
+       rep(1,dim(regmod$coefficient)[2])))
   z.stat          p        angle
0.9909448  0.0160000  0.1346768
```

The two vectors are significantly different and we conclude that there are non-isometric relationships within the set of measurements. Note that the residual variation of regressed log-measurement variables corresponds to variation that cannot be explained by growth.

Jolicoeur [51] used regression for this purpose; there are, however, some problematic issues to this because residuals record measurement error as well [15, 103]. Since regression is degenerate, it is usually preferable to use major axis or the first principal component to approximate size.

By analogy with major axis methods for fitting regression, a third method consists of considering the first principal component (see the following chapter) of variation as a size-axis. This method has been explained in detail by Jolicoeur in several papers [49, 50]. As for the regression method, one should be careful when several groups are present in the analysis since the size estimate is not necessarily the same combination of initial variables. To examine differences between groups, it is possible to project the original data of one of the groups on the principal axes of the second group as in the original paper of Jolicoeur and Mosimann [50]. It is presented in the following Section dealing with multivariate morphometrics.

3.4 Multivariate Morphometrics

The description of geometric properties of objects usually requires more than two variables and, hence multivariate analysis. It is possible to produce scatterplots for three variables using some graphical way visualize volumes. An alternative possibility consists of using a triangle-plot visualization.

3.4.1 Visualization of More than Two Distance Measurements

One can simultaneously visualize every bivariate relationship between variables with the pairs function. Even if this tool is convenient for visualizing every relationship, it is not an easy way to examine the distribution of variables in the space they define; especially when there are more than than 10 variables, which is usual in multivariate morphometric analysis.

We have already seen functions and packages for representing data in a trivariate space. The scatterplot3d, persp, and rgl plotting functions are diverse alternatives for representing data in three dimensions.

One can produce triangle-plots using the triangle.plot function of the ade4 package (Fig. 3.10).

```
>data(iris)
>library(ade4)
>tp<-triangle.plot(iris[,c(1,2,4)],cpoint=0,show.position=F)
>points(tp, pch=c(1,2,8)[iris[,5]])
```

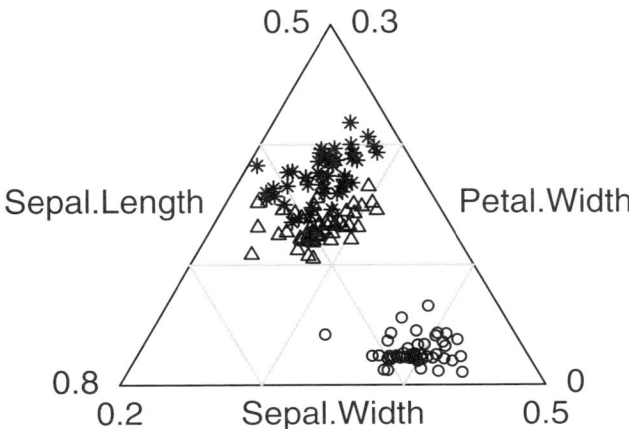

Fig. 3.10. Triangular plot for trivariate observations

This representation is not a direct representation of raw data, because it arranges data so that the three variables present in one observation sum up to one. However, it can be convenient if we consider the sum of variables to be a proxy for size. In this case, this representation shows shape variation in the data.

When more than three variables are present, it becomes impossible to easily visualize the space described by the variables in a simple graph. If one examines bivariate plots between possible variables in a population sample, one usually notices that variables form both oblique and elongated scatters, and that most relationships between pairs of variables behave similarly. It is thus trivial to understand that in a trivariate space, one will produce an elongated ellipsoid. One can examine directions of maximum variation along the axis as an exploration of possible causes for relationships between variables. In morphometry, these causes are likely to have a geometric interpretation, and the first principal axis reflects that all distance variables are proportional to each other. This axis is thus interpreted as a size-axis.

3.4.2 Principal Component Analysis

The principal component analysis (PCA) involves nothing more than moving the variable space and examining axes that reflect maximum of variation and covariation. PCA transforms the data to a new coordinate system such that the greatest variance of the data lies on the first transformed new variable (called the first principal component), the second greatest variance on the second transformed variable, and so on. The orthogonal axes of the PCA summarize variation decreasing in order and individuals observation are projected along axes. The score of a given observation on a given axis corresponds to the projection of the data on that axis. Examining variation on the first axes provides a way to reduce the variable space to dimensions that express most variation. Actually, each axis corresponds to a linear combination of original variables. The first corresponds to the main direction of the variance-covariance structure of individual observations.

The score of individuals ($y_{11 \rightarrow n}$) on the first axis corresponds to the linear combination of variables $x_1, x_2, x_3, \ldots x_p$ with coefficients $u_1, u_2, u_3, \ldots u_p$ so that

$$y_{11 \rightarrow n} = u_{11}x_1 + u_{12}x_2 + \ldots + u_{1p}x_p = \mathbf{U}'_1\mathbf{x} \ .$$

The variance of the scores $\sigma^2_{y_1}$ is maximized and equal to λ_1. $\mathbf{U_1}$ is the first eigenvector, λ_1 is called the first eigenvalue and is estimated from the sample dispersion matrix ($\mathbf{S}$) and $\mathbf{U_1}$. λ_1 is defined by the relationship

$$\lambda_1 = \mathbf{U}'_1\mathbf{S}\mathbf{U_1} \ .$$

To find the solution, one solves the equation

$$(\mathbf{S} - \lambda_1 I)\mathbf{U_1} = \mathbf{0} \ ,$$

I being the identity matrix. We generalize to obtain the whole set of eigenvectors $\mathbf{U}$ and eigenvalues $\mathbf{D}$. We will note $\mathbf{D}$ as the diagonal matrix of eigenvalues. We have

$$\mathbf{SU} = \mathbf{UD}, \quad \text{with} \quad \mathbf{UU}' = I.$$

It follows that $\mathbf{S} = \mathbf{UDU}'$.

R makes this calculation using the `eigen` function in the **base** package. Interpreting the signs and magnitude of the relative contribution of each variable on eigenvectors is equivalent to understanding the shape and size variation along each axis. Imagine a series of body measurements and one measurement of eye size. If an axis shows that all body measurements have a positive contribution while eye size has a negative contribution, the interpretation is that variation along the axis corresponds to the variation between two extremes: large individuals with small eyes and small individuals with large eyes.

The spectral decomposition has some interesting properties. Notice that the sum of eigenvalues equals the sum of variances of each original variable. The relative contribution of a given principal axis for explaining the overall variation corresponds then to the ratio between its corresponding eigenvalue and the sum of eigenvalues.

Usually only a few components are necessary for describing most of the variation. How to choose the optimal number of components is a rather subjective matter. For example, the analyst may select the first n principal axes so that they summarize more than 99% of the overall variance. Some other criteria have been published [52], but there is no true rule to follow.

Although PCA is nothing else than changing the coordinate system and manipulating matrices, several packages are available to perform PCA. The reader can find much of the theory, many applications, and further developments in Jolliffe [52]. The `prcomp` and `princomp` functions perform PCAs. These functions differ in the way that they work on variance-covariance matrices or correlation matrices. The first performs a singular-value decomposition, while the second one performs a spectral decomposition.

Since in morphometrics, we perform the analysis for data that are all expressed in the same unit, there is no reason to scale the data.

```
>library(Flury)
>data(turtles)
>prcomp(turtles[1:24,2:4])
Standard deviations:
[1] 13.974070  1.920563  1.050634

Rotation:
               PC1          PC2          PC3
Length 0.8401219 -0.48810477 -0.23653541
Width  0.4919082  0.86938426 -0.04687583
Height 0.2285205 -0.07697229  0.97049145
>prcomp(turtles[1:24,2:4])$x
           PC1          PC2         PC3
1  -24.9951007 -2.19437639  1.8904369
2  -22.6443870  0.94900046 -0.4745848
3  -19.9803270  1.71155944 -1.0414072
4  -12.8980031  2.44068345  1.4703782
```

```
5  -11.7944935   2.89893523   0.2164755
...
```

The `prcomp` function returns the standard deviation of each principal component. Squared standard deviations are equal to singular values. The `Rotation` value contains the contribution of the original variables on principal components. Note that each column of this matrix corresponds to a vector of unit size. For projecting the observations on eigenvectors, the original data are matrix post-multiplied by the rotation matrix, or are more simply extracted using the `x` item (of the `matrix` class) of the `prcomp` object. By analyzing contributions, one can interpret the meaning of each principal component. The first PC is positively related to all measurements; it corresponds, in addition, to the major axis of variation and can be interpreted as a size axis. The second PC contrasts width, and both length and height, and thus corresponds to the relative width of the turtle carapace; the last corresponds to the relative height and opposes high and low carapaces for individuals of similar size. The `summary.prcomp` function returns the percent of variation explained by each principal component.

```
>summary(prcomp(turtles[1:24,2:4]))
Importance of components:
                         PC1     PC2     PC3
Standard deviation      13.974  1.9206 1.05063
Proportion of Variance   0.976  0.0184 0.00552
Cumulative Proportion    0.976  0.9945 1.00000
```

The first axis represents 97.6% of the total variation, which is slightly more than when we used the regression model for filtering size (even when ignoring the intercept). This is normal, since like the major axis method, the PCA finds the fit that minimizes net distance to the axes. We have seen earlier that we can interpret the first component as a size axis. In contrast, further axes correspond to shape axes. By using the principal component approach to decompose morphological variation, the size axis becomes a function of the original variables.

The `biplot` function is convenient for producing a biplot graph where both observations and variables are projected in the same graph, making the interpretation easier (Fig. 3.11). `biplot` summarizes the results of the previous analysis for two PCs (in the example below: the first two PCs of the `turtles` dataset). The left and bottom axes use the unit for observations, while the top and right axes are graduated according to the contributions of original variables.

```
>biplot(prcomp(turtles[1:24,2:4]),xlim=c(-0.4,0.5),col=1)
```

Although the first axis represents a size-axis when one works on raw measurements, we know that growth in organisms is not always isometric. The relationships between measurement distances follow the allometry equation of Huxley (see Section 3.2.6). Transforming raw data into their logarithms allows allometry to be recorded by the first PC. The first component of a PCA of logged variables corresponds to a growth-axis, recording both isometric and allometric variations. The two

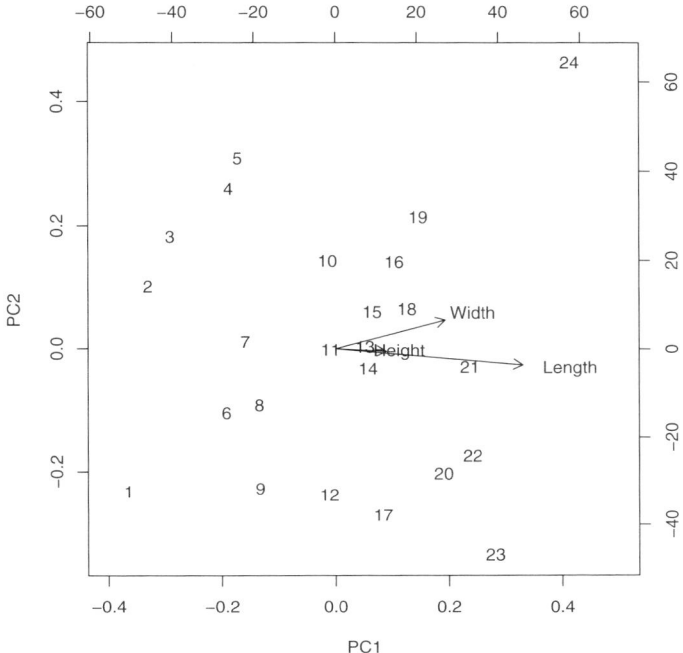

Fig. 3.11. Biplot graph for the PCA applied to the measurements of the males from the `turtles` dataset

remaining components correspond to variation not explained by isometry or allometry. In other words, they correspond to variation not explained by general growth. One can estimate whether growth follows an isometric model by comparing the first axis with an hypothetical isometric axis as demonstrated in Jolicoeur [49]. Jolicoeur suggests using a Chi-square test on the value

$$(n-1)(\lambda_1 V_1 \mathbf{S}^{-1} V_1' + \frac{1}{\lambda_1} V_1 \mathbf{S} V_1' - 2) ,$$

with n being the number of observations, V_1, the theoretical eigenvector under the hypothesis of isometry (i.e., all components equal to $\sqrt{\frac{1}{p}}$), with λ_1 being the first eigenvalue (variance on the first PC), $\mathbf{S}$ the variance-covariance matrix, and p being the number of log-transformed variables. The `isojoli` function performs this new multivariate test for isometry on a matrix of measurements.

Function 3.3. `isojoli`

Argument:
  `mat`: *Matrix of* `n` *observations and* `p` *variables.*
Values:
  `Chisp`: *Observed statistic for testing isometry.*
  `p`: *p-value.*

```
1  isojoli<-function(mat)
2  {n<-dim(mat)[1]; p<-dim(mat)[2]
3  S<-var(log(mat))
4  V1<-rep(sqrt(1/p),p)
```

Compute the first singular value.

```
5  L1<-svd(S)$d[1]
6  chiobs<-(n-1)*(L1*t(V1)%*%ginv(S)%*%V1+(1/L1)
7  +       *t(V1)%*%S%*%V1-2)
8  unlist(list(Chisq=chiobs, p=pchisq(chiobs,p-1,
9  +         lower.tail=F)))}
```

This new function is applied on the `turtles` dataset to discover whether growth is iso- or allometric.

```
>isojoli(turtles[1:24,2:4]
        Chisq              p
2.662345e+01 1.654976e-06
```

The test tells that variation because of size is not fully isometric.

To test for allometry, one can alternatively compute the angle between the theoretical axis of isometry and that of allometry, and compare this angle with a null distribution of angles obtained from random vectors.

```
>unlist(coli(prcomp(log(turtles[1:24,2:4]))[[2]][,1],
+     rep(sqrt(1/3),3)))
 z.stat          p        angle
0.9906529 0.0180000 0.1368338
```

Understanding shape differences between groups is fundamental in morphometrics. In our dataset, we wish to evaluate whether sex dimorphism is present. For a given population, the first principal axis represents the size variable. However, one cannot *a priori* think that males and females have a similar first PC. One can nonetheless project the whole data on the PCs defined by one group as in Jolicoeur [49], and examine the distribution of observations on the second and third axes of shape (Fig. 3.12).

```
>pca<-prcomp(turtles[1:24,2:4])
>proj<-turtles[,2:4]%*%pca[[2]]
>plot(proj[,2:3], xlab="PC2", ylab="PC3",pch=21,
```

```
>        bg=c("black","white")[turtles[,1]],asp=1)
>lines(ELLI(proj[1:24,2], proj[1:24,3]))
>lines(ELLI(proj[25:48,2], proj[25:48,3]))
```

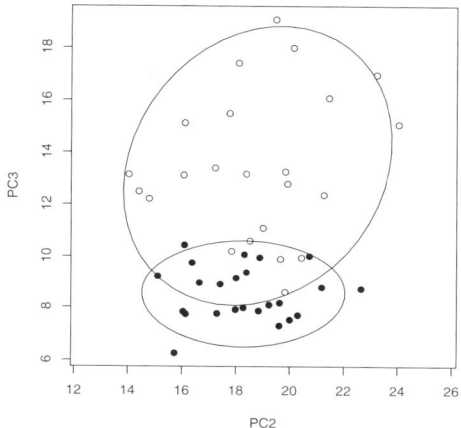

Fig. 3.12. Projection of the male and female observation on the second and third principal axes defined by variation of male turtles

In our example, males and females differentiate on the third axis, which is related to relative height variation (Fig. 3.12). We can interpret the PCA in terms of shape difference between sexes. However, PCA is not the preferred way to explore shape differences.

3.4.3 Analyzing Several Groups with Several Variables

Two questions can be addressed when several variables are measured on several groups.

- What are the differences between groups? If one can characterize differences between groups, it means that one has some reference for identifying unknown observations as belonging to one or the other group. Characterizing differences between groups is the goal of discriminant analysis.
- Are groups significantly different? This is the goal of multivariate analysis of variance.

For addressing both questions, the total variation $\mathbf{T}$ has to be decomposed in a between-group variation $\mathbf{B}$ and in a within-group variation $\mathbf{W}$. Let's say that we have b groups, each containing $n_{1 \to b}$ observations, and that each observation is depicted by p variables. $\mathbf{X}$ is The total matrix of observations and contains $\sum n_i = n$ observations. The total sum of squares and cross-products is calculated as

$$\mathbf{T} = (\mathbf{X} - \bar{\mathbf{X}})'(\mathbf{X} - \bar{\mathbf{X}}) ,$$

where $\bar{\mathbf{X}}$ is the matrix of n rows with p columns with values equal to the mean of corresponding columns of $\mathbf{X}$. This total sum of squares and cross-products has $(\sum n_i) - 1$ degrees of freedom and can be partitioned into a within-group and between-group sums of squares.

The within-group sum of squares and cross-products corresponds to the sum of within-group sum of squares which has $(\sum n_i) - b$ degrees of freedom. It is obtained as

$$\mathbf{W} = \sum_{1}^{b} \mathbf{W_i} .$$

The between-group sum of squares and cross-products is calculated from differences between total and within-group variation.

$$\mathbf{B} = \mathbf{T} - \mathbf{W} ;$$

$\mathbf{B}$ has $b - 1$ degrees of freedom. Discriminant analysis and MANOVA work basically with $\mathbf{B}$ and $\mathbf{W}$.

Linear Discriminant Analysis

Discriminant analysis finds linear combinations of variables that describe intergroup differences. These combinations define linear discriminant functions. The linear discriminant coefficients are defined from the non-null eigenvectors of the between-group variance-covariance "scaled" by the within-group variance-covariance.

The variance-covariance matrix is first computed $\mathbf{V_B} = \mathbf{B}/(b - 1)$ and $\mathbf{V_W} = \mathbf{W}/(n - b)$. Then $\mathbf{V_B}$ is premultiplied by the inverse of $\mathbf{V_W}$ to obtain the $\mathbf{V_{B/W}}$ matrix, such that

$$\mathbf{V_{B/W}} = \mathbf{V_W^{-1}} \mathbf{V_B} .$$

The $\mathbf{V_{B/W}}$ matrix has $k = b - 1$ non-null eigenvalues. The corresponding eigenvectors $\mathbf{U_k}$ are conserved to calculate scaled eigenvectors $\mathbf{C}$. These correspond to the linear discriminant coefficients an are obtained such as

$$\mathbf{C} = \mathbf{U}(\mathbf{U}'\mathbf{V_W}\mathbf{U})^{-0.5} .$$

The original data are projected onto the functions defined by the standardized linear discriminant coefficients to obtain individual scores.

The lda function of the **MASS** package performs all these operations. We use the raw data of Anderson iris dataset to check it. The first argument is the matrix of observations, the second is the grouping factor.

```
>library(MASS); data(iris)
>miris<-as.matrix(iris[,1:4])
>lda(miris,as.factor(iris[,5]))
Call:
lda(miris, as.factor(iris[, 5]))

Prior probabilities of groups:
    setosa versicolor  virginica
 0.3333333  0.3333333  0.3333333

Group means:
           Sepal.Length Sepal.Width Petal.Length Petal.Width
setosa            5.006       3.428        1.462       0.246
versicolor        5.936       2.770        4.260       1.326
virginica         6.588       2.974        5.552       2.026

Coefficients of linear discriminants:
                     LD1         LD2
Sepal.Length   0.8293776  0.02410215
Sepal.Width    1.5344731  2.16452123
Petal.Length  -2.2012117 -0.93192121
Petal.Width   -2.8104603  2.83918785

Proportion of trace:
   LD1    LD2
0.9912 0.0088
```

The function returns the group mean, the probability for each observation to belong to a given group, the group means, the coefficients of the linear discriminant, and the proportion of the trace explained by the discriminant functions (here two, since we have three groups). The first discriminant function explains most of the between-group variation. It opposes flowers with small petals and large sepals with flowers with large petals and small sepals.

The plot.lda function projects the observations on linear discriminant and displays a plot. One can alternatively calculate the projection by post-multiplying the observations by the coefficients of the linear discriminant functions. The discriminant analysis has been performed on the whole morphology, taking into account size and shape, but it can be applied to shape variables or log-shape ratios as well to investigate shape differences. We compare both analyses. Both plots are illustrated in Fig. 3.13.

```
>size<-apply(miris,1,prod)^(1/(dim(miris)[2]))
>shapeiris<-miris/size
>formlda<-lda(miris, as.factor(iris[,5]))
>shapelda<-lda(shapeiris, as.factor(iris[,5]))
>proj1<-miris%*%formlda$scaling
>proj2<-shapeiris%*%shapelda$scaling
>layout(matrix(c(1,2),1,2))
```

```
>plot(proj1,pch=(1:3)[as.factor(iris[,5])],asp=1,
+      cex=0.6,xlab="FD1",ylab="FD2",main="Form FDA")
>plot(proj2,pch=(1:3)[as.factor(iris[,5])],asp=1,
+      cex=0.6,xlab="FD1",ylab="FD2",main="Shape FDA")
```

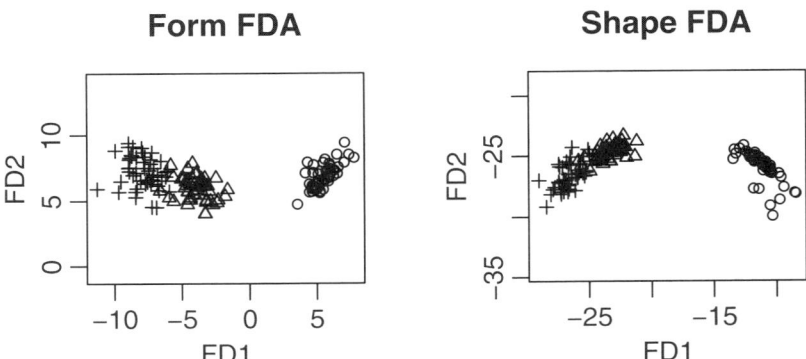

Fig. 3.13. Plot of linear discriminant analyses applied to the `iris` dataset considering form and shape variables respectively. The three species are represented by different symbols (*Iris setosa*: circles, *Iris virginica*: crosses, *Iris versicolor*: triangles)

The first linear discriminant scores returned by both analyses show that there is an important differentiation between *I. setosa* and both *I. virginica* and *I. versicolor*. The second axis shows a slight differentiation between *virginica* and the two other species. There is, however, considerable overlap on the second axis (Fig. 3.13).

An important quantity related to discriminant analysis is the Mahalanobis distance d_m. It is calculated from group means and from a variance-covariance matrix. This distance is a measure of distance between group means and is scaled by the pooled within-group covariance $\mathbf{V}_W$. This measure is meaningful, if the within-group variance-covariance matrices are similar enough between groups to be pooled (if not, things become much more difficult to estimate in practice, but solutions are available such as the quadratic discriminant analysis that is computed with the `qda` function of the **MASS** package). Given two mean vectors $\bar{X}_i$ and $\bar{X}_j$ defining the centroid of two groups, the Mahalanobis distance d_m between these groups is given by the relationship

$$d_m = \sqrt{(\bar{X}_i - \bar{X}_j)'\mathbf{V}_W^{-1}(\bar{X}_i - \bar{X}_j)} \, ,$$

where d_m is the distance separating groups according to the discriminant axes. We check this with a few lines of code:

First, compute distances between mean groups in the space defined by discriminant functions.

```
>meangroup<-formlda$mean
>meanproj<-meangroup%*%formula$scaling
>dist(meanproj)
               setosa versicolor
versicolor   9.479672
virginica   13.393458    4.147417
```

Compute the Mahalanobis distance between the first two groups "by hand."

```
>W<-var(miris[1:50,])*49+var(miris[51:100,])*49+
+      var(miris[101:150,])*49
>VCVw<-W/(150-3)
>sqrt(diff(meangroup[1:2,])%*%solve(VCVw)
+             %*%t(diff(meangroup[1:2,])))
            versicolor
versicolor   9.479672
```

Or, for simplicity, use the `predict` function.

```
>dist(predict(formlda, meangroup)$x)
               setosa versicolor
versicolor   9.479672
virginica   13.393458    4.147417
```

Since d_m is a measure of distance between groups, taking into account within-group variation, it is a very useful quantity for multivariate statistics. Consider the problem of deciding whether an observation belongs to one group. One needs not only to define the distance of the point to the centroid of the group, but also to know the variation in the scatter of points. The closer the point is to the centroid of the group, the more difficult it is to distinguish it from any other point of the group. Although one can assume that variation around the centroid is multinormal, it may not be equal in all directions. To resolve this problem, one must consider the matrix of intragroup variance-covariance that measures variation in all directions. For a unique observation, one can understand the Mahalanobis distance as the distance of the point from the center of the group divided by the width of the ellipsoid in the direction of the tested point. This distance measures the distance between a point and the group mean, taking into account the within-group covariance-variance matrix ($\mathbf{V_W}$), such as

$$d_m(X_i) = \sqrt{(X_i - \bar{X})'\mathbf{V}_W^{-1}(X_i - \bar{X})} \, .$$

This equation corresponds to the generalization of the Mahalanobis distance defined from two groups to the Mahalanobis distance between a group mean and a given observation. It can be used for predicting the probability that any observation belongs to a given group. The `predict` function is useful for allocating an observation to a given group. Indeed, it returns the probability of one observation belonging to any given group. Given any new observation, the discriminant analysis provides a diagnostic tool whose accuracy depends on sampling effort.

```
>model<-lda(iris[1:148,1:4],as.factor(iris[1:148,5]))
```

```
>predict(model, iris[149:150,1:4])
$class
[1] virginica virginica
Levels: setosa versicolor virginica

$posterior
          setosa    versicolor virginica
149 3.525273e-40 1.546625e-05 0.9999845
150 6.312764e-33 2.167781e-02 0.9783222

$x
          LD1       LD2
149 -5.918934 2.3626043
150 -4.715782 0.3282361
```

The model has allocated the two unknown observations to the species *virginica*; this is reliable. The function returns the posterior probabilities of these observations as a member of one of the three groups, and returns their position (scores) in the discriminant space. If allocation of observations is always exact, one can infer that differences between groups are significant.

MANOVA

Multivariate analysis of variance tests whether groups are similar. It is based on a multivariate extension of the F-test. The Mahalanobis distance is very close to the Hotelling t^2 statistic which corresponds to a multivariate generalization of the Student test for hypothesis testing. This generalization also holds for testing any given multivariate linear model. The Hotelling t^2 statistic is obtained such that

$$t^2 = n_i(\bar{X}_i - \bar{X})'\mathbf{V}_{W_i}^{-1}(\bar{X}_i - \bar{X}) ,$$

n_i being the number of observation in the sample, $\bar{X}_i$ being a sample mean, $\bar{X}$ the population mean, and $\mathbf{V}_{W_i}$ the sample variance-covariance matrix.

t^2 follows a distribution T^2 (Hotelling T-square distribution) of parameters, p (number of variables) and n_i (number of individuals in the sample). If X_i is a random multivariate observation of p elements following a multivariate Gaussian distribution of parameters $\sim N_p(\bar{X}, \mathbf{V})$, and $\mathbf{V}_{W_i} \sim \mathbf{V}_{W_p}(m, \mathbf{V})$ follows a Wishart distribution with the same variance-covariance $\mathbf{V}$ then

$$\frac{m-p+1}{pm}T^2 \sim F_{p,m-p+1} .$$

This property allows hypothesis testing. The Hotelling's two-sample t^2-statistic is calculated as

$$t^2 = \frac{n_1 n_2}{n_1 + n_2}(\bar{X}_1 - \bar{X}_2)'\mathbf{V}_W^{-1}(\bar{X}_1 - \bar{X}_2) \sim T^2(p, n_1 + n_2 - 2) ,$$

and can be related to the F-distribution by the equation

$$\frac{n_1 + n_2 - p - 1}{(n_1 + n_2 - 2)p} t^2 \sim F(p, n_1 + n_2 - 1 - p) ,$$

$\bar{X}_1$ and $\bar{X}_2$ are the multivariate group means, and $\mathbf{V}_W$ is the intragroup pooled variance-covariance matrix. In the two-group case, $\mathbf{V_W}$ is obtained as follows

$$\mathbf{V}_W = \frac{\sum_{i=1}^{n_1}(X_{1i} - \bar{X}_1)(X_{1i} - \bar{X}_1)' + \sum_{i=1}^{n_2}(X_{2i} - \bar{X}_2)(X_{2i} - \bar{X}_2)'}{n_1 + n_2 - 2} .$$

The Hotelling-Lawley trace T_{HL}^2 is an extension of the Hotelling two-sample t^2-statistics for comparing multivariate variances. One can use it in multivariate analysis of variance. It is given by

$$t_{HL}^2 = \text{trace}(\mathbf{BW}^{-1}) ,$$

where $\mathbf{B}$ is the effect sum of squares and cross-products, and $\mathbf{W}$ is the error sum of squares and cross-products. One can approximate the Hotelling trace T_{HL}^2 by the F-distribution. The approximation is different whether the number of dimensions is smaller or greater than the number of degrees of freedom for the error term. For the following, N is the total number of observations, k is the degrees of freedom for the effect term (factor levels - 1), $w = n - k - 1$ is the degrees of freedom of the error term, and p is the number of variables (or space dimensions). To calculate the F-approximation, first define $m = \frac{w-p-1}{2}$. Then, the approximation is given by the relationship

$$\frac{2(sm + 1)}{s^2(2t + s + 1)} T_{HL}^2 \sim F(s(2t + s + 1), 2(sm + 1)) ,$$

with

$$s = \min(p, k) ,$$

and

$$t = \frac{|p - k| - 1}{2} .$$

If $m > 0$, Mckeon [73] gives an alternative approximation:

$$\frac{4 + \frac{pk+2}{b-1}}{pk} \frac{T_{HL}^2}{c} \sim F(pk, 4 + \frac{pk + 2}{b - 1}) ,$$

with

$$b = \frac{(p + 2m)(k + 2m)}{(2m + 1)(2m - 2)} ,$$

and

$$c = \frac{2 + \frac{pk+2}{b-1}}{2m} .$$

Mckeon [73] showed that this approximation is closer to the actual estimate.

Other multivariate tests are available such as Pillai, Wilks lambda, etc. The `manova` and `summary.manova` functions return the results of the tests and different degrees of freedom. Users must adapt the method whenever they consider interaction should be taken into account as the error term. The resulting table for the `iris` dataset is produced as follows:

```
>summary(manova(miris~as.factor(iris[,5])),test="Hotelling")
                  Df Hotelling-Lawley approx F
as.factor(iris[, 5])    2               32.48    580.53
Residuals             147

                  num Df den Df     Pr(>F)
as.factor(iris[, 5])        8     286 < 2.2e-16 ***
Residuals
---
Signif. codes:   0 *** 0.001 ** 0.01 * 0.05 . 0.1   1
```

The first column of the table summarizes degrees of freedom of the effect and residual variance-covariance matrices, the second gives the Hotelling-Lawley value, and the third to fifth give the transposition of the multivariate test in terms of an F-test with respective degrees of freedom. The last cell is the probability for accepting the null hypothesis of equality between groups. In the example, the interspecific variance is larger than the intraspecific one.

The same rules of ANOVA concerning factor type and marginality apply to MANOVA; users have to be careful when performing two-way MANOVA to understand the assumptions of the test and determine whether they may violate marginality principles.

Burnaby's Approach

Some special extensions for morphometric data have been developed in multivariate statistics. The approach of Burnaby [16, 103] allows growth invariant discriminant functions to be defined. The originality of the approach is to project the original data onto a subspace orthogonal to the growth vectors, or to any other nuisance factors (for example, an ecological one). For projecting data on a given vector or a given space, we post-multiply the data by the eigenvectors or the linear discriminant function defining a new base. One projects the data orthogonally to a given base G by post-multiplying them by $I - G(G'G)^{-1}G'$, with I being the identity matrix, G being the matrix of $p \times k$ rows and columns, k being the number of nuisance vectors, p the number of variables.

Other methods with similar aims have been developed (for example, see [44]) but most of them have drawbacks (see [103]). We will use the method of Burnaby for generating a growth independent dataset with the `crabs` dataset for the first species. The idea is to project data orthogonally to both of the growth vectors defining each sex. Since we are removing two growth functions, we expect to find *in-fine* only $5 - 2$ dimensions in the final invariant growth space.

```
>data(crabs)
>crab<-as.matrix(log(crabs[1:100,4:8]))
>G<-cbind(prcomp(crab[1:50,])[[2]][,1],
+         prcomp(crab[51:100,])[[2]][,1])
>I<-diag(1,5)
>ortho<-I-G%*%ginv(t(G)%*%G)%*%t(G)
>newdata<-crab%*%prcomp(crab%*%ortho)[[2]][,1:3]
>summary(manova(newdata~ as.factor(crabs[1:100,2])))
                          Df  Pillai approx F num Df
as.factor(crabs[1:100, 2]) 1  0.2603  11.2628      3
Residuals                 98

                         den Df    Pr(>F)
as.factor(crabs[1:100, 2])   96 2.138e-06 ***
Residuals
---
Signif. codes:  0 *** 0.001 ** 0.01 * 0.05 . 0.1   1
```

Sex dimorphism in crabs does not alter only size or growth patterns but the test indicates that relative body depth is different for females and males of similar size. Further observations of bivariate plots, especially when examining body depths and rear widths, show that the body is deeper and the rear body is narrower for males in comparison to females of similar size.

Clustering

Grouping factors are not always known and the user may wish to see whether the data have a grouping structure. R offers many methods for examining similarities between observations and for possibly inferring a grouping structure between observation in morphometric datasets. Clustering, k-means, or Gaussian mixture models are among the most current in use, although their applications to morphometrics are still uncommon in the literature. There are several related functions and packages in R (see Section 3.8).

There is two groups of methods for identifying whether data are structured: hierarchical clustering methods and partitional clustering methods. Hierarchical clustering builds the hierarchy from the individual elements by progressively merging clusters, while partitional clustering assigns an observation to a given group. The first method provides the relationships between all observations, but does not define groups *a posteriori*, while the second classifies observations into a given number of groups. Here we have to determine whether it is possible to identify three groups from the iris dataset using both methods.

All hierarchical clustering methods usually work on a dissimilarity matrix computed either by the function or supplied by the user. One can calculate dissimilarity matrices from the data using several functions, but the more common is dist. The dist function calculates a dissimilarity matrix from a set of observations and variables according to several distance metrics. For measurements, it is typical to use the Euclidean distance that is the default method of the function.

Table 3.8. Some packages for exploration of grouping structures

Function	Method	Package Name
hclust	Hierarchical clustering	stats
agnes	Hierarchical clustering	cluster
nj	Neighbour joining tree estimation	ape
mst	Minimum spanning tree	ape
spantree	Minimum spanning tree	vegan
kmeans	k-means clustering	stats
clara	Defines k cluster from a dataset	cluster
fanny	Determines spherical cluster by fuzzy clustering	cluster
pam	partition the data into k cluster	cluster
Mclust	find the optimal model for partioning the data	mclust

Hierarchical clustering returns a tree or dendrogram that is a representation of similarity and dissimilarity between individuals. There are many different algorithms for clustering the observations, and the final tree nodes and branch length depends on the method used for clustering the observations. If the initial aim is to define clusters from the data without trying to understand the relationship between data, the "ward" method will have the advantage of finding rather spherical clusters; but if the idea is to appraise the structure of the data in the form space, the "average" method can be more easily understood.

```
>data(iris)
>rownames(iris)<-paste(toupper(substr(iris[,5],1,2)),
+       rownames(iris), sep="")
>bb<-hclust(dist(iris[,1:4]), method="ave")
>dend<-as.dendrogram(bb)
```

Note the use of the toupper *function for capitalizing fonts of the extracted strings.* plot *plots directly the* hclust *object, however, the* dendrogram *function combined with* plot *provides more possibilities with the graphic device. For displaying different colors and symbols for tips of the dendrogram, we modify the class of the dendrogram object with a local function placed as argument of the* dendrapply *function.*

```
>local({
+   colLab <<- function(n) {
+        if(is.leaf(n)) {
+          a <- attributes(n)
+          i <<- i+1
+          attr(n, "nodePar") <-
+            c(a$nodePar, list(lab.col = mycols[i],
+            pch=mysymbols[i],col=mycols[i],
+            lab.cex=0.5, cex=0.5 ))}
+      n }
+    mycols <- c("blue","red","green")[as.factor
+             (substr(labels(dend),1,2))]
+    mysymbols<-c(15,17,1)[as.factor(substr(labels(dend),
```

```
+                                                      1,2))]
+       i <- 0})
>b <- dendrapply(dend, colLab)
>plot(b, main="UPGMA on the iris data set")
```

Notice some new commands in the `colLab` *function. There is the* `substr` *function for extracting a string within a vector. The* `attributes` *and* `attr` *functions access the attributes and a specific attribute of an object respectively.*

Since there is a large number of species in the set, we may prefer to display species on a circular tree, rather than on a rectangular one. We can take advantage of functions that are actually used for plotting phylogenies. The `radial.phylog` *function of the* **ade4** *package plots a radial tree and returns a graphical display easier to visualize. First, we have to transform the hierarchical clustering into a tree with the* `newick2phylog` *function.*

```
>library(ade4)
>kk<-hclust2phylog(bb, FALSE)
>radial.phylog(kk,clabel.l=0.5,cleaves=0,circle=1.7)
>points(0,0,pch=21,cex=2,bg="grey")
```

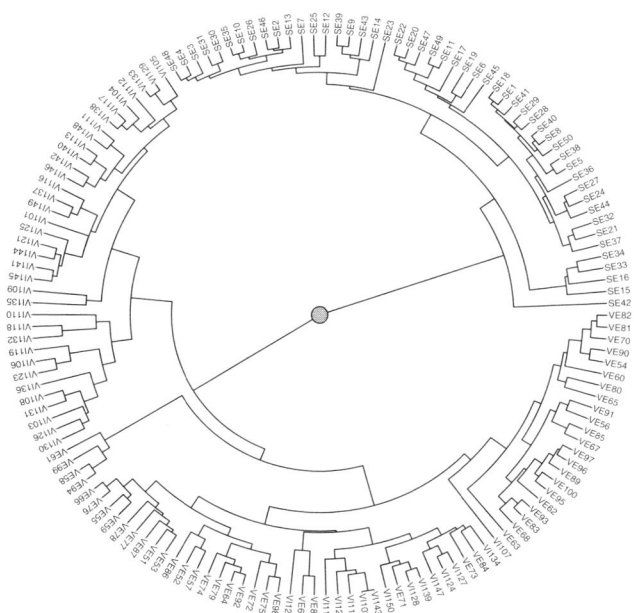

Fig. 3.14. Hierarchical clustering organized in a circular dendrogram on observations of the `iris` dataset. The three species are represented by abbreviations VI, VE, SE followed by numbers indicating their rank in the dataset

The plot (Fig. 3.14) shows that *Iris setosa* forms a clear cluster while the two other species are clustered together. In this second cluster, some members of the *virginica* group are clustered with *versicolor*.

The `kmeans`, `pam` or `clara` functions are some of the numerous partitional clustering methods implemented in R; they need, in addition to the data file, a specified numbers of clusters in their arguments. Users can let the function calculate the dissimilarity matrix or they can specify it with functions such as `dist` or `daisy`.

```
> pam(iris[,1:4],3, stand=F)$clustering
  [1] 1 1 1 1 1 1 1 1 1 1 1 1 1 1 1 1 1 1 1 1 1 1 1 1 1
 [26] 1 1 1 1 1 1 1 1 1 1 1 1 1 1 1 1 1 1 1 1 1 1 1 1 1
 [51] 2 2 3 2 2 2 2 2 2 2 2 2 2 2 2 2 2 2 2 2 2 2 2 2 2
 [76] 2 2 3 2 2 2 2 2 2 2 2 2 2 2 2 2 2 2 2 2 2 2 2 2 2
[101] 3 2 3 3 3 3 2 3 3 3 3 3 3 2 2 3 3 3 3 2 3 2 3 2 3
[126] 3 2 2 3 3 3 3 3 2 3 3 3 3 2 3 3 3 2 3 3 3 2 3 3 2
>palette(c("black", "grey50"))
>par(mar=c(5,4,1,1))
>plot(pam(x=iris[,1:4],k=3), main="",col.p="black")
```

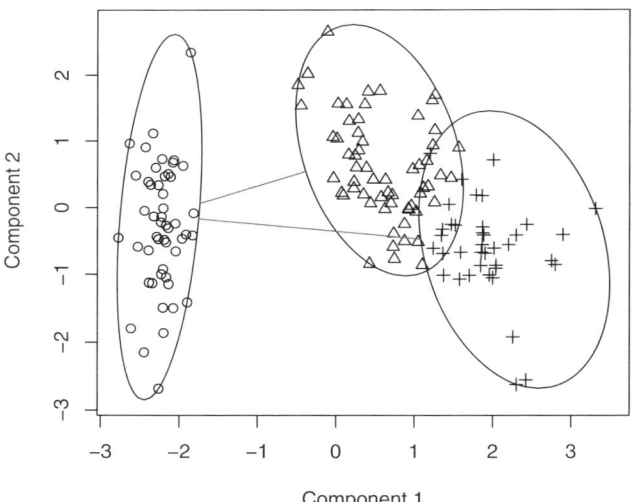

Component 1
These two components explain 95.81 % of the point variability.

Fig. 3.15. Partitional clustering plot on the `iris` dataset. The three groups are represented by different symbols and ellipses are drawn around groups

The `pam` function returns two plots when combined with `plot`. The second plot (Fig. 3.15) is a principal component on the correlation matrix with the group as identified by the algorithm. When we used the `pam` function on raw measurements, all individuals of the first group "*setosa*" have been correctly identified, there were two misidentified *virginica* as *versicolor*, and 14 misidentifications of *versicolor* as *virginica*.

The ideal number of clusters is not known in advance. There is a graphical way to select the best number of clusters called the *elbow criterion*. In progressing on the graph from the left to the right, and in successively removing dots of the curve, the last point permitting a convex elbow corresponds to the number of clusters. For obtaining the number of clusters, we first need to plot the evolution of the ratio of within-group variances by the total variance. The best number of clusters is appraised when adding a group does not improve the explained variance more than expected. This method is arbitrary but sensible. We will see how to implement this protocol for the k-means method (remember, however, that the method of k-means is not an exploration of the best partition; here, it deserves an example and partitions should be reiterated).

```
>totv<-sum(diag(var(iris[,1:4])))*149
>SSratio<-numeric(10)
>for (i in 2:10)
+      {mod<-kmeans(iris[,1:4],i)
+       SSratio[i]<-(totv-sum(mod$withinss))/totv}
>plot(1:10, SSratio, type="l",xlab="number of clusters",
+        ylab="% of explained variance" )
>points(1:10, c(0,SSratio[2:10]), pch=3)
```

By examining the plot produced in Fig. 3.16, the optimal number of clusters for the iris *dataset is two or three with the k-means method.*

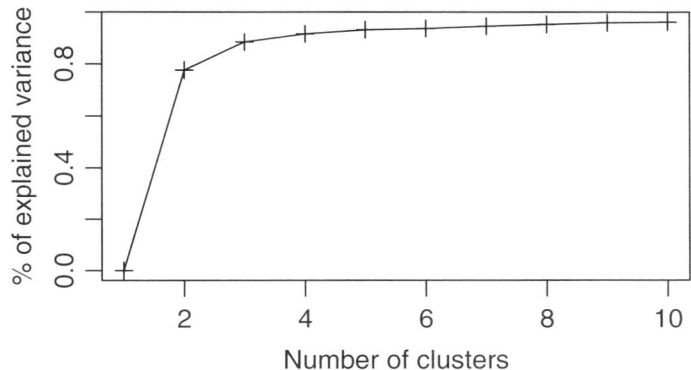

Fig. 3.16. The elbow method for selecting the number of clusters. Here it is applied to the iris dataset with the k-means method

Once clusters are defined with hierarchical or partitional clustering methods, you can check whether they correspond to a taxonomic entity.

It may be convenient to select not only the optimal number of clusters, but also the method for partitioning the data. For this goal, the mclust package provides many possibilities and has functions for selecting the best model. For other functions performing partitional clustering, examples provided with the functions in R can serve as good tutorials.

Note that clustering can work not only on observations, but also on variables. One can use clustering methods to identify whether some variables are closer together. In this respect, the variable clustering should be estimated on the basis of the correlation matrix rather than on the covariance.

3.4.4 Analyzing Relationships Between Different Sets of Variables

One possible aim of morphometrics is to compare morphometric measurements simultaneously with a set of ecological variables, with correspondence tables, or with one other set of morphometric measurements performed on the same individuals.

The `iris` dataset actually has two categories of measurements: two on petals and two on sepals. This dataset allows the study of relationships not only between two measurements but between two structures: sepal and petal. Here we want to discover whether there is a relationship between sepal measurements and petal measurements; and we want characterize this relationship. Several strategies are available: canonical correlation analysis or two-block partial least-squares.

Rohlf and Corti [105] give other reasons for preferring the two-block partial least-squares approach with examples taken from morphometrics. The two-block partial least-squares is a simple methodology for exploring the relationships between two sets of variables. Given n observations with p variables forming a dataset $\mathbf{M}$, one can decompose this dataset into two subsets $\mathbf{M_1}$ and $\mathbf{M_2}$, each of size $n \times p_1$ and $n \times p_2$ with $p_1 + p_2 = p$. One can partition the total variance-covariance matrix ($\mathbf{V}$) or correlation matrix ($\mathbf{R}$) of $\mathbf{M}$ such that

$$\mathbf{V} = \begin{pmatrix} \mathbf{V_{11}} & \mathbf{V_{12}} \\ \mathbf{V_{21}} & \mathbf{V_{22}} \end{pmatrix} ;$$

$\mathbf{V_{11}}$ and $\mathbf{V_{22}}$ are the variance-covariance matrices of each set $\mathbf{M_1}$ and $\mathbf{M_2}$, while $\mathbf{V_{12}}$ or its transpose $\mathbf{V_{21}}$ contains the covariance between the two sets.

The approach decomposes $\mathbf{V_{12}}$ in $\mathbf{F_1DF'_2}$ by singular-value decomposition. The diagonal matrix ($\mathbf{D}$) corresponds to the singular values, while the columns of $\mathbf{F_1}$ are the contributions for the linear combination of the variables of the first set and $\mathbf{F_2}$ are those for the second set. The first columns of $\mathbf{F_1}$ and $\mathbf{F_2}$ give the best least-squares approximation of the covariance between both sets. This relationship is measured by the ratio of the sum of the first singular values by the total sum of the singular values of the dataset. The sum of singular values is a measure of the total covariance between the two sets. If, rather than using the covariation matrix, one works on the correlation matrix, the ratio of the sum of eigenvalues by $p_1 \times p_2$ provides a measure of the overall squared covariance between the two sets of variables. It is possible to project the observations onto the singular vectors for constructing different plots. The projection is achieved by matrix multiplication. To interpret as principal component axis, the final projection should be centered (for example, with the `scale` function, and `scale=F` argument). The observations are projected on $\mathbf{F}_1$ by post-multiplying original data by $\mathbf{F}_1 \mathbf{M}_1 \times \mathbf{F}_1$.

There are several measures of association between the two sets of variables. We would recommend using the R_v coefficient defined by Escoufier [29] because it works directly on covariance and variance rather than on correlations. This coefficient is a measure of correlation between both sets. Its squared value is analogous to the R^2 coefficient in the bivariate case. It is the percentage of covariance of one set predicted by the other set. We compute it such that

$$R_v = \frac{\mathrm{trace}(\mathbf{V}_{21}\mathbf{V}_{12})}{\sqrt{\mathrm{trace}\mathbf{V}_{11}{}^2 \mathrm{trace}\mathbf{V}_{22}{}^2}} \; .$$

If variables in one set have different metrics, then one must work on correlation rather than on covariance matrices. The computations are similar.

Here, we write the `pls` function that computes the singular values, singular vectors, and the Rv coefficient from two subsets of variables organized as two matrices. Working on correlation rather than on covariance is similar to using the scaled original values.

Function 3.4. `pls`

Arguments:
   M1: *First variable subset arranged in a matrix of* n *observations and of* p1 *variables.*
   M2: *Second variable subset arranged in a matrix of* n *observations and of* p2 *variables.*
Values:
   Rv: *Rv coefficient.*
   F1: *Singular vectors for the first set.*
   F2: *Singular vectors for the second set.*
   D: *Singular values.*

```
1  pls<-function(M1, M2)
2  {p1<-dim(M1)[2]; p2<-dim(M2)[2]; n<-dim(M1)[1]
3  sM12<-svd(var(cbind(M1,M2))[1:p1, (p1+1):(p1+p2)])
4  vM12<-var(cbind(M1,M2))[1:p1, (p1+1):(p1+p2)]
5  vM21<-var(cbind(M1,M2))[(p1+1):(p1+p2), 1:p1]
6  v11<-var(M1)
7  v22<-var(M2)
8  D<-sM12$d; F1<-sM12$u; F2<-sM12$v
9  Rv<-sum(diag(vM12%*%vM21))/sqrt(sum(diag(v11%*%v11))*
10          sum(diag(v22%*%v22)))
11 list(Rv=Rv, F1=F1, F2=F2, D=D)}
```

The covariation between petal and sepal forms of the `iris` dataset is investigated for the first species of the set.

```
>pls1<-pls(iris[1:50,1:2], iris[1:50,3:4]); pls1
$Rv
```

```
[1] 0.07605671

$F1
            [,1]        [,2]
[1,] -0.7918234 -0.6107501
[2,] -0.6107501  0.7918234


$F2
            [,1]        [,2]
[1,]  0.8232088 -0.5677388
[2,] -0.5677388 -0.8232088


$D
[1] 0.024410433 0.001279043
```

The covariation between sepal and petal morphologies is rather low ($Rv < 0.1$).
When we examine singular values, we notice that most of the covariation for shape
or form is concentrated on the first dimension of covariation. In terms of morphology, variables are all similarly signed on the first axes, indicating that covariation
is primarily explained by size (big petals are found with big sepals). The second
axis shows a second pattern of covariation with longer than wide petals related with
longer than wide sepals. In our case, one can interpret the first axis as a covariation
axis because of isometric growth, and the second as a major axis of shape covariation.
It is possible to reiterate the procedure for the three species, one will find similar covariation patterns in all. These covariations are stronger for the second species. This
shows that petal and sepal morphologies are more coordinated in the second species
than in the others. One can bootstrap variables among observations to obtain a confidence interval for the Rv coefficient, and to compare different Rv coefficients.

The CCA package provides other functions for computing canonical correlation
analysis that permits similar studies. The cancor function of the stats package is
also performing canonical correlation analysis.

One can appraise relationships between sets of variables using variable clustering.
The Hmisc package provides the varclus function for this purpose. The distances
between variables are based by default on correlations, but alternative methods are
available. You can also have a look at Qannari et al. [86] to appraise other distances
between quantitative variables. For each species of the iris dataset, the variables
are clustered according to the UPGMA method. After the analysis, one can compare
the branching patterns between species in trees for interpreting changes of covariation (Fig. 3.17).

```
>library(Hmisc)
>plot(varclus(as.matrix((iris[1:50,1:4])),method="ave"))
>title("setosa")
>plot(varclus(as.matrix((iris[51:100,1:4])),method="ave"))
>title("versicolor")
>plot(varclus(as.matrix((iris[101:150,1:4])),method="ave"))
>title("virginica")
```

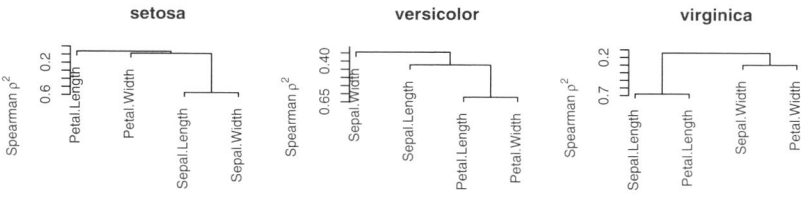

Fig. 3.17. Variable clustering among the different *Iris* species

Measurements are clustered in *Iris setosa* and *Iris versicolor* according to structures (sepal or petal) while the measurements are clustered according analogous shape measurements (length and width) in *Iris virginica* (Fig. 3.17).

One can appraise the significance of the relationships between two sets of variables with the Mantel or Procrustes tests.

The Mantel test aims to determine whether dissimilarities or covariation matrix are similar. For instance, the Mantel test [70] is used to determine whether there is a relationship between geographical distances versus morphological Euclidean distances, or ecological data versus morphological data. One can use the Mantel test to investigate whether the position of observations is similar between shape spaces or form spaces. The `dist` function produces a dissimilarity matrix according to several kinds of distance. One can compare the distance matrices using the `mantel` function of **vegan**, `mantel.test` of **ape**, or `mantel.rtest` of **ade4**.

```
>library(ape)
>unlist(mantel.test(as.matrix(dist(iris[1:50,1:2])),
        as.matrix(dist(iris[1:50,3:4]))))
  z.stat          p
188.9105    0.2530
```

We can compare the relative positions of *Iris* observations in the space defined by original variables between two sets M_1 and M_2 with a Procrustes test (see [84]). Actually, we can interpret each individual observation as a location in the p_1 and p_2 dimensioned spaces. If the relative position of individuals is similar between both spaces, it means that the first and second variable sets are related. Applying a PCA on the data can reduce the space dimensions of the two sets to make them comparable. The x first components of both spaces are extracted, and the test determines whether the position of observations depict a similar configuration in each variable set. To determine whether the relative position of each observation is comparable, whatever the variables used, one uses the Procrustes test with the `protest` function of the **vegan** package. The test performs scaling, rotations, translations and eventually reflections for finding the best match between configurations. Since the test allows reflection, it looks at the geometry of the space rather than at functional relationships between its components. The statistic used is the Procrustes distance between both configurations (see the following chapter), and the test works in permuting rows of one of the configurations to appraise the statistical distribution.

```
>library(vegan)
>protest(iris[1:50, 1:2], iris[1:50,3:4])
Call:
protest(X = iris[1:50, 1:2], Y = iris[1:50, 3:4])

Correlation in a symmetric Procrustes rotation:   0.2443
Significance:   0.083
Based on 1000 permutations.
```

The `protest` function returns an object which can be plotted with `plot`. This provides a graphical display for examining the fit.

Neither Procrustes nor Mantel tests find significant similarity between distances obtained from the sepal and petal measurements for the first species of Iris (*Iris setosa*). The relationships between morphologies are nonetheless significant for the two other species. A last alternative test could be used to compute a χ^2 statistic measure of association between the two covariance matrices; however, the degrees of freedom are uncertain because of the correlative nature of the data.

3.4.5 Comparing Covariation or Dissimilarity Patterns Between Two Groups

Similarly to the previous section, the data matrix is partitioned into different groups of observations. One has p variables and each group has n_i observations. One can characterize each group by its proper covariation or correlation matrix that contains information needed for making between-group comparisons. Since one can cluster variables, this can be a first tool for analyzing qualitatively whether two or more groups differ in their covariation patterns.

One can compare the difference between covariance matrices using a Mantel test, or using a nonparametric test on the singular vectors of the variance-covariance matrix (for example, comparing angles between eigenvectors and angles between random vectors). To test for similarity between covariation matrices with the Mantel test, one should include the diagonal, which contains variances. There is the possibility to modify a function in the R environment with the `fix` function. Writing `fix(myfunction)` edits the function code; users can thus use the editor to modify the function. In the function for the Mantel test, we must specify to take into account the lower triangle matrix, by including the diagonal. In typing `fix(mantel.test)`, we notice that several other internal functions are needed for running the test, especially those for computing the observed and theoretical values. This function, called `mant.zstat`, works with the function `lower.triang`. In this latter function, we have to replace "<" by "<=" to select the lower triangle that includes the diagonal term. Once done, one can close the editor, and the `mantel.test` function will be modified in the R environment. This will not be saved for a future session if the environment is not saved.

Whether covariation between measurements is similar between the first two species of *Iris* is assessed with the help of this modified Mantel test.

```
>unlist(mantel.test(var(iris[1:50,1:4]),
+     var(iris[51:100,1:4])))
   z.stat          p
0.06815982 0.22100000
>unlist(mantel.test(var(iris[1:50,1:4]),
+     var(iris[101:150,1:4])))
   z.stat          p
0.09155193 0.22200000
>unlist(mantel.test(var(iris[51:100,1:4]),
+     var(iris[101:150,1:4])))
  z.stat           p
0.2658072 0.0000000
```

Covariation patterns are different between *Iris setosa* and *Iris versicolor* and between *Iris setosa* and textitIris virginica but similar between *versicolor* and *virginica*.

The `coli` function, programmed in Section 3.3 of this chapter, can investigate similarity between singular or eigen vectors. Singular vectors or eigenvectors of the different covariance matrices represent the main axes of covariation; this tests whether the principal axes are collinear. We can perform every possible comparison between singular vectors, and store them in a matrix. If one uses this test repeatedly, the values must be corrected by Bonferroni adjustment.

```
>res<-matrix(NA,4,4)
>for (i in 1:4)
+    {for (j in 1:i)
+       {res[i,j]<-coli(svd(var(iris[1:50,1:4]))$u[,i],
+        svd(var(iris[51:100,1:4]))$u[,j])$p}}
      [,1]   [,2]   [,3]   [,4]
[1,] 0.136    NA     NA     NA
[2,] 0.336  0.364    NA     NA
[3,] 0.552  0.092  0.470    NA
[4,] 0.982  0.638  0.918  0.012
```

Contrary to the Mantel test, we find that most singular vectors are similar. The difference is because the Mantel test incorporates information about the overall variance-covariance contained in the data, while the test for collinearity only considers the directions of principal axes of variation (without considering the amount of variance they carry).

Problems

3.1. Using lmer

Load the nlme package, and perform an analysis of variance on the `musdom.txt` dataset with the `lmer` function. Are there any major differences in the presentation of results when compared with the traditional ANOVA approach? Can you directly obtain the results with `aov` by specifying the error term in the formula?

3.2. Using car
Construct an unbalanced two-way ANOVA design. Change the order of entrance of each factor. What do you notice with the standard `summary.aov` command? Use the `Anova` function of the car package. How can we calculate sums of squares of type II, writing code using the base and stats packages?

3.3. Using diagnostic tools for examining residuals
Load the lmtest package and perform diagnostics on the linear models that have been applied in this section. Do they all meet required conditions?

3.4. Phylogenetic comparative methods
Load the `carniherbi49` dataset of the ade4 package. Plot the first phylogeny; using the `pic` function, calculate phylogenetic independent contrasts for body Mass, hind limb measurement, and running speed. Is the computed relationship between variables influenced by taking into account the phylogeny?

Using the cubic root of mass as an approximation for size, do you find any significant relationships with body size? Divide running speed by body length to appraise running speed relative to body size. Do heavier animals move faster or slower than light ones? Do you have to transform the variables before assuming a linear relationship?

3.5. Standardized discriminant vectors
We have used the `lda` function to compute standardized linear discriminant vectors. Using the formula in the book, develop a function using linear models for appraising variance-covariance matrices in the `iris` dataset, and calculate the eigenvalues and the standardized linear coefficients. To appraise the inverse square root of a matrix, one must diagonalize the desired symmetric matrix, and then obtain the singular or eigenvectors **U** and eigenvalues **D**. The square root is given by

$$\mathbf{U}\mathbf{D}^{0.5}\mathbf{U}' .$$

The square root of a diagonal matrix corresponds to the matrix of the square root of its diagonal elements.

3.6. Isometry and allometry
Test whether there is isometry in each sex of the `turtles` dataset. If so, test whether allometry is expressed the same way in males and females. Develop appropriate linear models and use the `anova` function to investigate the effect of sex on allometric growth.

3.7. Clustering
By applying the partional clustering method of the `pam` function to shape variation in the `iris` dataset, check whether groups are identical to those defined on form variation.

3.8. Clustering

Using the elbow method and the `pam` function, estimate the number of groups that are present in the `iris` dataset on the basis of shape and form variation.

3.9. Two-block partial least-squares

Write a function to perform a test on the significance of the correlation coefficient between two sets of variables. Use this function to test the significance of covariation between sepal and petal morphologies.

4

Modern Morphometrics Based on Configurations of Landmarks

Most modern morphometric methods deal directly with the geometric information contained in the configuration rather than on a collections of ad-hoc measured distances. Traditional collection of distances have, indeed, few chances to extract exhaustively the geometric information contained in landmarks (see reference [116] for a critical view of traditional morphometrics). Furthermore, new techniques based on configurations have the enormous advantages of offering a visualization and a geometrical interpretation of variation or change of the whole configuration (in contrast to multivariate morphometrics for which the analyst must examine loadings of transformed variables on principal, discriminant or canonical axes).

I explain and detail how to use geometrical and modern techniques in the following two chapters; statistical aspects are exposed in Chapter 6. This chapter addresses the application of modern morphometrics to sets of points whose relative positions have anatomical and homologous grounds basis. Chapter 5 mainly deals with configurations made of sets of points (pseudolandmarks) describing an outlines.

Considerable advances have been made in the last 30 years for taking into consideration as much geometric information as possible from a set of landmarks. Many of the landmark-based methods that have been developed in the second half of the last century have isolated the shape component in removing a scale factor, translation and rotation effects on the configuration (see Sections 4.2, 4.3). However, some other approaches developed in the 1990s have undertaken a very different protocol, using methods that were not dependent on mathematical procedures performing translation, rotation and scaling (Sections 4.4, 4.5) which confer on them some advantages. The former methods, however, have the enormous advantage of having undergone more statistical and methodological developments concerning visualization and decomposition of shape variation along several components.

4.1 The Truss Network Approach of Strauss and Bookstein

The truss network approach described in Strauss and Bookstein [116] is one of the first protocols described for visualizing and understanding shape changes and variation from a given set of configurations. The approach consists of selecting distances of objects objectively rather than randomly, and it allows reconstruction of landmark locations from the set of distance measurements. The method based on the truss network approach, however, has been less successful and less used than superimposition methods; and even if location of landmarks can be obtained using a limited number of measurements, this method should be avoided [10].

Here I briefly present some R implementations, from the selection of landmarks to the reconstruction of the shape. Although not commonly used, it provides a link between multivariate morphometrics and geometric morphometrics. In exhaustively registering information concerning landmark positions on the shape by using distance measurements, this approach is considered to be less biased than classic methods, which tend to accentuate the effect of certain landmarks without intending to do so. The truss network approach allows coordinates of landmarks to be estimated from a minimum set of measurements when no digitizing device is available. It thus constitutes a link between data acquisition and geometric morphometric analysis.

In the truss network approach, the morphology is systematically covered by a set of landmarks organized in quadrilaterals each having two diagonals (that can be considered as tetrahedrons too). The truss is easy to define uniquely for elongated structures, where landmarks lie close to the outline. However, the landmark connection becomes less objective when landmarks occupy both the outline and span the inner part of the object. Defining the set of quadrilaterals is easy for a 2-multiple number of landmarks (Fig. 4.1). It is more difficult if there is a landmark left over at one end. Nevertheless, its position can be defined by its orientation relative to the last quadrilateral.

For starting, one must find some automatic definition of a truss network. For convenience, we will first need to align the main axis of the M configuration along the x-axis. One achieves this by multiplying the coordinates by the eigenvectors of the covariation matrix estimated from the coordinates, and by checking for eventual final reflection of the configuration.

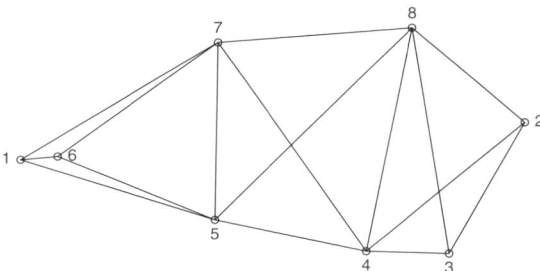

Fig. 4.1. The distances of the truss network for the dataset gorf.dat of the shapes package

```
>library(shapes)
>M<-gorf.dat[,,3]
```

Orient the longer axis of the configuration parallel to the x-axis.

```
>Ma<-M%*%svd(var(M))$u
```

Compare the sign of the first angle of the original configuration with the one of the rotated con-figuration to detect for reflection with the `angle2d` *function defined in the previous chapter (see Section 2.5.2).*

```
>if(round(angle2d(M[1,]-M[2,],M[3,]-M[2,]),3)!=
+    round(angle2d(Ma[1,]-Ma[2,],Ma[3,]-Ma[2,]),3))
+    {Ma[,1]=-Ma[,1]}
```

We need to define the landmarks at the vertices of each quadrilateral, and to store them in an object. In the example developed below, I used an object of the `list` class, each element of the `list` being a vector containing the indices of landmarks involved in the quadrilateral. First quadrilateral vertices are chosen as the set of four landmarks, which coordinates are the more on the left. Further quadrilateral vertices are successively selected from the left to the right along the x-axis.

I followed the approach of Carpenter et al. [18] for reconstructing the configuration. For this approach, we need all landmarks to be involved in at least three inter-landmark distances. If we have an odd number of landmarks, the landmark left over at an end is registered with one of the vertex of the closest quadrilateral. It results that each element of the `list` is obligatory in a set of indices for a quadrilateral.

```
>Ma1<-Ma
>truss<-list()
>rownames(Ma1)<-1:nrow(Ma1)
>a<-1
>while (nrow(Ma1)>4)
+    {truss[[a]]<-NA
+    vert1<-as.numeric(rownames(Ma1)[which.min(Ma1)])
+    truss[[a]][1]<-vert1; Ma1<-Ma1[-which.min(Ma1),]
+    vert1<-as.numeric(rownames(Ma1)[which.min(Ma1)])
+    truss[[a]][2]<-vert1; Ma1<-Ma1[-which.min(Ma1),]
+    vert2<-as.numeric(rownames(Ma1)[which.min(Ma1)])
+    truss[[a]][3]<-vert2; Ma2<-Ma1[-which.min(Ma1),]
+    vert2<-as.numeric(rownames(Ma2)[which.min(Ma2)])
+    truss[[a]][4]<-vert2; Ma2<-Ma2[-which.min(Ma2),]
+    a<-a+1}
>truss[[a]]<-as.numeric(rownames(Ma1))
>if(length(truss[[a]])==3)
+    {truss[[a]]<-c(truss[[a]],truss[[a-1]][2])}
>truss
[[1]]
[1] 1  6  7  5
[[2]]
[1] 7  5  4  8
```

```
[[3]]
[1] 2 3 4 8
```

For drawing the truss as in Fig. 4.1, we apply a triple loop.

```
>plot(Ma,asp=1,axes=F,xlab="", ylab="")
>nq<-length(truss)
>for (i in 1:nq)
+    {for (j in 1:length(truss[[i]])){
+        for (k in 1:j){
+        segments(Ma[truss[[i]][k],1],Ma[truss[[i]][k],2],
+        Ma[truss[[i]][j],1],Ma[truss[[i]][j],2])}}}
```

Reconstructing the relative positions of landmarks from the distances of the network is less obvious [18, 116]. The approach of Carpenter et al. [18] estimates the location of the series of landmarks starting from a prototype and weighting for known interlandmark distances. For the truss not to fold onto itself or not be reflected, it is recommended to start with an initial approximate prototype. I tried the algorithm with a prototype of random landmark positions, that nonetheless, most of the time, finds the correct initial configuration. The algorithm works with different matrices and vector $\mathbf{X}$, $\mathbf{Y}$, and β, so that

$$\mathbf{X} = \begin{pmatrix} x_1 \; y_1 \; 1 \\ \vdots \; \vdots \; \vdots \\ x_p \; y_p \; 1 \end{pmatrix} \; ; \; \beta = \begin{pmatrix} -2x_i \\ -2y_i \\ r_i^2 \end{pmatrix} \; ; \; \text{and } \mathbf{Y} = \begin{pmatrix} d_{i1}^2 - r_1^2 \\ \vdots \\ d_{ip}^2 - r_p^2 \end{pmatrix} \; ,$$

where d_{ij} is the distance between landmarks i and j, and where r_i is the distance of the landmark i to the origin. The above vectors and matrices satisfy the equation

$$\mathbf{X}\beta = \mathbf{Y} \; .$$

One can calculate β using the equation

$$\beta = (\mathbf{X}'I\mathbf{X})^{-1}(\mathbf{X}'I\mathbf{Y}) \; ,$$

where I is the identity matrix. Then the identity matrix is exchanged by a diagonal matrix named W where w_{ij} diagonal elements equal to 1 if the d_{ij} distance is available and to 0 otherwise, which allows us to weight for known distances. Then the above equation is used by replacing I with W for finding the new locations for all landmarks, starting from the first to the last. The newly calculated configuration replaces the prototype and one iterates the procedure until convergence. Carpenter et al. [18] suggest translating the coordinates of the centroid of successive configurations to the origin for insuring convergence, and to limit the number of iterations to 50.

In the example, the second configuration of the verb+gorf.dat+ dataset is used as prototype shape. I divided its coordinates by 10 to show that the algorithm works well even if the first estimated distances are far from those in the final shape.

```
>Xap<-gorf.dat[,,2]/10
```

Dista, *the matrix of available distances, is obtained using the truss network approach (as defined in the previous code). Nonavailable distances appear as* NA *in the matrix.*

```
>Dista<-matrix(NA,8,8)
> for (i in 1:nq){for (j in 1:4){for (k in 1:4){
+      a<-truss[[i]][k]; b<-truss[[i]][j]
+      Dista[a,b]<-sqrt(sum((Ma[a,]-Ma[b,])^2))}}}
>DD<-Dista
```

In the code below, inc *corresponds to the indices of missing interlandmark distances including the* i[th] *landmark and any other.*

```
>a<-1
>b<-10
>while (a<50 & b>0.01){
+      X1<-Xap
+      for (i in 1:8){
+        inc<-which(is.na(Dista[,i]))
+        d1<-dim(Xap[inc,])[1]
+        d2<-dim(Xap[inc,])[2]
+        Dista[inc,i]<-sqrt(apply((Xap[inc,]-
+        matrix(Xap[i,],d1,d2,byrow=T))^2,1,sum))
+        Ra<-apply(Xap^2,1,sum)
+        Y<-Dista[,i]^2-Ra
+        W<-diag(1,8); diag(W)[inc]<-0
```

The solve *function returns the inverse of a matrix.*

```
+        coord<-solve((t(cbind(Xap,1))%*%W%*%
+         cbind(Xap,1)))%*%(t(cbind(Xap,1))%*%W%*%Y)
+        Xap[i,1]<--coord[1]/2
+        Xap[i,2]<--coord[2]/2}
```

Translate the centroid of the new configuration on the origin.

```
+        Xap[,1]<-Xap[,1]-mean(Xap[,1])
+        Xap[,2]<-Xap[,2]-mean(Xap[,2])
+        b<-sum(abs(dist(Xap)-dist(X1)))
+        Dista<-DD
+        a<-a+1}
>points(Xap,asp=1)
```

There are other ways to select a minimal number of distances for covering information on the object; for example, the selection of distances can be based on a Delaunay triangulation returned by the delaunayn function of the **geometry** package. The delaunayn function returns the vertices of triangles involved in the triangulation – then stored as a matrix. The trimesh function displays the triangle mesh. We type the following code to produce the Fig. 4.2. Note, however, that unlike the truss network method, reconstructing the relative position of the landmarks can be ambiguous.

```
>plot(Ma,asp=1,axes=F,xlab="",ylab="")
>dd<-delaunayn(Ma)
>dd
      [,1] [,2] [,3]
[1,]    4    3    2
[2,]    4    8    2
[3,]    4    7    8
[4,]    4    7    5
[5,]    6    5    1
[6,]    6    7    1
[7,]    6    7    5
>trimesh(dd,Ma,add=T)
```

Note the add=T argument, used for displaying the Delaunay triangles on the configuration.

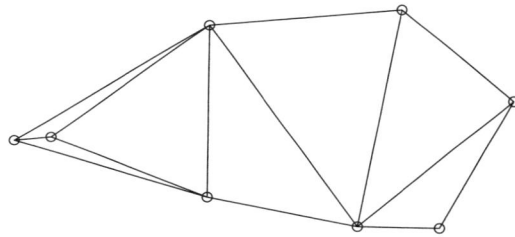

Fig. 4.2. A Delaunay triangulation on the third configuration of the gorf.dat dataset

When measurements are selected, it is theoretically possible to perform any kind of multivariate analysis and to interpret changes by reconstructing morphologies by using the truss network estimated on PC, LD or canonical axes.

4.2 Superimposition Methods

Although the idea of superimposing configurations to quantify their form or shape difference seems rather trivial, the first application of superimposition methods in the field of morphometrics is rather young. This first application may be that of Sneath [115] who used the Procrustes method developed by Green [41] for multivariate statistics. The term Procrustes for statistical purposes was first coined by Hurley and Cattel [45]. In fact "Procrustes" is an analogy from the Greek mythology. Procrustes was a bandit who trapped travelers in Attica. He later invited his victims to lie down on an iron bed and compared their size with the bed size. Limbs of too-tall victims were amputated while too-short victims were stretched (or flattened) to fit the bed size. Like Procrustes did with his victims, superimposition methods fit and scale one configuration (the victim) onto a reference (the bed).

4.2.1 Removing the Size Effect

Superimposition methods first attempt to remove the size, orientation and position information from the form information contained in the configuration (note this is true for the methods exposed in the three following sections but not the last one). Although the definition of size is not trivial, it is an important feature that can differ within a set of configurations, and that can drive shape variation (allometry). One must therefore estimate size from the set of configurations, and test factors that could influence size. Size is considered as a univariate measure, calculated by applying a function of size measure to the original coordinates. A $g(M)$ function of size measure must be real valued and should satisfy

$$g(aM) = ag(M) \qquad \text{for any positive scalar } a \text{ .}$$

The most commonly used measure of size for a configuration is the centroid size (e.g., [8, 38, 54, 27]) . It is defined as the square root of the sum of squared distances from each landmark to the centroid of the configuration. It is independent of position or orientation of the configuration. One computes centroid size directly from raw coordinates. The distances of landmarks to the centroid are calculated as the square root of the sum of squares of the difference between their coordinates and those of the centroid. Let M be a configuration matrix with p landmarks (in rows) and k coordinates (in columns) and sc_M be the centroid size of M, then

$$sc_M = \sqrt{\sum_{i=1}^{p}\sum_{j=1}^{k}(M_{ij} - \overline{M_{.j}})^2} \text{ .}$$

The centroid coordinates M_c of the configuration are the arithmetic mean for each coordinate dimension:

$$M_c = \frac{1}{p}\sum_{i}^{p} M_p \text{ .}$$

If the configuration is arranged in a matrix, one can calculate centroid coordinates with the `apply` function of R that applies the same function to the margin of a matrix. The following `centcoord` function calculates the coordinates of the centroid of the `M` configuration.

Function 4.1. `centcoord`

Argument:
    `M`: *Configuration matrix.*
Value:
    Vector of centroid coordinates.

```
centcoord<-function(M){apply(M,2,mean)}
```

We can use it for plotting a configuration and its centroid.

```
>M<-matrix (c(2,0,1,1,0,0,0,-1,2,-2),5,2, byrow=T)
>centcoord(M)
>plot(M)
>polygon(M)
>points(t(centcoord(M))
```

The coordinates of the scaled configuration (*Ms*) are calculated by dividing the original coordinates of the configuration by the centroid size, such as

$$Ms = \frac{M}{sc_M} \ .$$

We program a small function, `centsiz`, for scaling the configuration to unit centroid size, and for returning both the scaled configuration and the centroid size. The configuration passed as argument of the function is organized as a matrix object (`M`) of p rows and k columns. We could have used the `centcoord` function, which would then become an internal function of `centsiz`. However, it is also simpler to derive the centroid size directly from the variance of each x, y or z-dimension of the configuration.

Function 4.2. `centsiz`

Argument:
    `M`: *Configuration matrix.*
Values:
    `centroid_size`: *Centroid size.*
    `scaled`: *Configuration matrix, scaled to centroid size.*

```
centsiz<-function(M)
{p<-dim(M)[1]
size<-sqrt(sum(apply(M, 2,var))*(p-1))
list("centroid_size" = size,"scaled" = M/size)}
```

The `csize` function of the **Rmorph** package calculates the centroid size of a set of configurations. One can pass this set into argument of the function under various types of object (vector (of size $k \times p$), matrix (of dimensions n and $k \times p$), or array of dimensions p, k, and n). The second argument of that function indicates the number of dimensions.

Dryden and Mardia [27] have proposed a normalized form of the centroid size ($sc_M/\sqrt{p}$ or $sc_M/\sqrt{p \times k}$) for comparing configurations with different numbers of landmark.

The distance between two landmarks $d_{p_1 p_2}$ defining a baseline has also been proposed as a measure of the size scalar [33] and was later used by Bookstein [7, 10] as geometric registration for extracting shape information of 2D configurations. We write the `basesiz` function. It returns the baseline size and the scaled configuration and accepts an `M` argument of the `matrix` class. The function should take into account the `p1` and `p2` indices of the baseline landmarks as well.

Function 4.3. `basesiz`

> *Arguments:*
> > `M`: *Configuration matrix.*
> > `p1`: *Index of the first baseline landmark.*
> > `p2`: *Index of the second baseline landmark.*
> *Value:*
> > *Baseline size.*

```
1  basesiz<-function(M, p1, p2)
2  {sqrt(sum((M[p1,]-M[p2,])^2))}
```

4.2.2 Baseline Registration and Bookstein Coordinates

For 2D data, Bookstein [7, 10] suggests removing the effect of similarity transformation by sending the two landmarks M_{p_1} and M_{p_2} defining the baseline to a fixed position of respective coordinates $(-1/2, 0)$ and $(1/2, 0)$. These Bookstein coordinates Mb_{i1} and Mb_{i2} are defined so that

$$Mb_{i1} = (M_{p_2 1} - M_{p_1 1})(M_{p_i 1} - M_{p_1 1}) + (M_{p_2 2} - M_{p_1 2})(M_{p_i 2} - M_{p_1 2})/d_{p_1 p_2}^2 - \frac{1}{2},$$

and

$$Mb_{i2} = (M_{p_2 1} - M_{p_1 1})(M_{p_i 2} - M_{p_1 2}) - (M_{p_2 2} - M_{p_1 2})(M_{p_i 1} - M_{p_1 1})/d_{p_1 p_2}^2,$$

where $d_{p_1 p_2}$ is the baseline size.

The `booksteinM` function is written to compute the coordinates of superimposed configurations using the baseline registration. We will write it for a configuration matrix passed as the `M` argument of dimensions `p` and `k`. The function calls `basesiz`, which we have developed above. The function rotates, translates and scales `M` onto the baseline defined by the `p1` and `p2` landmark indices.

Function 4.4. `booksteinM`

Arguments:

    `M`: *Configuration matrix.*

    `p1`: *Index of the first baseline landmark.*

    `p2`: *Index of the second baseline landmark.*

Value:

    Scaled configuration matrix aligned on the baseline of coordinates $(-0.5, 0)$ and
        $(0.5, 0)$.

Required function: `basesiz`.

```
  booksteinM<-function(M, p1, p2)
  {D<-basesiz(M, p1, p2)
  m<-matrix(NA, nrow(M), ncol(M))
  p1<-M[p1,]
  p2<-M[p2,]
  m[,1]<-(((p2[1]-p1[1])*(M[,1]-p1[1])+(p2[2]-p1[2])
  +      *(M[,2]-p1[2]))/(D^2))-0.5
  m[,2]<-((p2[1]-p1[1])*(M[,2]-p1[2])-(p2[2]-p1[2])
  +      *(M[,1]-p1[1]))/(D^2)
  m}
```

We use this function for performing the same task on a set of configurations organized in an `array` object using a single loop. This is implemented within the `booksteinA` function.

Function 4.5. `booksteinA`

Arguments:

    `A`: *Array containing configuration matrices.*

    `p1`: *Index of the first baseline landmark.*

    `p2`: *Index of the second baseline landmark.*

Value:

    Array of scaled configuration matrices aligned on the baseline of coordinates $(-0.5, 0)$
        and $(0.5, 0)$.

Required functions: `booksteinM, basesiz`.

```
  booksteinA<-function(A, p1, p2)
  {B<-array(NA, dim=c(dim(A)[1],dim(A)[2],dim(A)[3]))
  for (i in 1: dim(A)[3])
      {B[,,i]<-booksteinM(A[,,i], p1, p2)}
  B}
```

We can appraise a mean shape using the Bookstein coordinates as the configuration of coordinates corresponding to the mean of all individual coordinates. We develop the `mbshape` function to perform this task. The function returns the mean shape matrix and works with the same arguments as in `booksteinA`.

Function 4.6. `mbshape`

Arguments:
    A: *Array containing configuration matrices.*
    p1: *Index of the first baseline landmark.*
    p2: *Index of the second baseline landmark.*
Value:
    Matrix of mean shape coordinates.
Required functions: `booksteinA, booksteinM, basesiz.`

```
1  mbshape<-function(A,p1, p2)
2  {B<-booksteinA(A, p1, p2)
3  k<-dim(A)[2]
4  mbshape<-matrix(NA, dim(A)[1], dim(A)[2])
5  for (i in 1:k)
6      {mbshape[,i]<-apply(B[,i,], 1, mean)}
7  mbshape}
```

The `bookstein2d` function of the **shapes** package performs a baseline registration for 2D data. We plot the resulting superimposition with the `plotshapes` function of the same package (see Fig. 4.3).

```
>library(shapes)
>data(gorf.dat)
>B<-bookstein2d(gorf.dat)
>plotshapes(B$bshpv)
```

We can produce more customized graphs with our own functions. Here is an example of code and the corresponding graph (see Fig. 4.4).

```
>layout(matrix(c(1,2),1,2))
>data(gorm.dat)
>plot(mbshape(gorf.dat,1,2),pch=18,asp=1,
+      xlab="",ylab="",axes=F)
>points(mbshape(gorm.dat,1,2), pch=22)
>lines(mbshape(gorm.dat,1,2)[c(1,6,7,8,2,3,4,5,1),])
>lines(mbshape(gorf.dat,1,2)[c(1,6,7,8,2,3,4,5,1),],lty=2)
>Fe<-booksteinA(gorf.dat, 1, 2)
>Ma<-booksteinA(gorm.dat, 1, 2)
>plot(Fe[,1,],Fe[,2,],asp=1,axes=F, xlab="",ylab="",
+      cex=0.5,pch=18)
>points(Ma[,1,],Ma[,2,], cex=0.5,pch=22)
>segments(-0.5,0,0.5,0, lw=2)
```

The baseline corresponds to the location of two landmarks for the 2D registration; as a consequence the set of all possible shapes can be expressed in a $p \times (2-4)k$ space, k being the number of dimensions of the configuration. It is possible to check it by computing the singular values of the variance-covariance matrix: the last four singular values should be null. The `svd` function returns a `list` containing singular vectors and singular values of a rectangular matrix.

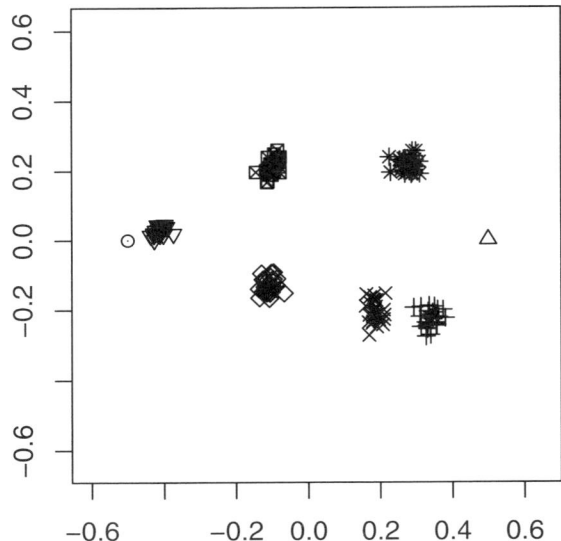

Fig. 4.3. A visualization of the Bookstein registration obtained with functions of the shapes package: configurations corresponding to eight landmarks digitized on the midline section for 30 female gorilla skulls are registered onto the first two landmarks

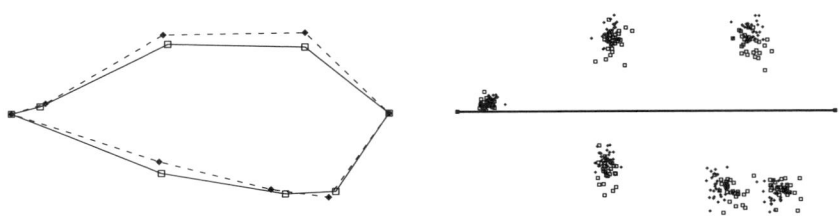

Fig. 4.4. Customized visualization of the Bookstein registration, on the left: mean skull shapes for male and female gorilla (from the `gorf.data` dataset); on the right: superimposed configurations of the 30 and 28 individuals for both sexes, with the baseline plotted as a full segment

```
>a<-mbshape(gorf.dat,1,2)
```

Reorganize data so that each configuration corresponds to a row with coordinates of landmarks in columns.

```
>ma<-matrix(Fe, 30, 16, byrow=T)
```

Compute the variance-covariance matrix.

```
>vma<-var(ma)
```

Compute the singular values.

```
>round(svd(vma)$d,6)
 [1] 0.002213 0.000753 0.000600 0.000228 0.000170 0.000101
 [7] 0.000093 0.000087 0.000071 0.000047 0.000028 0.000021
[13] 0.000000 0.000000 0.000000 0.000000
```

As expected the four last singular values are null.

Bookstein registration is possible for 3D data. In this case, two landmarks, M_{p_1} and M_{p_2}, are set at coordinates $(-1/2, 0, 0)$ and $(1/2, 0, 0)$. The third landmark, M_{p_3}, used for the registration, is set on the x, y-plane. The transformed coordinates of M_{p3} are then: $(Mb_{p_3 1}, Mb_{p_3 2}, 0)$. The $Mb_{p_3 2}$ coordinate is enforced to be positive in order to avoid reflection. Registration needs three translations, 1 scaling, and three rotations for removing location, orientation and scale effects. Bookstein coordinates vary thus in a space of $3 \times p - 7$ dimensions. Producing the registered configurations follows three steps. The first step translates the configurations so that the mid point between M_{p_1} and M_{p_2} corresponds to the origin of the system. The coordinates of the mid point are thus removed from the coordinates of the whole configuration (M) for each dimension k to obtain a translated configuration (Mt), such as

$$Mt_{.k} = M_{.k} - 1 \frac{M_{p_1 k} + M_{p_2 k}}{2} \, ,$$

with **1** being a column vector of p 1, p being the number of landmarks in the configuration. In R code, the corresponding `tranb` function computes the transformation for a configuration matrix so that the middle of the baseline becomes of coordinates $(0, 0, 0)$.

Function 4.7. `tranb`

Arguments:
    M: *Configuration matrix.*
    p1: *Index of the first baseline landmark.*
    p2: *Index of the second baseline landmark.*
Value:
    Translated configuration matrix.

```
1  tranb<-function(M, p1, p2)
2  {M-matrix((M[p1,]+M[p2,])/2, nrow(M), ncol(M), byrow=T)}
```

Translated configurations (*Mt*) are later rescaled by dividing the coordinates by the baseline distance between the points M_{p_1} and M_{p_2}:

$$Mts = Mt/d_{p_1 p_2} \ .$$

The `transb` function translates and scales the M using a baseline registration defined by the landmarks `p1` and `p2`. It is developed below.

Function 4.8. `transb`

Arguments:
　　M: *Configuration matrix.*
　　p1: *Index of the first baseline landmark.*
　　p2: *Index of the second baseline landmark.*
Value:
　　Translated and scaled configuration matrix.
Required function: `basesiz`.

```
1  transb<-function(M, p1, p2)
2  {tranb(M, p1, p2)/basesiz(M, p1, p2)}
```

The third step is more difficult to implement since we have to operate a series of clockwise rotations around the x, y and z-axes through respective angles θ, ω, ϕ. One calculates these angles from the transformed coordinates of the three selected landmarks Mts_{pk} of the *Mts* matrix. We calculate them following three more steps.

1. Align the baseline on the (x, z)-plane, and compute the first angle θ. This angle is a rotation along the z-axis. It corresponds to the angle between the positive x-axis and the x and y-coordinates of the second landmark Mts_{p_2}. The configuration is rotated around the z-axis, using the $\mathbf{\Gamma_z}$ rotation matrix defined as

$$\mathbf{\Gamma_z} = \begin{pmatrix} \cos\theta & \sin\theta & 0 \\ -\sin\theta & \cos\theta & 0 \\ 0 & 0 & 1 \end{pmatrix} \ .$$

We calculate the rotated coordinates *Mtsr* so that

$$Mtsr = (\mathbf{\Gamma_z} Mts')' \ .$$

2. Align the baseline along the x-axis. One must rotate the new *Mtsr* to send the points $Mtsr_{p_1,p_2}$ to the coordinates $(-0.5, 0, 0)$ and $(0.5, 0, 0)$. This involves a rotation of ω around the y-axis. This angle is defined by the vectors $\overrightarrow{Mtsr_{p_2}, O}$, and $(0.5, 0, 0)$. Then rotate *Mtsr* around the y-axis using the $\mathbf{\Gamma_y}$ rotation matrix such that

$$\mathbf{\Gamma_y} = \begin{pmatrix} \cos\omega & 0 & -\sin\omega \\ 0 & 1 & 0 \\ \sin\omega & 0 & \cos\omega \end{pmatrix} \ .$$

We calculate the new rotated coordinates *Mtsrr* as

$$Mtsrr = (\mathbf{\Gamma_y} Mtsr')' \ .$$

3. Align the triangle $Mtsrr_{p_1, p_2, p_3}$ on the (x, y)-plane. Since $Mtsrr_{p_1, p_2}$ are already aligned on the x-axis, this corresponds to a rotation of ϕ along the x-axis. One must rotate the new $Mtsrr$ to send the points $Mtsrr_{p_3}$ to the coordinates $(x, y, 0)$. The rotation angle is defined by the vectors $(0, Mtsrr_{p_3 2}, 0)$ and $(0, Mtsrr_{p_3 2}, Mtsrr_{p_3 3})$. Then, one rotates $Mtsrr$ around the y-axis using the Γ_x rotation matrix defined as

$$\Gamma_x = \begin{pmatrix} \cos\phi & 0 & \sin\phi \\ 0 & 1 & 0 \\ -\sin\phi & 0 & \cos\phi \end{pmatrix},$$

The rotated matrix of coordinates (Mb) is obtained as

$$Mb = (\Gamma_x Mtsrr')'.$$

These coordinates are the 3D Bookstein coordinates Mb. They finally correspond to

$$Mb = (\Gamma_x \Gamma_y \Gamma_z Mts')'.$$

We write the `bookstein3d` function that performs a 3D registration of a given configuration according to a baseline and a reference plane defined by three landmarks `p1`, `p2` and `p3`.

Function 4.9. `bookstein3d`

Arguments:
  M: *Configuration matrix.*
  p1: *Index of the first baseline landmark.*
  p2: *Index of the second baseline landmark.*
  p3: *Index of the third landmark of the reference plane.*
Value:
  Scaled configuration matrix aligned on reference plane and on the baseline of coordinates $(-0.5, 0)$ *and* $(0.5, 0)$.
Required functions: `transb`, `basesiz`, `angle3`.

```
bookstein3d<-function(M, p1, p2, p3)
{m<-transb(M, p1, p2)
te1<-angle3(c(1,0,0),c(m[p2,1],m[p2,2],0))
Rz<-matrix(c(cos(te1),sin(te1),0,-sin(te1),cos(te1),0,
     0,0,1),3,3,byrow=T)
m1<-t(Rz%*%t(m))
om1<-angle3(c(m1[p2,1],0,0),c(m1[p2,1],0,m1[p2,3]))
Ry<-matrix(c(cos(om1),0,-sin(om1),0,1,0,
     sin(om1),0,cos(om1)),3,3,byrow=T)
m2<-t((Ry)%*%(Rz)%*%t(m))/basesiz(M,p1,p2)
ph1<-angle3(c(0,m2[p3,2],0),c(0,m2[p3,2:3]))
Rx<-matrix(c(1,0,0,0,cos(ph1),sin(ph1),
     0,-sin(ph1),cos(ph1)),3,3,byrow=T)
t(Rx%*%Ry%*%Rz%*%t(m))}
```

Since it is possible to define a mean shape with Bookstein coordinates, we can analyze shape variation. Some variability models have been developed around Bookstein coordinates (for example, see [27]; however, these registrations usually introduce spurious correlations because of the registration protocol. Moreover, the superimposition depends on the definition of the baseline (and for 3D on the definition of the reference plane).

4.2.3 Procrustes Methods and Kendall Coordinates

The choice of the baseline for registration methods is not always trivial, and this strategy introduces some further biases; it is therefore desirable to find a more objective method for inferring scale, translation and rotation from the whole dataset of landmarks. This is the purpose of Procrustean methods that have been developed throughout the 20th Century. An important part of the revolution of morphometrics announced by Rohlf and Marcus in 1993 has been done in the quest for optimally superimposing a set of configuration and inferring shape variation from these superimpositions. This quest began with the research of Kendall [54], Goodall [38] and Bookstein [7, 8]. Important theoretical work and statistics have been developed around the dimensionalty and geometry of this space during the last two decades, which gave rise to the science of shape statistics. Kendall [54] introduced the Procrustes distance as a metric of a shape space defined by a general superimposition procedure. This space later received the name of the Kendall shape space [114] for which I provide a summarized explanation in Section 4.2.4.

The Least Squares Approach

Procrustes superimpositions in their strict sense are least-squares methods to estimate superimposition parameters (scale, rotation, translation) between configurations. The protocol used for the Procrustes superimposition aims to minimize the sum of squared distances between similar landmarks of configurations by allowing size, rotation and translation to be adjusted. This sum is called the Procrustes distance $d_{F_{M1,M2}}$.

The superimposition of two configuration matrices, *M1* and *M2*, involves estimating three parameters α, Γ, and β to minimize the quantity $d_{F_{(M1,M2)}}$. Therefore,

$$d_{F_{(M1,M2)}} = \min \| M2 - \beta M1 \Gamma - \mathbf{1}_p \alpha' \| ,$$

where β is a scalar for the size parameter, Γ is a square rotation matrix of $k \times k$ dimensions for the orientation parameter, α is the location parameter corresponding to a vector of k values, and $\mathbf{1}_p$ is a column vector of p 1.

Removing Location

One can filter the location effect by removing the centroid coordinates $M1_c$ and $M2_c$ for *M1* and *M2* configurations. The coordinates of the centroid define the translation parameter α. We have already written the `centcoord` function for returning the coordinates of the centroid of the configuration when we have been defining the centroid size (Section 4.2.1). To perform the translation, one must remove the value of these coordinates from the ones of the original configuration. This yields the centered the configuration (noted X in the following section). The `transl` function translates a configuration so that its centroid is set at the origin.

Function 4.10. `transl`

Argument:
    M: *Configuration matrix.*
Value:
    Translated configuration matrix (so that the centroid is sent at the origin).
Required function: `centcoord`.

```
transl<-function(M)
{M - matrix(centcoord(M), nrow(M), ncol(M), byrow=T)}
```

The `scale` function with the `scale=F` argument performs the same task more rapidly. It is used in the `transl` function that faster computes the coordinate of the translated configuration.

Function 4.11. `transl`

Argument:
    M: *Configuration matrix.*
Value:
    Translated configuration matrix (so that the centroid is sent at the origin).

```
transl<-function(M){scale(M,scale=F)}
```

Here we check that our functions work properly: The centroid coordinates of the centered configuration should equal zero, and the centroid size of the centered configuration should be the same as the original one.

```
>test<-gorf.dat[,,5]
>ntest<-transl(test)
>apply(ntest, 2, mean)
[1] 0 0
>centsiz(test)[[1]]
[1] 229.2441
>centsiz(ntest)[[1]]
[1] 229.2441
```

Alternatively one can obtain the centered configurations (noted X) from original configurations (M) by premultiplying the configuration matrix (M) by the centering matrix. The centering matrix has a diagonal equal to $1 - 1/p$ and lower and upper triangle cells equal to $-1/p$. This multiplication is easy to compute in R. In our example, we must type the code:

```
>p<-nrow(test)
>cm1<-diag(1,p)-1/p
>cm1%*%test
          [,1]   [,2]
[1,] -10.125 111.5
[2,]  21.875 -98.5
[3,] ...
```

More interestingly, one can calculate the centering matrix by premultiplying the sub Helmert matrix by its transpose [27]. The Helmert matrix is a square $p \times p$ orthogonal matrix with the first row of elements equal to $1/\sqrt{p}$, then the ith row has $i - 1$ elements equal to $-1/\sqrt{i(i-1)}$, followed by one element equal to $(i-1) \times 1/\sqrt{i(i-1)}$, and $p - i$ zeros. The sub-Helmert matrix is a Helmert matrix with its first row dropped. The helmert function allows us to compute the Helmert matrix of size $p \times p$:

Function 4.12. `helmert`

Argument:
  p: *Number of landmarks.*
Value:
  Helmert matrix.

```
1  helmert<-function(p)
2  {H<-matrix(0, p, p)
3  diag(H)<--(0:(p-1)) * (-((0:(p-1))*((0:(p-1))+1))^(-0.5))
4  for (i in 2:p){H[i,1:(i-1)]<- -((i-1)*(i))^(-0.5)}
5  H[1,]<-1/sqrt(p)
6  H}
```

We check that the premultiplication of the sub-Helmert matrix by its transpose is equal to the centering matrix:

```
>t(helmert(4)[-1,])%*%helmert(4)[-1,]
          [,1]   [,2]   [,3]   [,4]
[1,]  0.75 -0.25 -0.25 -0.25
[2,] -0.25  0.75 -0.25 -0.25
[3,] -0.25 -0.25  0.75 -0.25
[4,] -0.25 -0.25 -0.25  0.75
```

By premultiplying the configuration by the sub-Helmert matrix, one removes location considering the contrasts of the data. This is one step for obtaining Kendall coordinates [27] – Kendall coordinates are similar to Bookstein coordinates but location, rotation and scaling are not computed the same way. In removing location using the sub-Helmert matrix, we obtain a configuration known as the Helmertized configuration (*Mh*), a terminology that we find in several publications (e.g., [27]). This configuration has $(p \times k)$ - k coordinates. The `helmertm` function returns the

Helmertized configuration of a raw configuration matrix.

Function 4.13. `helmertm`

Argument:
    M: *Configuration matrix.*
Value:
    Helmertized configuration matrix.
Required function: `hermert`.

```
helmertm<-function(M)
{helmert(nrow(M))[-1,]%*%M}
```

Filtering Size

We can also remove size information by scaling all configurations so that they all have a similar size (according to the definition of size we adopt). Removing size is done by dividing coordinates by the size of the object. Scaling the centered configuration and the Helmertized configuration, respectively, give the preshape (Z) and the centered preshape (Zc). One can plot centered preshapes but they define a space that loses $k + 1$ dimensions by comparison with the raw data (configurations occupying a space of $p \times k$ dimensions). Preshape configurations are nearly of full rank (They have $kp - k$ coordinates and only one dimension is lost during scaling).

The preshape space is the space that represents all possible preshapes for configurations of p landmarks and k coordinates. This terminology was first introduced by Kendall [54]. If we scale all configurations to unit size, the preshape space can be considered as a hypersphere of radius 1 in $(p - 1) \times k$ dimensions. Rotation nuisance is still not filtered in this space.

Removing Orientation

By comparison to the preshape space, the shape space is the set of of all possible shapes for given number of landmarks and coordinates. Shapes are invariant to rotation in the shape space, in addition to being invariant to scale and translation (as in the preshape space). Since there are less dimensions in the shape and preshape spaces than there are geometric coordinates, these spaces are non Euclidean. In the preshape space all possibilities for rotations of a given shape are organized along an orbit called a fiber [54, 27]. A fiber in the preshape space corresponds thus to a shape in the shape space. Finding the rotation parameters to superimpose *M1* on *M2* is equivalent to finding the shortest distance between both fibers in the preshape space.

If we allow *M2* to adjust its size in order to decrease interlandmark differences with *M1*, then *Z2* will move onto the radius of the hypersphere until the distance with *Z2* to be minimized. The best match is achieved when the full Procrustes distance is found. I followed [27] to illustrate the purpose (Fig. 4.5).

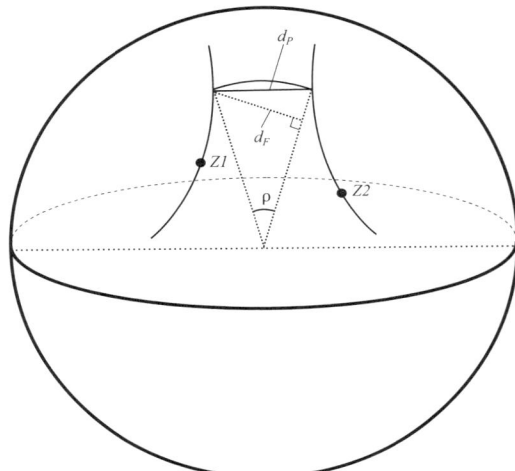

Fig. 4.5. A schematic illustration of the preshape space with the preshape $Z1$ and $Z2$ on their fibers. The chordal distance d_P between both fibers is the partial Procrustes distance. ρ is the smallest angle between $Z1$ and $Z2$ over rotation of $Z1$ and $Z2$. The shortest distance between $Z1$ and the radius of $Z2$ is the full Procrutes distance d_F

Kendall coordinates are invariant to location, orientation and scaling. One computes Kendall coordinates (fide [27]) using complex algebra and Helmertized configurations. For 2D data, the x and y-dimensions of the Helmertized configuration can be passed to the real and imaginary part of complex numbers. Kendall coordinates Mk are produced by dividing the complex coordinates of Mh by the first complex coordinate. They are thus independent of size, rotation and location. Besides, Kendall coordinates are interesting for understanding the geometry of shape spaces (see Section 4.2.4). We return them with the `kendall2d` function.

Function 4.14. `kendall2d`

Argument:
    M: *Configuration matrix.*
Value:
    Matrix of Kendall coordinates.
Required functions: `herlmertm`, `helmert`.

```
kendall2d<-function(M)
 {Mh<-helmertm(M)
 mhc<-complex(nrow(Mh),Mh[,1], Mh[,2])
 (mhc/mhc[1])[-1]}
```

For 2D configurations, one can also use this complex transformation on Bookstein coordinates. Then, one can ignore the coordinates of the baseline, because they are invariant from configuration to configuration. It is easy to pass from complex Bookstein coordinates Mb to complex Kendall coordinates Mk or from Kendall co-ordinates to Bookstein coordinates applying the following equations:

$$Mk = \sqrt{2}\mathbf{H}_1 Mb \;,$$

$\mathbf{H}_1$ being the lower right $(p-2) \times (p-2)$ partition matrix of the Helmert matrix. and

$$Mb = (\mathbf{H}_1'\mathbf{H}_1)^{-1}\mathbf{H}_1' Mk/\sqrt{2} \;.$$

The second possibility is appraising the Γ rotation matrix directly from centered configurations or from centered preshapes by matrix operations. For achieving this, one must remember that the full Procrustes distance d_F between $M1$ and $M2$ is a measure of shape difference between these configurations, and must satisfy the following equation:

$$d_F = \min \|Z2c - \beta 1 Z1c\Gamma\| \;,$$

where $Z2c$ and $Z1c$ are the centered configurations of $M1$ and $M2$ scaled to unit centroid size. $\beta 1$, Γ being scaling and rotation parameters. Alternatively one can write

$$d_F = \min \|X2 - \beta X1\Gamma\| \;,$$

where $X2$ and $X1$ are the centered configurations of $M1$ and $M2$. β, **Gamma** being scaling and rotation parameters.

Note that even though one works on centered and scaled configurations, there is still a scale parameter; although configurations have been scaled to identical size, one can adjust the size of one of the configurations to optimize the minimizing of interlandmark distances. We have

$$\Gamma = \mathbf{UV}' \;,$$

where $\mathbf{U}$ and $\mathbf{V}$ are matrices of singular vectors coming from the singular-value decomposition of the product of the transpose of $X2$ by $X1$. This is written

$$X2'X1 = \mathbf{V\Delta U}' \;.$$

The trace of $\mathbf{\Delta}$ (sum of λ_k elements) contains information for computing the scaling parameter for fitting the $X1$ configuration onto $X2$. One can obtain the full Procrustes distance using the relation

$$d_F = \sqrt{\left(1 - (\sum_{i=1}^{k} \lambda_i)^2\right)} \;.$$

The parameter β for best fit is computed as

$$\beta = \frac{\text{trace}(X2'X1\Gamma)}{\text{trace}(X1'X1)} \;.$$

Note in this equation that the centroid size of *M1* is the square root of the sum of the k eigenvalues of $X1'X1$.

The full Procrustes distance between *M1* and *M2* is given by

$$d_F = \sqrt{\operatorname{trace}\big((X2 - \beta X1\boldsymbol{\Gamma})'(X2 - \beta X1\boldsymbol{\Gamma})\big)} \, .$$

When one considers orientation, one must specify whether or not one considers reflection. The possible reflection can be taken into account by paying attention on the sign of the determinant of $X2'X1$. The singular-value decomposition considers both orientation and reflection. When there is a reflection to improve the fit, the determinant is negative. One can remove this possible reflection by checking the sign of the determinant, and by inverting the sign of the last column of **V**. In this case, the smallest value λ_k is set as negative $\iff \det(X2'X1) < 0$.

One can perform the same operations on the centered configurations scaled to unit centroid size (centered preshapes) rather than to work on centered configurations. In this case, one computes $\beta 1$.

We write the `fPsup` function to perform the full Procrustes superimposition of *M1* onto *M2*. For convenience, configurations are scaled to centroid size in the computation.

Function 4.15. `fPsup`

Arguments:
  M1: *Configuration matrix to be superimposed onto the centered preshape of* M2.
  M2: *Reference configuration matrix.*
Values:
  Mp1: *Superimposed centered preshape of* M1 *onto the centered preshape of* M2.
  Mp2: *Centered preshape of* M2.
  `rotation`: *Rotation matrix.*
  `scale`: *Scale parameter.*
  DF: *Full Procrustes distance between* M1 *and* M2.
Required functions: `centsiz`, `trans1`.

```
 1  fPsup<-function(M1, M2)
 2  {k<-ncol(M1)
 3  Z1<-trans1(centsiz(M1)[[2]])
 4  Z2<-trans1(centsiz(M2)[[2]])
 5  sv<-svd(t(Z2)%*%Z1)
 6  U<-sv$v; V<-sv$u; Delt<-sv$d
 7  sig<-sign(det(t(Z2)%*%Z1))
 8  Delt[k]<-sig*abs(Delt[k]) ; V[,k]<-sig * V[,k]
 9  Gam<-U%*%t(V)
10  beta<-sum(Delt)
11  list(Mp1=beta*Z1%*%Gam,Mp2=Z2,rotation=Gam,scale=beta,
12      DF=sqrt(1-beta^2))}
```

Remember that the full Procrustes distance is also equal to the square root of the sum of the squared distances between homologous coordinates of superimposed configurations (previously scaled to unit size). To check this relationship, we first write the `ild2` function that calculates the p interlandmark distances between two configurations.

Function 4.16. `ild2`

Arguments:
   `M1`: *First configuration matrix of k dimensions and p landmarks.*
   `M2`: *Second configuration matrix of k dimensions and p landmarks.*
Value:
   Vector of interlandmark distances between configurations.

```
ild2<-function(M1, M2){sqrt(apply((M1-M2)^2, 1, sum))}
```

```
>test<-fPsup(gorf.dat[,,1], gorf.dat[,,2])
>test$DF
[1] 0.0643504
>sqrt(sum(ild2(test$Mp1, test$Mp2)^2))
[1] 0.0643504
```

One can match the two translated and scaled shapes without optimizing the scaling transformation. In this case, both configurations are kept to unit centroid size, and the superimposition is called a partial Procrustes superimposition. The optimal rotation is the same, whether or not scaling is included in the minimization. The partial Procrustes distance d_P is equal to the square root of the sum of the squared distances between homologous coordinates of the superimposed configurations [27]. However, one ignores the revaluation of the parameter β once centered preshape configurations are rotated onto each other. The partial Procrustes distance d_P corresponds to the smallest distance between *Z1* and *Z2* on the preshape hypersphere . We program the partial Procrustes superimposition between two configurations under the name `pPsup`.

Function 4.17. `pPsup`

Arguments:
   `M1`: *Configuration matrix to be superimposed onto the centered preshape of* `M2`.
   `M2`: *Reference configuration matrix.*
Values:
   `Mp1`: *Superimposed centered preshape of* `M1` *onto the centered preshape of* `M2`.
   `Mp2`: *Centered preshape of* `M2`.
   `rotation`: *Rotation matrix*
   `DP`: *Partial Procrustes distance between* `M1` *and* `M2` *configurations.*
   `rho`: *Trigonometric Procrustes distance.*
Required functions: `centsiz, trans1`.

```
1  pPsup<-function(M1,M2)
2  {k<-ncol(M1)
3  Z1<-transl(centsiz(M1)[[2]])
4  Z2<-transl(centsiz(M2)[[2]])
5  sv<-svd(t(Z2)%*%Z1)
6  U<-sv$v; V<-sv$u; Delt<-sv$d
7  sig<-sign(det(t(Z1)%*%Z2))
8  Delt[k]<-sig*abs(Delt[k]) ; V[,k]<-sig * V[,k]
9  Gam<-U%*%t(V)
10 beta<-sum(Delt)
11 list(Mp1=Z1%*%phi,Mp2=Z2, rotation=Gam,
12  DP=sqrt(sum(ild2(Z1%*%phi, Z2)^2)),rho=acos(beta))}
```

Since one can include or not include scaling in the superimposition procedure, it is necessary to define a measure of shape difference not concerned with the scaling option. This measure is the trigonometric Procrustes distance ρ. It is the smallest curvilinear length between preshapes on the hypersphere or the angle between superimposed configurations and the origin of the hypersphere. ρ is defined as the arccosine of the sum of λ_i:

$$\rho = \arccos \sum_{i=1}^{k} \lambda_i \ .$$

As illustrated in Fig. 4.5, there are trigonometric relationships between the Procrustes, the full Procrustes and the partial Procrustes distances:

$$d_P = \frac{\sin \rho}{\cos (\rho/2)} = 2 \sin (\rho/2) \ ; \quad d_F = \sin \rho \ .$$

When the differences between shapes are small, full Procrustes distances, partial Procrustes distances and trigonometric Procrustes distances are nearly similar.

Procrustes Superimposition for More than Two Configurations

I have presented two kinds of Procrustes superimposition for matching two objects: respectively, the full ordinary Procrustes analysis and the partial Procrustes analysis. Writing functions for superimposing more than two shapes is less trivial since we have to find a general procedure for superimposing several configurations. For this goal, one must define an objective reference for the superimposition. The basic idea is to define a mean shape as a reference for allowing some assessment of the variability of the shape sample. The mean shape must be a parameter of central tendency of the shape distribution. The generalized Procrustes analysis (GPA) is a method that searches the average shape whose sum of pairwise squared coordinates with other rotated configurations is minimized. Several algorithms have been described for finding the best overall fit [39, 107]. These algorithms iteratively rotate configurations with a trial average shape. The average shape is re-estimated from the superimposed coordinates, and the operation is iterated until the algorithm converges. The superimposition can consider a posterior scaling between configurations and the mean shape (full GPA) for improving the fit, or it can keep all configurations to unit size (partial GPA). In any case, the mean shape is estimated from the whole set of configurations, and is constrained to unit size.

We first develop a general function called (mshape) that computes an averaged shape from an *A* array object of p, k and n dimensions.

Function 4.18. mshape

Argument:
   A: *Array containing configuration matrices.*
Value:
   Averaged configuration matrix.

```
mshape<-function(A){
 apply(A, c(1,2), mean)}
```

The full general Procrustes superimposition looks at minimizing the sum of squared norms of pairwise differences between all shapes in the sample such that

$$Q = \min(\frac{1}{n}\sum_{i=1}^{n}\sum_{j=1+i}^{n}\|(\beta_i M i \mathbf{\Gamma}_i + 1_p \alpha_i) - (\beta_j M j \mathbf{\Gamma}_j + 1_p \alpha_j)\|^2) ,$$

where β, $\mathbf{\Gamma}$ and α are respectively the scalar for scale, the rotation matrix and the translation vector of k values. The algorithm for computing the full generalized Procrustes analysis involves three steps:

1. Compute centered preshapes removing location effects (translation) and scaling all objects to unit centroid size.
2. Rotate and scale the centered preshape configuration (*Zic*) onto the average configuration appraised from all other centered preshape configurations (rotated and not yet rotated).
3. Iterate step 2 until the quantity Q cannot be reduced anymore.

We write the fgpa function that performs the full GPA.

Function 4.19. fgpa

Argument:
   A: *Array containing configuration matrices.*
Values:
   rotated: *Array of superimposed configurations.*
   iterationnumber: *Number of iterations.*
   Q: *Convergence criterion.*
   interproc.dist: *Minimal sum of squared norms of pairwise differences between all shapes in the superimposed sample.*

mshape: *Mean shape configuration.*
cent.size: *Vector of centroid sizes.*
Required functions: trans1, centsiz, mshape, fPsup.

```
1  fgpa<-function(A){
```

Extract information about the size of the array.

```
2  p<-dim(A)[1]; k<-dim(A)[2]; n<-dim(A)[3]
```

Create an empty array for storing scaled and rotated configurations, and an initial zero vector for storing centroid size.

```
3  temp2<-temp1<-array(NA, dim=c(p,k,n))
4  Siz<-numeric(n)
```

Translate and scale configurations to unit size.

```
5  for(i in 1:n)
6      {Acs<-centsiz(A[,,i])
7       Siz[i]<-Acs[[1]]
8       temp1[,,i]<-trans1(Acs[[2]])}
```

Define the quantity Qm *that must be minimized. Here* Qm *is the sum of Procrustes distances between configurations.*

```
9   Qm1<-dist(t(matrix(temp1,k*p,n)))
10  Q<-sum(Qm1); iter<-0
```

Loop until differences between shapes do not decrease anymore.

```
11  while (abs(Q)>0.00001)
12     {for (i in 1:n){
```

*Define the mean shape (*M*) ignoring the configuration that is going to be rotated.*

```
13      M<-mshape(temp1[,,-i])
```

Perform a full Procrustes superimposition between the mean shape and the ith *configuration.*

```
14      temp2[,,i]<-fPsup(temp1[,,i],M)[[1]]}
15     Qm2<-dist(t(matrix(temp2,k*p,n)))
16     Q<-sum(Qm1)-sum(Qm2)
17     Qm1<-Qm2
18     iter=iter+1
19     temp1<-temp2}
20
21  list(rotated=temp2,iterationnumber=iter,Q=Q,
22      interproc.dist=Qm2,mshape=centsiz(mshape(
23      temp2))[[2]],cent.size=Siz)}
```

Notice that the mean of rotated configurations has a relatively smaller size than the mean shape used as a reference (the mean shape has unit size); however, it has the same shape. This is because the Procrustes adjustment considers both size and rotation for minimizing interlandmark distances.

```
>centsiz(fgpa(gorf.dat)$mshape)[[1]]
[1] 1
>centsiz(mshape(fgpa(gorf.dat)$rotated))[[1]]
[1] 0.9980163
```

Gower [39] and Rohlf and Slice [107] give a solution for ensuring that the sum of centroid sizes for all configurations reaches the number of configurations. Here the scale parameters β_i are rescaled at every iteration so that $\sum_i^n \text{trace}(Mip_\star Mip'_\star) = n$, with $Mip_\star$ being the ith rotated and centered configuration during one iteration process. The Rohlf and Slice [107] algorithm involves eight steps:

1. Compute centered preshape configurations for removing translation effects and scaling all objects to unit centroid size.
2. Use the first configuration ($Z1c$) as the first reference, and rotate and scale the preshape configurations (Zic) onto this reference.
3. Compute the consensus of the superimposed configurations.
4. Compute the residual sum of squares.
5. Set the individual scale factor β_i to 1.
6. Rotate each of the $Mp_\star$ superimposed configurations with their new Y consensus using the scale factor previously calculated, and rescale each newly obtained configuration by $\beta_\star/\beta$. The scale factors are computed as

$$\frac{\beta_\star}{\beta} = \sqrt{\frac{\text{trace}(Mip_\star Y'_\star)}{\text{trace}(MipMip'_\star) \times \text{trace}(Y_\star Y'_\star)}} .$$

7. Compute a new consensus, new scale factors and a new sum of squares.
8. Reiterate steps 6 and 7 until the difference between the sums of squares are not changing above a given tolerance level.

The `fgpa2` function performs all eight steps.

Function 4.20. `fgpa2`

Argument:
    `A`: *Array containing configuration matrices.*
Values:
    `rotated`: *Array of superimposed configurations.*
    `iterationnumber`: *Number of iterations.*
    `Q`: *Convergence criterion.*
    `intereuclidean.dist`: *Minimal sum of squared norms of pairwise differences between all shapes in the superimposed sample.*
    `mshape`: *Mean shape configuration.*
    `cent.size`: *Vector of centroid sizes.*

Required functions: `trans1, centsiz, mshape, fPsup`.

```
1 | fgpa2<-function(A)
```

Start as for the `fgpa` *function.*
Extract information about the size of the array.

```
2 | {p<-dim(A)[1]; k<-dim(A)[2]; n<-dim(A)[3]
3 | temp2<-temp1<-array(NA, dim=c(p,k,n))
4 | Siz<-numeric(n)
```

Step 1: Translate and scale to unit size.

```
5 | for (i in 1:n)
6 |      {Acs<-centsiz(A[,,i])
7 |       Siz[i]<-Acs[[1]]
8 |       temp1[,,i]<-trans1(Acs[[2]])}
```

Initialize and set the type of objects that are going to be used for the iteration.

```
9 | iter<-0; sf<-NA
```

Step 2: Use the first configuration as reference for the first superimposition.

```
10 | M<-temp1[,,1]
11 | for (i in 1:n)
12 |    {temp1[,,i]<-fPsup(temp1[,,i],M)[[1]]}
```

Step 3: Define a new consensus.

```
13 | M<-mshape(temp1)
```

Step 4: Calculate the square root of the sum of paired residual squares differences.

```
14 | Qm1<-dist(t(matrix(temp1,k*p,n)))
15 | Q<-sum(Qm1); iter<-0
```

Step 5: Set the scaling factor to 1.

```
16 | sc<-rep(1,n)
```

Start the loop.

```
17 | while (abs(Q)>0.00001){
```

Step 6: Rotate and scale the configuration to the current consensus.

```
18 |        for (i in 1:n){
19 |          Z1<-temp1[,,i]
20 |          sv<-svd(t(M)%*%Z1)
21 |          U<-sv$v; V<-sv$u; Delt<-sv$d
22 |          sig<-sign(det(t(Z1)%*%M))
23 |          Delt[k]<-sig*abs(Delt[k])
24 |          V[,k]<-sig*V[,k]
```

```
25      phi<-U%*%t(V)
26      beta<-sum(Delt)
27      temp1[,,i]<-X<-sc[i]*Z1%*%phi}
```

Step 6: Define a new consensus.

```
28      M<-mshape(temp1)
```

Step 6: Compute the rescaling factor and rescale superimposed configurations.

```
29      for (i in 1:n)
30        {sf[i]<-sqrt(sum(diag(temp1[,,i]%*%t(M)))
31          /(sum(diag(M%*%t(M)))*sum(diag(temp1[,,i]
32          %*%t(temp1[,,i]))))))
33        temp2[,,i]<-sf[i]*temp1[,,i]}
```

Step 7: Compute a new consensus, new scale factors, and the difference between the square roots of sum of paired squared differences.

```
34      M<-mshape(temp2)
35      sc<-sf*sc
36      Qm2<-dist(t(matrix(temp2,k*p,n)))
```

Step 8: Until Q is not below the tolerance, reiterate steps 6 and 7.

```
37      Q<-sum(Qm1)-sum(Qm2)
38      Qm1<-Qm2
39      iter=iter+1
40      temp1<-temp2}
41    list(rotated=temp2,iterationnumber=iter,Q=Q,
42      intereuclidean.dist=Qm2, mshape=
43      centsiz(mshape(temp2))[[2]], cent.size=Siz)}
```

The procGPA function of the **shapes** package nearly performs the same tasks, except it is optimized to run more rapidly. It also uses complex algebra for 2D data. One obtains the rotated configurations by selecting the procGPA(object) $rotated element. Although different algorithms are used, results in terms of variation and mean shape are very close each other.

The plotshapes function of the **shapes** package can display the plot of rotated configurations (see Fig. 4.6).

```
>plotshapes(procGPA(gorf.dat)$rotated,
    joinline=c(1,6,7,8,2,3,4,5,1))
```

We can adopt a more customized code with the common functions of R and our own functions. This confers some more graphical possibilities (see Fig. 4.7). Segments linking landmarks are drawn with the lines function and a logical indexing (in our examples, indices are stored in the vector joinline). These links gives an idea of the morphology of the objects.

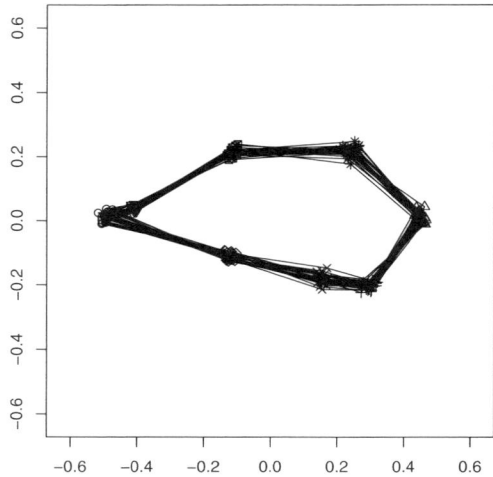

Fig. 4.6. Full Procrustes superimposition of the configurations corresponding to the skulls of female gorilla of the `gorf.dat` dataset. We have produced the plot with the `plotshapes` function of the paishapes package

```
>layout(matrix(c(1,2),1,2))
>par(mar=c(0.5,0.5,0.5,0.5))
>rot<-procGPA(gorf.dat)$rotated
>ms<-procGPA(gorf.dat)$mshape
>plot(rot[,1,],rot[,2,],axes=F,asp=1,
-      cex=0.6,xlab="",ylab="")
>plot(rot[,1,],rot[,2,],axes=F,asp=1,
-      cex=1,xlab="",ylab="")
>joinline<-c(1,6:8,2:5,1)
>for (i in 1:30)
-          {lines(rot[joinline,,i],col="grey")}
>points(rot[,1,],rot[,2,],cex=0.7)
>lines(ms[joinline,],col="black")
```

Partial generalized Procrustes superimposition follows the same methodology, except that it does not rescale the size of centered preshape configurations for optimizing the superimposition fit [107]. The `pgpa` function is very close to those we have written before, and should result in theory in the definition of a similar mean shape.

Fig. 4.7. Full Procrustes superimposition of the configurations corresponding to the skulls of female gorilla of the `gorf.dat` dataset. The graph is produced with the usual functions of R. Left: scatterplot of the superimposed coordinates; right: scatterplot with links between landmarks of the mean shape configuration

Function 4.21. `pgpa`

Argument:
  `A`: *Array containing configuration matrices.*
Values:
  `rotated`: *Array of superimposed configurations.*
  `it.number`: *Number of iterations.*
  `Q`: *Convergence criterion.*
  `intereucl.dist`: *Minimal sum of squared norms of pairwise differences between all shapes in the superimposed sample.*
  `mshape`: *Mean shape configuration.*
  `cent.size`: *Vector of centroid sizes.*
Required functions: `trans1, centsiz, mshape, pPsup`.

```
1  pgpa<-function(A)
```

Extract the number of landmarks, coordinate dimensions, and number of configurations contained in the array.

```
2  {p<-dim(A)[1];k<-dim(A)[2];n<-dim(A)[3]
```

Create an empty array for storing scaled and rotated configurations.

```
3  temp2<-temp1<-array(NA, dim=c(p,k,n)); Siz<-numeric(n)
```

Translate every configuration by aligning their centroid with the origin, and scale them to unit size.

```
4  for (i in 1:n)
5      {Acs<-centsiz(A[,,i])
6       Siz[i]<-Acs[[1]]
7       temp1[,,i]<-trans1(Acs[[2]])) }
```

Define the quantity Qm *that should be minimized.*

```
8   Qm1<-dist(t(matrix(temp1,k*p,n)))
9   Q<-sum(Qm1); iter<-0
```

Loop until differences between shape coordinates do not decrease anymore.

```
10  while (abs(Q)>0.00001)
11     {for(i in 1:n){
```

Define the mean shape ignoring the configuration that is going to be rotated.

```
12         M<-mshape(temp1[,,-i])
```

Perform a partial Procrustes superimposition between the mean shape and each configuration.

```
13         temp2[,,i]<-pPsup(temp1[,,i],M)[[1]]}
14     Qm2<-dist(t(matrix(temp2,k*p,n)))
15     Q<-sum(Qm1)-sum(Qm2)
16     Qm1<-Qm2
17     iter=iter+1
18     temp1<-temp2}
19  list("rotated"=temp2,"it.number"=iter,"Q"=Q,"intereucl.dist"=
20     Qm2,"mshape"=centsiz(mshape(temp2))[[2]],"cent.size"=Siz)}
```

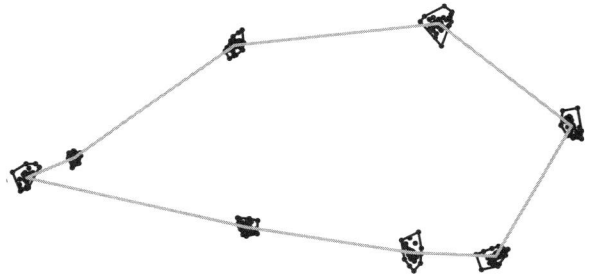

Fig. 4.8. Plot of the partial generalized Procrustes superimposition of the configurations corresponding to the skulls of female gorilla of the `gorf.dat` dataset, with the `plot.gpa` function of the **Rmorph** package. The full gray link lines correspond to the mean shape

The `gpa` function of the **Rmorph** package returns the same results and is optimized to run more rapidly. When shape variation is small, full Procrustes GPA and partial Procrustes GPA return very similar fits. The plots of superimposed configurations are nearly identical. Depending on arguments, the `plot.gpa` function provides different kinds of graph using an argument of the `gpa` class. It works in a similar way as the `plot.opa` function with an argument of the `opa` class. Default options display a scatterplot of superimposed configurations together with the mean shape (Fig. 4.8). It can draw 99% confidence ellipses for each landmark for 2D data. One can use the `ellipse` function for other tasks. Below is a practical example using the **Rmorph** package. Before starting the superimposition, we rotate configurations along their major axes, checking for eventual reflection. This step is not necessary, but it is convenient for graphical purposes. To align the configurations along their first principal axes, I write a small function called `aligne`.

Function 4.22. `aligne`

Argument:
    A: *Array containing configuration matrices.*
Value:
    Array of configurations aligned on their first principal axis.
Required functions: `angle2d, angle3.`

```
aligne<-function(A)
{B<-A
 n<-dim(A)[3]; k<-dim(A)[2]
 for (i in 1:n)
  {sv<-svd(var(A[,,i]))
   M<-A[,,i]%*%sv$u
   v1<-A[2,,i]-A[1,,i]; v2<-A[3,,i]-A[1,,i]
   V1<-M[2,]-M[1,]; V2<-M[3,]-M[1,]

   if (k ==2)
       {if (round(angle2d(v1,v2),3)!=
             round(angle2d(V1,V2),3))
                   {M[,1]=-M[,1]}}
   if (k ==3)
       {if (round(angle3(v1,v2),3)!=
        round(angle2d(V1,V2),3))
                   {M[,1]=-M[,1]}}
   B[,,i]<-M}
 B}
```

After aligning configuration on their first principal axes, we re-superimpose configurations with the `gpa` function.

```
>GORF<-aligne(gorf.dat)
>ji<-gpa(GORF, links=c(1, 6:8, 2:5,1))
>plot.gpa(ji, what="res")
-----------------------------------------------------
Rmorph(gpa):  partial  Procrustes superimposition
              threshold=  1e-04
              tangent space projection=  orthogonal
              reflection=  TRUE
              data dimensions: Npoints =  8
                  dim =  2
```

```
                           Nobj =   30
    iterations             Convergence (delta)
    1                      0.05734024
    2                      2.477402e-06
    done
    elapsed time=  0.16 seconds
```

To compare two samples of configurations, it is necessary to compute a GPA in-volving individuals of both samples and calculating an average shape from the whole set of configurations. In the following example, I gather two datasets to perform a sin-gle partial Procrustes GPA. Then, I plot the superimposed configurations belonging to the different groups on a single graph that can be used for a preliminary analysis (see Fig. 4.9).

Concatenate the gorf.dat *and* gorm.dat *datasets into the* gor *array to perform a unique Procrustes superimposition.*

```
>gor<-array(c(gorf.dat, gorm.dat), dim=c(8,2,59))
>go<-pgpa(aligne(gor))
>fe<-go$rotated[,,1:30]
>ma<-go$rotated[,,31:59]
>plot(fe[,1,],fe[,2,],asp=1,xlab="",ylab="",
+       cex=0.8,axes=F,pch=20,col="grey65")
>points(ma[,1,], ma[,2,],pch=4,cex=0.8)
>for (i in 1:8)
+    {lines(ELLI(ma[i,1,],ma[i,2,]))}
>for (i in 1:8)
+    {lines(ELLI(fe[i,1,],fe[i,2,]),lwd=2,col="grey65")}
>FE<-mshape(fe)
>MA<-mshape(ma)
```

joinline *are indices that permit us to draw lines between desired landmarks. The line can be interrupted using* NA *instead of an index.*

```
>joinline<-c(1,6:8,2:5,1)
>lines(FE[joinline,],col="grey65",lwd=2)
>lines(MA[joinline,])
```

One can adapt the algorithms and previous functions to compute a weighted av-erage. This can be useful later for statistical issues (see Problems of this chapter).

4.2.4 The Kendall Shape Space and the Tangent Euclidean Shape Space

We can better understand the geometry of shape space by learning from the sim-plest case for triangles. For triangles in the plane, there are six coordinates for a configuration, which means that raw data is organized in a six-dimensional space. After computing the preshape coordinates (removal of the scale and the 2 transla-tion parameters), the preshape space corresponds to the surface of a hypershpere of $k(p-1)$ dimensions and of unit size radius. The shape space corresponds to a space

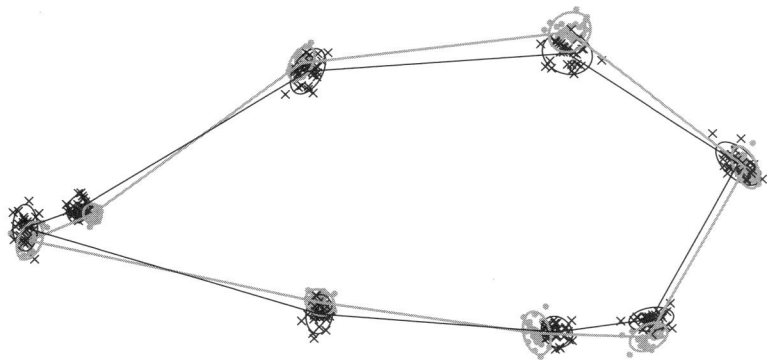

Fig. 4.9. Partial generalized Procrustes superimposition of the configurations corresponding to the gorilla skulls of the datasets `gorf.dat` and `gorm.dat`. 95% confidence ellipses for the variation of each landmark according to sex. Females correspond to gray dots, gray consensus links and 95% ellipses, while males correspond to black-color and thin links and ellipses

of $k \times p - k - 1 - k(k-1)/2$ dimensions as demonstrated by Kendall [54] (in the case of triangles, this corresponds to a spherical surface). However, this surface is not a plane, and for triangle configurations, the space corresponds to the surface of a sphere of radius $r = 1/2$ [53, 27]. The mapping of Kendall coordinates for triangles onto the sphere is given by

$$x = \frac{1 - r^2}{2(1 + r^2)} \; ; \; y = \frac{Mk_{.1}}{1 + r^2} \; ; \; z = \frac{Mk_{.2}}{1 + r^2} \; .$$

One can calculate Kendall spherical coordinates (θ and ϕ) using the equations:

$$x = \frac{1}{2}\sin\theta\cos\phi \; ; \; y = \frac{1}{2}\sin\theta\sin\phi \; ; \; z = \frac{1}{2}\cos\theta \; .$$

For shapes with more landmarks, the geometry of the shape space is more complicated and no longer corresponds to the surface of a sphere but to a more complex curved surface with more than three dimensions (namely a manifold). More practically, Rohlf [98] shows that after a GPA, aligned specimens on the reference (the mean shape) can be represented on the surface of a hemisphere of radius 1 with the pole defined as the mean shape. For shapes more complex than triangles, this corresponds to a hyperhemishpere of radius 1 but with more dimensions. Rohlf [98] calls it the hemisphere of aligned preshapes on the reference (see Fig. 4.10).

Although shape spaces defined by superimposition methods have less dimensions than raw data or nonredundant measurements, they are nonEuclidean and correspond to a curved surface. Nobody will recommend applying traditional statistics directly in this space because traditional statistics relies on the Euclidean metric, which is not the same as the Procrustes one. To perform usual statistical methods, one must

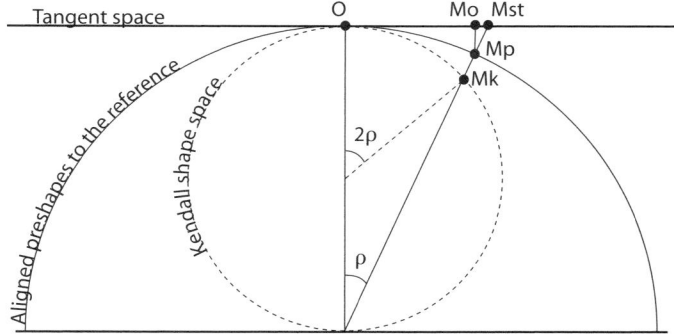

Fig. 4.10. Illustration of a section of the tangent shape, Kendall shape, and aligned preshape space for the triangle case. **O** corresponds to the mean reference shape, while **Mk** corresponds of Kendall position of a shape in the sphere. **Mo** and **Ms** are, respectively, the orthogonal and stereographic projections of the shape of **M** onto the tangent shape space

first project the surface of the hyperhemisphere onto a "flat" tangent space where the Euclidean metrics allows us to use Euclidean statistics. The data are projected on a tangent shape space (also called Kendall tangent space or Kent tangent space). The contact between spaces is chosen as the mean shape. Working on variation in the tangent space is a rather perilous estimation since the projection can introduce distortion for the largest distances. However, provided that variation is small, one can assume that the portion of the shape hyperhemisphere and tangent space are nearly flat and nearly confused.

The projection onto a Euclidean space can be orthogonal or stereographic (see Fig. 4.10). Note that both projections will introduce biases for shapes being very different from the mean shape: the orthogonal projection minimizes large differences while stereographic projection accentuates them.

The stereographic projection is produced by adjusting the size scale factor for the configuration to be projected onto the tangent space. To perform this projection, we use simple trigonometric relationships and divide the coordinates of the aligned configurations by the cosine of the Procrustes distance ρ between shapes and the mean shape. The `stp` function performs a stereographic projection of a configuration set, the mean shape being the pole.

Function 4.23. `stp`

Argument:
    A: *Array containing superimposed configuration matrices.*
Value:
    Array of projected configurations onto the euclidean shape space.
Required function: `mshape`.

```
1  stp<-function(A)
2  {p<-dim(A)[1];k<-dim(A)[2];n<-dim(A)[3]
3  Yn<-mshape(A)
4  B<-array(NA, dim=c(p,k,n))
5  for (i in 1:n)
6     {rho<-2*asin((sqrt(sum((A[,,i]-Yn)^2)))/2)
7      B[,,i]<-A[,,i]/(cos(rho))}
8  return(B)}
```

One performs the orthogonal projection using the coordinates of aligned configurations and those of the mean shape. Rohlf [98] gives the following equation to compute the projected configurations organized in a $\mathbf{X}^*$ matrix of n rows $\times$ kp columns:

$$\mathbf{X}^* = \mathbf{X}(I_{kp} - x'_m x_m) \,,$$

$\mathbf{X}$ being the n rows $\times$ kp columns of aligned centered preshapes on the reference, x_m being the vectorized form of the mean shape of unit centroid size, and I_{kp} being a $kp \times kp$ identity matrix. The `orp` function performs an orthogonal projection on a configuration dataset of the `array` class with $p \times k \times n$ dimensions.

Function 4.24. `orp`

Argument:
   A: *Array containing superimposed configuration matrices.*
Value:
   Array of projected configurations onto the euclidean shape space.
Required function: `mshape`.

```
1  orp<-function(A)
2  {p<-dim(A)[1];k<-dim(A)[2];n<-dim(A)[3]
3  Y1<-as.vector(centsiz(mshape(A))[[2]])
4  oo<-as.matrix(rep(1,n))%*%Y1
5  I<-diag(1,k*p)
6  mat<-matrix(NA, n, k*p)
7  for (i in 1:n){mat[i,]<-as.vector(A[,,i])}
8  Xp<-mat%*%(I-(Y1%*%t(Y1)))
9  Xp1<-Xp+oo
10 array(t(Xp1), dim=c(p, k, n))}
```

Checking whether these operations have introduced important biases corresponds to appraising whether variation is small. We can estimate them using the correlations between Euclidean distances in the tangent shape space with pair-wise Procrustes distance in the shape space. If variation is small, one can apply Euclidean statistics onto the tangent space coordinates. Euclidean distances in the tangent shape space correspond to the square root of the sum of the squared differences between the coordinates of projected aligned configurations. Intershape distances are redundant, and using a p-value for the correlation coefficient is nonsense. Instead of this, we can directly use the correlation coefficient; if it is very close to 1, we can be confident that variation is small enough. Since distances are redundant, and the distance matrices are symmetric, we compute the squared coefficient of correlation for the upper half triangle of the matrices only. Here is an example with the corresponding figure (Fig. 4.11).

Perform a full GPA on a dataset (here `gorf.dat`*).*

```
>go<-fgpa(gorf.dat)
>n<-dim(gorf.dat)[3]
```

Extract Procrustes distances between configurations.

```
>proc<-go$interproc.dist
```

Calculate Euclidean distances.

```
>go<-fgpa(gorf.dat)
>n<-dim(gorf.dat)[3]
>proc<-go$interproc.dist
>tango<-orp(go$rotated)
>tang<-matrix(NA, n, n)
>for (i in 1:n){
+    {for (j in i:n){
+        tang[i,j]<-sqrt(sum(ild2(tango[,,i],
+                    tango[,,j])^2))}}
>proc.dist<-asin(proc)
>euc.dist<-as.dist(t(tang))
>plot(euc.dist, proc.dist,xlab=Euc.dist,
+                    ylab=expression(rho))
>abline(0,1, col="grey50")
>(cor(proc.dist, euc.dist))^2
[1] 0.9999968
```

Here Procrustes and Euclidean distances are very well correlated, indicating that variation is small.

Rohlf ([98, 99, 101]) and Slice [113] have shown that the generalized Procrustes superimposition was one of the less unbiased approaches among the newly developed morphometric methods to estimate variation in the shape space (however, see [93] for a different point of view).

4.2.5 Resistant-fit Superimposition

Other optimization criteria have been developed for superimposing the coordinates of configurations on each other, and some have interesting advantages. Indeed, since the optimization criterion in ordinary and general Procrustes analysis is to reduce interlandmark distances between configurations, this averages and allocates the change in shape to all landmarks rather than to really influential landmarks. Siegel and Benson [110] have noticed that the least-squares method resulted in a lack of fit

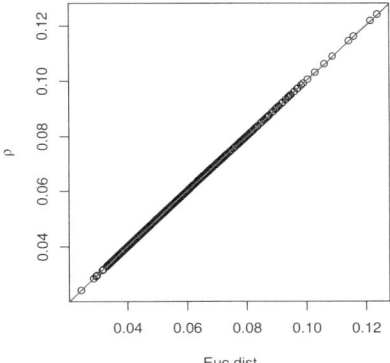

Fig. 4.11. Relationship between the Procrustes distances of the shape space and the Euclidean distances in the tangent shape space in the `gorf.dat` dataset

between configurations. If two configurations differ only in the position of a single landmark, the ordinary Procrustes method will allocate difference in shape to all landmarks rather than localizing the landmark that has changed in position. Siegel and Benson [110] thus developed a superimposition procedure based on robust regression using repeated medians. In the literature, these methods are referred to as resistant-fit methods. Rohlf and Slice [107] published algorithms for ordinary and generalized resistant-fit methods for 2D data.

Ordinary Resistant-Fit

For 2D data, the aim of the procedure is to fit an *M1* configuration onto an *M2* configuration by minimizing the number of mismatching landmarks between configurations. As with the least-squares procedure we start using the equation:

$$M1^\star = \mathbf{1_p}\alpha' + \beta M1\mathbf{\Gamma} \ ,$$

where α is a vector of translation of k elements, β is a scaling factor, and $\mathbf{\Gamma}$ is a $k \times k$ rotation matrix.

The procedure uses four steps:

1. Perform a first estimate of the superimposition of *M1* and *M2* using the full ordinary procedure. This approximation facilitates the estimation of the rotation parameter [107], and yields the *M1p* and *M2p* configuration matrices on which the resistant-fit parameters will be estimated.
2. Appraise the scale factor by resistant-fit. The scale factor corresponds to the repeated median of the ratios of corresponding interlandmark distances. The `dist` function can compute the interlandmark distances in the configuration, and one can coerce it into a `matrix` object with the code `as.matrix(dist(M))`.

The `medsize` function computes the repeated median size of a configuration; it uses the `median` function.

Function 4.25. `medsize`
Argument:
   M: *Configuration matrix.*
Value:
   Repeated median size of the configuration.

```
1  medsize<-function(M)
2  {mat<-as.matrix(dist(M))
3  median(apply(mat, 2, median, na.rm=TRUE))}
```

3. Estimate the rotation matrix. The rotation angle is appraised in a similar way, and the rotation matrix $\mathbf{\Gamma}$ is computed from the angle ϕ:

$$\mathbf{\Gamma} = \begin{pmatrix} \cos\phi & \sin\phi \\ -\sin\phi & \cos\phi \end{pmatrix} .$$

To obtain ϕ, all angles defined by homologous vectors in both configurations are computed, and the median angle-value is estimated. Finding the median value for an angle is problematic since angles are circular data with a period of 2π. The problem is nearly solved if configurations have been first superimposed according to the ordinary Procrustes superimposition. Indeed, one can expect that most adjustments of rotation will be close to 0. Siegel and Benson [110] suggest using values of ϕ ranging between $-\pi$ and π. For 2D data, we can take advantage of vectors written in their complex form as the combination of arguments and modulus. One must define a function for finding the arguments between all similar complex vectors of both configurations. The `argallvec` function computes the angles between homologous vectors of both configurations.

Function 4.26. `argallvec`
Arguments:
   X1: *First configuration matrix.*
   X2: *Second configuration matrix.*
Value:
   $p \times p$ *matrix of angles between homologous vectors of both configurations.*

```
1  argallvec<-function(X1, X2)
2  {p<-dim(X1)[1]
3  m<-m<-matrix(NA, p, p)
4  for (i in 1:p){ for (j in 1:p){
5      m[i,j]<-Arg(complex(1,X2[i,1],X2[i,2])
6          -complex(1,X2[j,1],X2[j,2]))-
7          Arg(complex(1,X1[i,1],X1[i,2])
8          -complex(1,X1[j,1],X1[j,2]))  }}
9  ((m+pi)%%(2*pi))-pi}
```

4. Perform the final translation. One appraises the tanslation vector (α) using the median of the differences in landmark position between the *X2* and *X1* matrices for each dimension.

The oPsup function performs an orthogonal resistant-fit between two configuration matrices.

Function 4.27. oPsup

Arguments:

M1: *Configuration matrix to be superimposed onto the centered preshape of* M2.

M2: *Reference configuration matrix.*

Values:

Mo1: *Superimposed centered preshape of* M1 *onto* Mo2.

Mo2: *Centered preshape of* M2.

Required functions: argallvec,fPsup, centsiz,trans1.

```
1  oPsup<-function(M1, M2)
2  {p<-dim(M1)[1]; k<-dim(M1)[2]
```

Step 1.

```
3  ols<-fPsup(M1, M2)
4  X1<-ols[[1]]; X2<-ols[[2]]
```

Step 2.

```
5  M<-as.matrix(dist(X2))/as.matrix(dist(X1))
6  beta<-median(apply(M, 2, median, na.rm=TRUE))
```

Step 3.

```
7  ARG<-argallvec(X1,X2)
8  phi<-median(apply(ARG, 2,median, na.rm=TRUE))
9  Gam<-matrix(c(cos(phi),-sin(phi),sin(phi),cos(phi)),2,2)
```

Step 4.

```
10  alpha<-X2-beta*X1%*%Gam
11  alpha<-matrix(apply(alpha,2,median), p, k, byrow=T)
12  list("Mo1"=beta*X1%*%Gam + alpha,"Mo2"=X2)}
```

We apply this method with an outlier introduced in one of the two configurations. The resulting superimposition is compared with the ordinary Procrustes superimposition in Fig. 4.12.

```
>layout(matrix(c(1,2),1,2))
>par(mar=c(1,1,1,1))
>M1<-gorf.dat[,,1]; M2<-gorf.dat[,,2]
>M1[1,2]<-250
>lsq<-fPsup(M1, M2)
>resf<-oPsup(M1, M2)
```

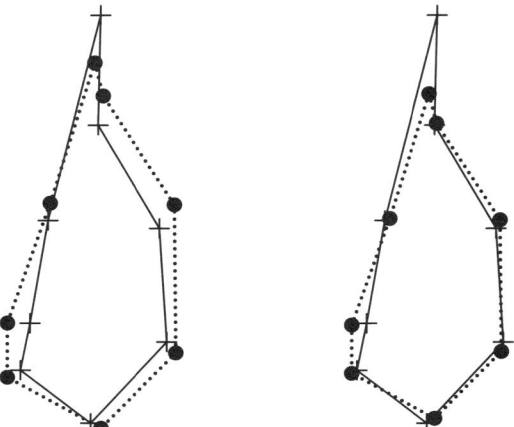

Fig. 4.12. Difference of behavior between the resistant-fit and the full-Procrustes-fit superimpositions when one outlier is present in one of the configurations. Left: ordinary-full-Procrustes superimposition; right: resistant-fit superimposition

Plot the ordinary-full-Procrustes superimposition.

```
>plot(lsq[[1]], asp=1,axes=F, xlab="",ylab="")
>points(lsq[[2]], pch=3)
>joinline<-c(1,6:8,2:5,1)
>lines(lsq[[1]][joinline,])
>lines(lsq[[2]][joinline,], lty=3)
```

Plot the resistant-fit superimposition.

```
>plot(resf[[1]], asp=1,axes=F, xlab="",ylab="")
>points(resf[[2]], pch=3)
>lines(resf[[1]][joinline,])
>lines(resf[[2]][joinline,], lty=3)
```

For 3D data, we use the same approach, taking into account the third coordinate. A difficulty remains for calculating the rotation matrix ($\mathbf{\Gamma}$), which does not depends on a single but on several angles. Siegel and Pinkerton [111] and Slice [112] have provided the formulae to compute $\mathbf{\Gamma}$ that depends on angles ϕ, ω, and θ. These angles, respectively, perform rotations around the x, y, and z-axes. We have

$$\mathbf{\Gamma}=\begin{pmatrix} \cos\theta\cos\omega + \sin\theta\sin\phi\sin\omega & \cos\theta\sin\phi\sin\omega - \sin\theta\cos\omega & \cos\phi\sin\omega \\ \sin\theta\cos\phi & \cos\theta\cos\phi & -\sin\phi \\ \sin\theta\sin\phi\cos\omega - \cos\theta\sin\omega & \sin\theta\sin\omega + \cos\theta\sin\phi\cos\omega & \cos\phi\cos\omega \end{pmatrix}.$$

One must first estimate a list of vectors to compute from the *M1* data matrix and the *M2* reference matrix in order to obtain $\mathbf{\Gamma}$. The rotation matrix is based on all sets of triplets for landmarks i, j and k. Let $M1_i$, $M1_j$, $M1_k$, $M2_i$, $M2_j$, $M2_k$ be the vectors containing the coordinates of the landmarks of the data and reference matrix for each dimension. Then, these vectors are used to determine sets of orthogonal vectors $M1_{ij}$, $M1_{ijk}$, $M1_{ijk\star}$, $M2_{ij}$, $M2_{ijk}$, $M2_{ijk\star}$ as follows:

$$M1_{ij} = \frac{M1_j - M1_i}{\|M1_j - M1_i\|} \quad ; \quad M2_{ij} = \frac{M2_j - M2_i}{\|M2_j - M2_i\|} \, ,$$

$$M1_{ijk} = \frac{(M1_k - M1_i) - [(M1_k - M1_i).M1_{ij}]M1_{ij}}{\|(M1_k - M1_i) - [(M1_k - M1_i).M1_{ij}]M1_{ij}\|} \, ,$$

$$M2_{ijk} = \frac{(M2_k - M2_i) - [(M2_k - M2_i).M2_{ij}]M2_{ij}}{\|(M2_k - M2_i) - [(M2_k - M2_i).M2_{ij}]M2_{ij}\|} \, ,$$

and

$$M1_{ijk\star} = M1_{ij} \times M1_{ijk} \, ; \quad M2_{ijk\star} = M1_{ij} \times M1_{ijk} \, .$$

The sets of orthonormal vectors are stored in $Y1$ and $Y2$ matrices for each configuration. To rotate a triplet of landmarks onto the other, one calculates Γ as follows:

$$\Gamma = Y1'Y2 \, .$$

One must appraise a median matrix from all possible triplets of points. The not-so-obvious solution is to transform all Γ_{ijk} rotation matrices into S_{ijk} skew-symmetric matrices (see [112]). Indeed, element-wise repeated medians of skew-symmetric matrices yield a skew-symmetric matrix (noted S), which in turn is associated with a unique orthogonal rotation matrix (Γ) by the relationships

$$S_{ijk} = (\Gamma_{ijk} + I_k)^{-1}(\Gamma_{ijk} - I_k) \, ,$$

I_k being a $k \times k$ identity matrix;

$$S = \text{median}_i^p(\text{median}_j^{p \neq i}(\text{median}_k^{p \neq i,j}(S_{ijk}))) \, ;$$

$$\Gamma = (I_k + S)(I_k - S)^{-1} \, .$$

We write the corresponding `rrot` function for achieving the rotation of the $M2$ configuration onto $M1$ following a resistant-fit approach:

Function 4.28. `rrot`

Arguments:
    `M1`: *Configuration matrix to be aligned onto* `M2`.
    `M2`: *Reference configuration matrix.*
Values:
    `ref`: *Configuration rotated onto the reference.*
    `targ`: *Reference configuration.*
    `Gamma`: *Rotation matrix.*

```
1   rrot<-function(M1, M2)
2   {p<-dim(M1)[1]; k<-dim(M1)[2]
3   P<-1:p
4   I<-diag(1,3)
5   A<-array(NA, dim=c(p,p,p,3,3))
6   for (i in P){
7     for (j in P[-i]){
8       for (k in P[-c(i,j)]){
9         M1i<-matrix(M1[i,])
10        M1j<-matrix(M1[j,])
11        M1k<-matrix(M1[k,])
12        M2i<-matrix(M2[i,])
13        M2j<-matrix(M2[j,])
14        M2k<-matrix(M2[k,])
15        M1ij<-(M1j-M1i)/sqrt(sum((M1j-M1i)^2))
16        M2ij<-(M2j-M2i)/sqrt(sum((M2j-M2i)^2))
17        M1ijk<-((M1k-M1i)-as.numeric(t(M1k-M1i)
18          %*%M1ij)*M1ij)/ sqrt(sum(((M1k-M1i)-
19          as.numeric(t(M1k-M1i)%*%M1ij)*M1ij)^2))
20        M2ijk<-((M2k-M2i)-as.numeric(t(M2k-M2i)
21          %*%M2ij)*M2ij)/ sqrt(sum(((M2k-M2i)-
22          as.numeric(t(M2k-M2i)%*%M2ij)*M2ij)^2))
23        M1s<-matrix(c(M1ij[2]*M1ijk[3]-M1ij[3]*M1ijk[2],
24          M1ij[3]*M1ijk[1]-M1ij[1]*M1ijk[3],M1ij[1]*
25          M1ijk[2]-M1ij[2]*M1ijk[1]))
26        M2s<-matrix(c(M2ij[2]*M2ijk[3]-M2ij[3]*M2ijk[2],
27          M2ij[3]*M2ijk[1]-M2ij[1]*M2ijk[3],M2ij[1]*
28          M2ijk[2]-M2ij[2]*M2ijk[1]))
29        X<-cbind(M1ij, M1ijk, M1s)
30        Y<-cbind(M2ij, M2ijk, M2s)
31        H<-t(X)%*%Y
32        A[i,j,k,,]<-S<-solve(H+I)%*%(H-I)}}}
33  S1<-apply(apply(apply(A,c(1,2,4,5),median,na.rm=T),
34    c(1,3,4),median,na.rm=T),2:3,median,na.rm=T)
35  H<-(I+S1)%*%solve(I-S1)
36  list("ref"=M2, "targ"=M1%*%t(H), "Gamma"=H)}
```

Once done, the orthogonal resistant-fit procedure is similar to the one for 2D data. The r3sup function performs a 3D resistant-fit for two configurations.

Function 4.29. r3sup

Arguments:
    M1: *3D configuration matrix to be superimposed onto the centered preshape of* M2.
    M2: *3D reference configuration matrix.*
Values:
    Mo1: *Superimposed centered preshape of* M1 *onto* Mo2.
    Mo2: *Centered preshape of* M2.

Required functions: fPsup, centsiz, trans1.

```
1   r3sup<-function(M1, M2)
2   {p<-dim(M1)[1]; k<-dim(M1)[2]
3   ols<-fPsup(M1, M2)
4   X1<-ols[[1]]; X2<-ols[[2]]
5   B<-as.matrix(dist(X2))/as.matrix(dist(X1))
6   M<-B; M<-t(M)
7   M[row(B)<col(B)]<-B[row(B)<col(B)]
8   M<-as.matrix(dist(X2))/as.matrix(dist(X1))
9   beta<-median(apply(M, 2, median, na.rm=TRUE))
10  Gam<-rrot(X1,X2)[[3]]
11  alpha<-X2-beta*X1%*%Gam
12  alpha<-matrix(apply(alpha,2,median),p,k, byrow=T)
13  list("Mo1"=beta*X1%*%Gam + alpha,"Mo2"=X2)}
```

Note that the procedure is consuming more time than the ordinary Procrustes full superimposition.

Generalized Resistant-Fit

Generalizing the above procedures for more than two shapes is not usual, since little or no statistical background about resistant-fit methods is available from the literature. This is inherent to several properties of the method and its philosophy:

- Unlike procedures minimizing the Procrustes distances, parameters are estimated using procedures that are based on a biased estimator of central tendency (the median), which makes it unclear whether one can interpret results statistically.
- If there are fewer points that closely fit between configurations, there can be ambiguity in the best fitting: indeed, several regions can fit, but which one to choose unambiguously?
- Resistant-fit methods are more time consuming than Procrustes methods.
- There is no symmetry by fitting M_1 onto M_2 and M_2 onto M_1 as for partial Procrustes superimpostion, and thus no similar distances between objects depending on the reference configuration.

However, the resistant approach has interesting properties for some datasets. At least they provide a different way o displaying and examining shape differences. Some datasets can display only one or a few landmarks that are known to exhibit large variation around the general shape of the configurations. Unlike least-squares distance optimization methods, resistant-fit approaches perform superimpositions that are not sensitive to the change of position of a single or few points. Analyzing variation of points around a landmark scatterplot produced by resistant-fit methods can allow us to understand and to localize shape difference with a different point of view that can be closer to biological reality. Rohlf and Slice [107] and Slice [112] published algorithms for fitting more than two shapes. They proceed in several steps as explained below:

1. First fit all objects using a generalized Procrustes superimposition.
2. Scale all configurations to have unit median size.
3. Compute a consensus that corresponds to the element-wise medians of all configurations and scale this consensus to unit median size.
4. Superimpose all configurations onto the consensus following the ordinary resistant-fit
5. Compute a new consensus estimate and measure the goodness-of-fit (a value that should be minimized). If the result is above a given level of tolerance, go back to step 3. Rohlf and Slice [107] suggest using the median difference between successive consensuses of two iterations, something I have adopted here.

The corresponding function for 2D data should use the previous functions we have programmed, but it should shortcut the ordinary resistant-fit by removing the former Procrustes alignment. We write `grf2` to compute a generalized resistant-fit for 2D configurations.

Function 4.30. `grf2`

Argument:
 A: *Array containing 2D configuration matrices.*
Values:
 `rotated`: *Array of superimposed configurations.*
 `iteration`: *Number of iterations*
 `limit`: *Convergence criterion*
 `medshape`: *Median shape configuration.*
Required functions: `pgpa`, `trans1`, `centsiz`, `mshape`, `pPsup`, `medsize`, `argallvec`.

```
1  grf2<-function(A)
2  {p<-dim(A)[1]; k<-dim(A)[2]; n<-dim(A)[3]
```

Step 1.

```
3  A<-pgpa(A)$rotated
4  D<-B<-array(NA, dim=c(p,k,n))
```

Step 2.

```
5  for (i in 1:n){
6      B[,,i]<-A[,,i]/medsize(A[,,i])}
```

Step 3.

```
7   Y<-apply(B, 1:2, median)
8   Y<-Y/medsize(Y)
9   A0<-10
10  iter<-1
```

Step 4.

```
11  while(A0>0.0005){
```

Ordinary resistant-fit shortcut.

```
12      for (i in 1:n){
13          M<-as.matrix(dist(Y))/as.matrix(dist(B[,,i]))
14          beta<-median(apply(M, 2, median, na.rm=TRUE))
15          ARG<-argallvec(B[,,i],Y)
16          phi<-median(apply(ARG, 2,median, na.rm=TRUE))
17          Gam<-matrix(c(cos(phi),-sin(phi),
18              sin(phi),cos(phi)),2,2)
19          alpha<-Y-beta*B[,,i]%*%Gam
20          D[,,i]<-beta*B[,,i]%*%Gam+matrix(apply(
21              alpha,2,median),p,k,byrow=T)}
22      Yb<-apply(D, 1:2, median)
```

Step 5.

```
23      A0<-median(sqrt(apply((Yb-Y)^2,1,sum)))
24      Y<-Yb<-Yb/medsize(Yb)
25      B<-D
26      iter<-iter+1}
27  list("rotated"=D,"limit"=A0,"iteration"=iter,"medshape"=Yb)}
```

The corresponding function for 3D data is very similar except for estimating the rotation. The `grf3` function performs a generalized resistant-fit for a set of 3D configurations.

Function 4.31. `grf3`

Argument:
   A: *Array containing 3D configuration matrices.*
Values:
   `rotated`: *Array of superimposed configurations.*
   `iteration`: *Number of iterations*
   `limit`: *Convergence criterion*
   `medshape`: *Median shape configuration.*
Required functions: pgpa, trans1, centsiz, mshape, pPsup, medsize, rrot.

```
1  grf3<-function(A)
2   {p<-dim(A)[1]; k<-dim(A)[2]; n<-dim(A)[3]
```

Step 1.

```
3  A<-pgpa(A)$rotated
4  D<-B<-array(NA, dim=c(p,k,n))
```

Step 2.

```
5      for (i in 1:n){
6      B[,,i]<-A[,,i]/medsize(A[,,i])}
```

Step 3.

```
 7  Y<-apply(B, 1:2, median)
 8  Y<-Y/medsize(Y)
 9  A0<-10
10  iter<-1
```

Step 4.

```
11  while(A0>0.005){
```

Ordinary resistant-fit shortcut.

```
12    for (i in 1:n){
13      M<-as.matrix(dist(Y))/as.matrix(dist(B[,,i]))
14      beta<-median(apply(M, 2, median, na.rm=TRUE))
15      Gam<-rrot(B[,,i],Y)[[3]]
16      alpha<-Y-beta*B[,,i]%*%Gam
17      D[,,i]<-beta*B[,,i]%*%Gam + matrix(apply(
18              alpha,2,median),p,k,byrow=T)}
19    Yb<-apply(D, 1:2, median)
```

Step 5.

```
20    A0<-median(sqrt(apply((Yb-Y)^2,1,sum)))
21    Y<-Yb<-Yb/medsize(Yb)
22    B<-D
23    iter<-iter+1}
24  list("rotated"=D,"limit"=A0,"iteration"=iter,"medshape"=Yb)}
```

The tolerance can be adjusted in the function and entered through one supplementary argument; I respectively set it at 0.0005 and 0.005 for 2D and 3D data. Slice [112] suggests limiting the number of iterations since there is no guarantee that the algorithm will return convergent results.

As demonstrated in Fig. 4.13, the generalized resistant-fit procedure can be more efficient than Procrustes superimposition for finding landmarks that bear more variation than others. This is mainly a matter of distribution of variation between landmarks. However, as stated before, statistical development around the resistant-fit procedure is nearly absent. Here is the code for producing Fig. 4.13.

```
>gls<-pgpa(gorm.dat)
>glsr<-gls$rotated
>grf<-grf2(gorm.dat)
>grfr<-grf$rotated
>layout(matrix(c(1,2),1,2))
>plot(glsr[,1,],glsr[,2,],pch=20,col="grey65",
+       asp=1,axes=F,xlab="", ylab="",main="GPA")
>points(gls$mshape,pch=3,cex=2,lwd=2)
>plot(grfr[,1,],grfr[,2,],pch=20, col="grey65",
+       asp=1,axes=F,xlab="", ylab="",main="GRF")
>points(grf$medshape,pch=3,cex=2,lwd=2)
```

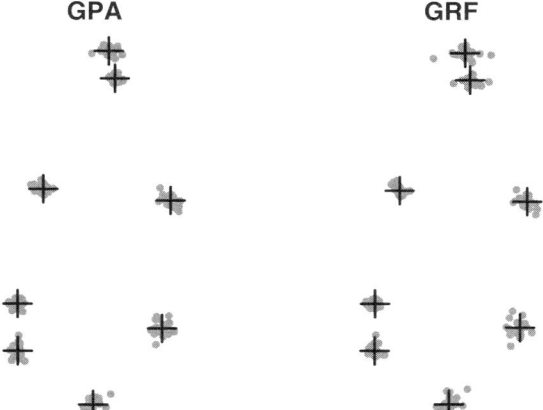

Fig. 4.13. Difference of behavior between generalized resistant and generalized Procrustes-fits. Generalized resistant-fit superimposition can be useful for investigating landmarks that affect the variation in shape (especially here with the landmarks at the apex of the configuration). Left: generalized partial Procrustes superimposition; right: generalized resistant-fit superimposition. Mean and median shapes correspond to crosses

You can find additional development around resistant-fit methods in [28] and references therein.

4.3 Thin-Plate Splines

It is easy to display differences between a couple of shapes with a field of arrows that goes from a reference shape to a target shape. The `arrows` function plots arrows and can help with this task as shown in the code example below. If shape change is too small, one can also amplify shape differences (see Fig. 4.14). Note, however, that when these vectors result from least-squares Procrustes superimposition, one cannot interpret them individually.

In the following example, shape differences are displayed between averaged sagittal skull sections for female and male gorillas.

Concatenate the `gorf.dat` *and* `gorm.dat` *datasets; perform a partial GPA; store rotated configurations in the* `fe` *and* `ma` *objects.*

```
>gor<-array(c(gorf.dat, gorm.dat), dim=c(8,2,59))
>go<-pgpa(aligne(gor))
>fe<-go$rotated[,,1:30]
>ma<-go$rotated[,,31:59]
```

Compute and plot mean shapes for females and males.

```
>FE<-mshape(fe)
>MA<-mshape(ma)
```

```
>layout(matrix(c(1,2),1,2))
>par(mar=c(1,1,1,1))
>plot(FE, asp=1, xlab="", ylab="", axes=F)
>joinline<-c(1,6:8,2:5,1)
>lines(FE[joinline,],col="grey50",lwd=2)
>lines(MA[joinline,],lty=3)
```

Draw arrows between corresponding landmarks of the mean shapes for males and females.

```
>arrows(FE[,1],FE[,2],MA[,1],MA[,2],
+        length=0.1,lwd=2)
>title("No amplification")
```

Amplify shape differences and arrows.

```
>mag<-3
>MA1<-MA+(MA-FE)*3
>plot(FE, asp=1, xlab="", ylab="", axes=F)
>lines(FE[joinline,], col="grey50", lwd=2)
>lines(MA[joinline,],lty=3)
>arrows(FE[,1],FE[,2],MA1[,1],MA1[,2],
+        length=0.1,lwd=2)
>title("Three-times amplification")
```

No amplification   **Three-times amplification**

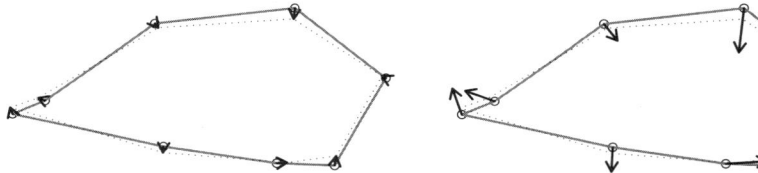

Fig. 4.14. Scatter of vectors depicting shape differences between superimposed skulls of mean female gorilla (reference) and mean male gorilla (target). On the right, vectors have been amplified 3 times

Rather than displaying only a field of vectors, it may be convenient to understand shape change in a more global way. For this task, one needs to interpolate how and which changes have affected the whole shape. Thin-plate splines were imported into the field of morphometrics by Bookstein at the end of the last century [10]. They mathematically express the deformation grids of D'Arcy Thompson [117] whose aim was to describe shape change with a few functions applied to mapping one specimen onto another. D'arcy Thompson grids were done by hand using an approximative and subjective way. The idea of deformation grids and thin-plate splines is that individual landmarks are not only to be affected by shape change, but that one can interpolate change in other regions of the shape being considered. Each landmark becomes a local indication of shape change, and one uses the whole set of landmarks for finding global and local shape change.

We can distinguish two types of shape change if we map an orthogonal grid onto the shape under consideration. The first will deform one object into the other leaving parallel lines of the grids parallel. In other words, they will transform squares of the reference grid into rectangles (compaction or dilatation) or into parallelograms (tilting) in the target grid.

Deformation of one triangle to another in two dimensions involves uniform transformations only. However, the situation changes when there are two more points than there are dimensions. Indeed, shape change does not correspond only to the effect of affine transformations, but to the combination of affine and nonaffine transformation. Uniform change does not take into account more local shape changes or possibilities of curvation (torsion) of the cartesian systems.

In order to find a mathematical solution for interpolating shape change in the whole cartesian system mapped onto a shape, Bookstein [9] imported mathematics that are used in continuum mechanics and applied them to the bending of a thin metal plate subject to physical constraints. The reader interested in the mathematical theory around thin-plate splines will find more information in [10]. I will present only how to estimate these pairs of thin-plate splines in the context of morphometrics.

For 2D data, thin-plate splines rely on the function $U(r) = r^2 \log r^2$ and the M configuration of p landmarks in two dimensions. From these, we define the three matrices: $\mathbf{P}$, $\mathbf{Q}$ and $\mathbf{L}$ as follows,

$$\mathbf{P} = \begin{pmatrix} 0 & U(r_{12}) & \dots & U(r_{1p}) \\ U(r_{21}) & 0 & \dots & U(r_{2p}) \\ \dots & \dots & \dots & \dots \\ U(r_{p1}) & U(r_{p2}) & \dots & 0 \end{pmatrix},$$

where r_{ij} is the distance between landmarks M_i and M_j.

$$\mathbf{Q} = \begin{pmatrix} 1 & M_{11} & M_{12} \\ 1 & M_{21} & M_{22} \\ 1 & \dots & \dots \\ 1 & M_{p1} & M_{p2} \end{pmatrix},$$

and

$$\mathbf{L} = \begin{pmatrix} \mathbf{P} & \mathbf{Q} \\ \mathbf{Q}' & \mathbf{0} \end{pmatrix}.$$

Then let N be the second configuration, and $V_x = (N_x\ 0\ 0\ 0)$ be a column vector; we use $\mathbf{L}^{-1}V_x = (w_1\ w_2 \dots w_p\ a_1\ a_x\ a_y)$ for defining the vector, $W_{1\cdot} = (w_1\ w_2 \dots w_p)$. The $f_x(x,y)$ function defines the transformed x-coordinate of a point Z of (x, y) coordinates everywhere in the plane, so that

$$f_x(x, y) = a_1 + a_x x + a_y y + \sum_{i=1}^{p} U(\|M_i Z\|),$$

where $\|M_i Z\|$ is the distance between a point Z of (x, y) coordinates in the plane and the M_i landmark. The same thing is repeated for the second dimension with V_y = $(N_y \ 0 \ 0 \ 0)$, W_2., $f_y(x, y)$.

The terms of the f functions can be grouped in two components: the first is an expression of the a coefficients that are the affine components and represent the behavior of f at infinity, and the second group is a sum of $U(r)$ functions. Furthermore, the quantity $I_f = \mathbf{W P W'}$ is proportional to the bending energy used for the deformation (= $8\pi \times$ the bending energy). The f functions minimize this quantity.

The bending energy matrix ($\mathbf{Be}$) corresponds to the $k \times k$ upper left submatrix of $\mathbf{L}^{-1}$. The decomposition of this matrix shows that three eigenvalues are equal to zero in two dimensions. The $p-3$ remaining non-zero eigenvectors are called the principal warp eigenvectors, while the eigenvalues are called the bending energies. Notice that we do not need the second target configuration to estimate these parameters.

The tps2d function written below returns the position of interpolated coordinates. The arguments of this function are the coordinates of points that are going to be interpolated in the target, the reference configuration matrix (matr), and the target configuration (matt). For convenience, the original coordinates to be mapped are written in a two-column matrix.

Function 4.32. tps2d

Arguments:
   M: Original coordinates to be mapped by TPS.
   matr: Reference configuration matrix.
   matt: Target configuration matrix.
Value:
   Interpolated coordinates arranged in a matrix object.

```
1   tps2d<-function(M, matr, matt)
2   {p<-dim(matr)[1]; q<-dim(M)[1]; n1<-p+3
```

Compute P, Q and L.

```
3    P<-matrix(NA, p, p)
4    for (i in 1:p)
5      {for (j in 1:p){
6          r2<-sum((matr[i,]-matr[j,])^2)
7          P[i,j]<- r2*log(r2)}}
8    P[which(is.na(P))]<-0
9    Q<-cbind(1, matr)
10   L<-rbind(cbind(P,Q), cbind(t(Q),matrix(0,3,3)))
```

Define the fx and fy functions.

```
11   m2<-rbind(matt, matrix(0, 3, 2))
12   coefx<-solve(L)%*%m2[,1]
13   coefy<-solve(L)%*%m2[,2]
14   fx<-function(matr, M, coef)
```

```
15    {Xn<-numeric(q)
16     for (i in 1:q)
17          {Z<-apply((matr-matrix(M[i,],
18              p, 2, byrow=T))^2, 1, sum)
19          Xn[i]<-coef[p+1]+coef[p+2]*M[i,1]+coef[p+3]*
20              M[i,2]+sum(coef[1:p]*(Z*log(Z)))}
21     Xn}
```

Calculate the interpolated coordinates.

```
22    matg<-matrix(NA, q, 2)
23    matg[,1]<-fx(matr, M, coefx)
24    matg[,2]<-fx(matr, M, coefy)
25    matg}
```

One can use this function for several purposes:

1. Interpolating the position of points of the target configuration that have been digitized only on the reference configuration (for instance, to estimate missing landmarks).
2. Interpolating change in other parts of the shape than at the position of the land-mark coordinates.
3. Constructing deformation grids.

Deformation grids are formalizations of D'arcy Thompson's idea. Once we have written the interpolating function, we can easily produce deformation grids with R programming. Here I provide a small function called tps, for which one must enter the number n of columns cells in its third argument. The grid should cover the target completely. For this purpose, we need to determine the minimal and maximal coordinates in order to appraise the range of the configuration on the x and y axes. For the graphical display to be slightly larger than the configuration, we enlarge the range of the graphic display by adding one fifth of the range of the target configuration on the left and right sides. To facilitate interpretation, the reference grid has square cells. Therefore, we adjust the number of row cells using the same interspace as raw columns. We write tps for producing grids of deformation according to our needs.

Function 4.33. tps

Arguments:
    matr: *Reference configuration matrix.*
    matt: *Target configuration matrix.*
    n: *Number of displayed column cells.*
Value:
    Deformation grid plot obtained by TPS interpolation.
Required function: tps2d.

```
1    tps<-function(matr, matt, n){
```

Define the range of the graph to estimate the size of the grid.

```
2    xm<-min(matt[,1])
3    ym<-min(matt[,2])
4    xM<-max(matt[,1])
5    yM<-max(matt[,2])
6    rX<-xM-xm; rY<-yM-ym
```

Calculate the coordinates of the intersections between the lines of the grid.

```
7    a<-seq(xm-1/5*rX, xM+1/5*rX, length=n)
8    b<-seq(ym-1/5*rX, yM+1/5*rX,
9            by=(xM-xm)*7/(5*(n-1)))
10   m<-round(0.5+(n-1)*(2/5*rX+ yM-ym)/
11           (2/5*rX+ xM-xm))
12   M<-as.matrix(expand.grid(a,b))
13   ngrid<-tps2d(M,matr,matt)
```

Plot the lines of the transformed grid.

```
14   plot(ngrid, cex=0.2,asp=1,axes=F,xlab="",ylab="")
15   for (i in 1:m){lines(ngrid[(1:n)+(i-1)*n,])}
16   for (i in 1:n){lines(ngrid[(1:m)*n-i+1,])}}
```

We compute the deformation grid between the mean configurations for the gorilla skulls belonging to females and males as:

```
> tps(FE, MA, 20)
> lines(FE[joinline,],col="grey50",lwd=2)
> lines(MA[joinline,],lty=3,lwd=2)
```

Alternatively to our function, the `tpsgrid` function of the **shapes** package displays deformation grids between the reference and the target configurations; users are invited to specify arguments concerning the width and location of the plot.

Note that deformation grids and TPS can map a form onto another, and they are independent of the scale factor.

We can amplify shape change in the target to magnify the deformation. This is the strategy that I used in the following example to produce the grids in Fig. 4.16.

```
>layout(matrix(c(1,2),1,2))
>mag1<-2; mag2<-4
>MA1<-MA+(MA-FE)*mag1
>tps(FE, MA1, 20)
>title("Two-times amplified deformation")
>MA2<-MA+(MA-FE)*mag2
>tps(FE, MA2, 20)
>title("Four-times amplified deformation")
```

Rather than displaying deformation grids, we can produce a field of vectors covering the whole shape and depicting shape change. The **sp** package offers a variety

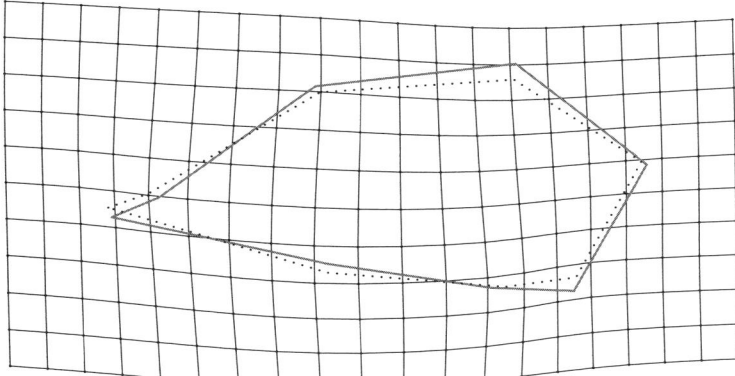

Fig. 4.15. Deformation grids produced on the skull configurations of the mean female gorilla (reference) and the mean male gorilla (target). The left (maxillary region) part is enlarged while the neurocranial part is smaller in the males in comparison to the females

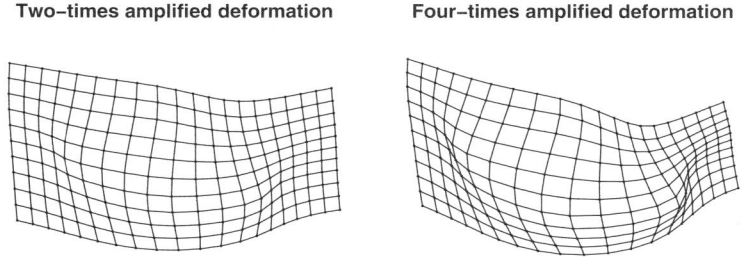

Fig. 4.16. Same grids as in Fig. 4.16 but with various amplification scalars

of functions for spatial data, and the `spsample` function can sample regularly or randomly spaced points on lines or polygons. This is convenient for exhaustively covering the shape surface. One can interpolate shape change everywhere in the considered shape displaying a vector field between the reference and the target. The plot (Fig. 4.17) and the following scripts offer some other possibilities for displaying deformation between two shapes.

```
>par(mar=c(1,1,1,1))
>plot(FE, asp=1,xlab="",ylab="",axes=F,pch=20,col="grey")
>lines(FE[joinline,],lwd=2,col="grey50")
>lines(MA[joinline,],lty=3,lwd=2)
```

We sample 200 points within the reference shape polygon. The `spsample` *function accepts objects of a class defined in the* sp *package. For creating a spatial polygon object, we use the* `Polygon` *function.*

```
>library(sp)
>sFE<-spsample(Polygon(FE[joinline,]),200,type="regular")
```

After defining TPS transformation, we draw arrows between the points of the interpolated target and the points sampled on the reference. The coordinates of the latter are found in the `@coords` *slot of the object returned by* `spsample`.

```
>sR<-sFE@coords
>sT<-tps2d(sR,FE,MA)
>for (i in 1:dim(sR)[1])
+    {arrows(sR[i,1],sR[i,2],sT[i,1],sT[i,2],length=0.05)}
```

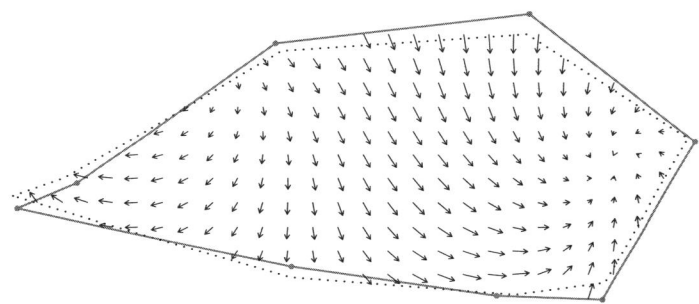

Fig. 4.17. Graphical display of TPS by way of a field of vectors. This field corresponds to deformations associated with differences between female and male gorilla skulls

We can also display the regions of the sagittal view of the skulls that show more deformation than others with a contour plot or an image plot. For this purpose, the z argument, passed in the `image` or `contour` functions, represents the amount of change at any given location – it is then is equivalent to the norm of the deformation vectors (Fig. 4.18).

```
>par(mar=c(1,1,1,1))
>layout(matrix(c(1,2),1,2))
>sFE1<-spsample(Polygon(FE[joinline,]),10000,type="regular")
>sR1<-sFE1@coords
>sT1<-tps2d(sR1,FE,MA)
>def<-sqrt(apply((sT1-sR1)^2,1,sum))
>xl<-length(unique(sR1[,1]))
>yl<-length(unique(sR1[,2]))
>im<-matrix(NA,xl,yl)
>xind<-(1:xl)[as.factor(rank(sR1[,1]))]
```

```
>yind<-(1:yl)[as.factor(rank(sR1[,2]))]
>n<-length(xind)
>for (i in 1:n){im[xind[i], yind[i]]<-def[i]}
>plot(FE, asp=1, xlab="", ylab="", axes=F)
>lines(FE[joinline,])
>contour(sort(unique(sR1[,1])),sort(unique(sR1[,2])),im,
+                     ,axes=F,frame=F,add=T)
```

Finally, we can superimpose the shape surface with vectors of deformations with an `image` plot (Fig. 4.18). We notice that deformation between both morphologies is nonuniform in Fig. 4.18.

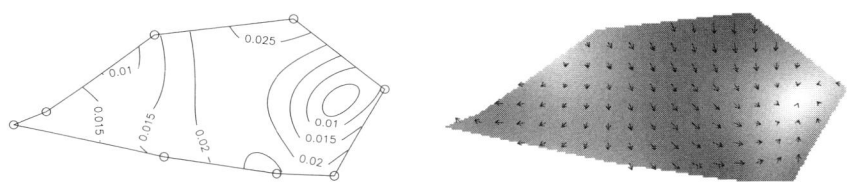

Fig. 4.18. Graphical display of TPS by the way of `contour` and `image` plot. Regions that show strong deformation are dark. These plots describe deformation associated with differences between averaged sagittal skull sections of female and male gorillas

```
>image(sort(unique(sR1[,1])),sort(unique(sR1[,2])),im,
+    col=gray((32:0)/32)[1:26],asp=T,axes=F,frame=F)
>sFE<-spsample(Polygon(FE[joinline,]),100, type="regular")
>sR<-sFE@coords
>sT<-tps2d(sR,FE,MA)
>for (i in 1:dim(sR)[1])
+    {arrows(sR[i,1],sR[i,2],sT[i,1],sT[i,2],length=0.04)}
```

You can find some other developments around TPS concepts in Dryden and Mardia [27]. These authors suggest modifying $\mathbf{P}$ by introducing a smoothing parameter γ (it should be post-multiplied by the identity matrix of the same size as $\mathbf{P}$). Bookstein [14] and Gunz et al. [42] provide further developments and future directions with TPS and splines (among which are presented several 3D applications).

4.4 Form and Euclidean Distance Matrix Analysis

Statistics must be adapted for coordinates of superimposed configurations because they are not independent data; the superimposition step introduces biases that theoretically prevent traditional statistics from being applied to them. In particular, when generalized superimposition aligns a series of shapes onto a reference, it introduces covariation between coordinates of superimposed configurations. Because of this, coordinates of superimposed configurations cannot be regarded as independent variables [62, 122]. Alternative methods were developed during the 1990s. Euclidean distance matrix analysis [59, 63, 92] is a coordinate-free approach relying on the set of all possible interlandmark distances that has been developed for avoiding problems with the superimposition method.

The EDMA approach works for form configurations and is invariant of translation and rotation. Moreover, several statistical applications have been developed around this approach, such as estimating the average form and the variance-covariance from a set of individuals, to allow further hypotheses to be tested [64, 60, 61] (position of influential landmarks, difference between sets of forms).

For any M configuration of p landmarks in k dimensions, there is an orthogonal a $p \times p$ matrix of interlandmark distances that one call the form matrix (**FM**). One can rapidly calculate this interlandmark distance matrix using the `dist` function (see Section 4.2.3). EDMA performs operations on the Euclidean matrix or on the vectorized upper or lower triangle of that matrix. Below is the code of the `FM` function that computes the full matrix of distances **FM**, and the code of the `fm` function that returns the vectorized form fm of the upper diagonal entry of $p(p-1)/2$ elements.

Functions 4.34. `FM` and `fm`

   Argument:
      `M`: *Configuration matrix.*
   Value:
      Full matrix of interlandmark distances in matrix form (for the `FM` *function).*
      Upper diagonal entry of the full matrix of interlandmark distances in vector form (for the `fm` *function).*

```
1   FM<-function(M){as.matrix(dist(M))}
2   fm<-function(M){mat<-FM(M);  mat[col(mat)<row(mat)]}
```

Thus, any object with p landmarks in k dimensions is represented by a point in a $p(p-1)/2$ dimensional Euclidean space. This space is called the form space. Not all sets of distances define forms because forms are constrained: the most trivial constraint is, for any kind of shape, no interlandmark distance can be greater than the sum of the others. The form space is thus a small part of the Euclidean space, and its geometry is constrained by limitations imposed by possible shapes. The observations are dependent in some extent in this limited space.

Two identical forms must have the same values for each element of the matrix, and correspond to the same point in the form space. If forms differ by their size only, the two points lie on a ray that passes through the origin. The form difference matrix (**FDM**) between the *M1* and *M2* configurations, corresponds to the pair-wise division of the elements of the matrices:

$$\mathbf{FDM}_{M1/M2} = \mathbf{FM1}/\mathbf{FM2} .$$

Two forms *M1* and *M2* have the same shape if the multiplication of **FM1** with a unique positive scalar ϕ equals **FM2**. Alternatively, it means that all elements of **FDM** are equal.

To examine which landmarks may have some influence on shape difference between configurations, Lele and Richtsmeier [64] suggest sorting the vector returned by the division of $fm(M1)$ by $fm(M2)$ and screening the landmarks that belong to distances that appear at the extreme of the ranked vector. One can achieve this with the `order` and `rank` functions, and by returning the name or the index of rows and columns of the matrix as in the following example.

```
>gorf.dat[1,2,1]
[1] 193
```

Introduce a Pinocchio effect in modifying the y-value of the first landmark of the first configuration.

```
>gorf.dat[1,2,1]<-250
>Ra1<-(rank(FM(gorf.dat[,,1])/FM(gorf.dat[,,2])
+       ,na.last="keep")+0.5)/2
>Ra1
       [,1] [,2] [,3] [,4] [,5] [,6] [,7] [,8]
[1,]   NaN   24   23   25   27   28   26   22
[2,]    24  NaN    6    7   16   17   14   21
[3,]    23    6  NaN   15    8   11    5   13
[4,]    25    7   15  NaN    4   10    2    3
[5,]    27   16    8    4  NaN   19   18   12
[6,]    28   17   11   10   19  NaN   20    9
[7,]    26   14    5    2   18   20  NaN    1
[8,]    22   21   13    3   12    9    1  NaN
>p<-dim(gorf.dat[,,1])[1]; Name<-NA
>for(i in 1:(p-1))
+     {Name[(sum(((p-1):1)[1:i])-p+i+1):(sum(((p-1):1)
+          [1:i]))]<-paste(i,(i+1):p,sep="-")}
>Ra2<-order(rank(fm(gorf.dat[,,1])/
+       fm(gorf.dat[,,2]),na.last=NA))
>Name[Ra2]
 [1] "7-8" "4-7" "4-8" "4-5" "3-7" "2-3" "2-4" "3-5" "6-8"
[10] "4-6" "3-6" "5-8" "3-8" "2-7" "3-4" "2-5" "2-6" "5-7"
[19] "5-6" "6-7" "2-8" "1-8" "1-3" "1-2" "1-4" "1-7" "1-5"
[28] "1-6"
```

Landmarks 1, 4 and 8 lie at the extreme of the distribution and influence difference in shape between configurations examined here.

Alternatively, one can remove the rank of median value from the vector difference, and keep the absolute values of the results. In ordering the distribution of the transformed distance ratios, distances including landmarks that have been altered will appear at the end of the vector and distances that have been less altered by shape change will appear first in the vector.

```
>fdm<-fm(gorf.dat[,,1])/fm(gorf.dat[,,2])
>Ra<-abs(fdm-median(fdm))
>Ra3<-order(rank(Ra, na.last=NA))
>Name[Ra3]
 [1] "2-7" "3-4" "3-8" "5-8" "3-6" "2-5" "2-6" "5-7" "5-6"
[10] "4-6" "6-8" "3-5" "2-4" "6-7" "2-3" "3-7" "4-5" "4-8"
[19] "4-7" "7-8" "2-8" "1-8" "1-3" "1-2" "1-4" "1-7" "1-5"
[28] "1-6"
```

A last solution consists of calculating the sum of divergences to the median value for each landmark, considering the whole **FDM** matrix. Landmarks of greatest influence will have the higher scores.

```
>FDM<-FM(gorf.dat[,,1])/FM(gorf.dat[,,2])
>rownames(FDM1)<-1:8
>FDM1<-abs(FDM-median(FDM, na.rm=T))
>round(apply(FDM1,2,sum,na.rm=T),2)
   1    2    3    4    5    6    7    8
4.93 0.61 0.44 0.65 0.82 2.82 0.92 0.71
```

As underlined by Cole [22] and Lele and Richtsmeier [64], the researcher should fully review and inspect the **FDM**. For a better visualization, Cole and Richtsmeier [22] have proposed graphically analyzing the scatter of ratios between interlandmark distances in the vertical axis and corresponding landmarks on the horizontal axis. The distribution of ratios involving each landmark provides information on their influence. Influential landmarks are characterized by a large range of related ratios, or by a skewed distribution of these ratios. I propose using landmark numbers as labels for the scatterplot; it makes interpretation easier (see Fig. 4.19).

```
>j<-cbind(rep(1:8,8), as.numeric(FDM))
>plot(j[-which(is.na(FDM)),],ylim=c(0,max(FDM,na.rm=T)),
+  cex=0,xlab="Landmark",ylab="Form difference",font.lab=2)
>text(j[-which(is.na(FDM)),],label=gl(8,7),cex=0.7)
>abline(h=median(FDM,na.rm=T),lty=3)
```

For statistical purposes, we have to define an average matrix form and variance-covariance for a set of several configurations. For a set of n individual configurations $(A_{i \rightarrow n})$, the mean of the squared Euclidean distances must be computed between landmarks for each pair of landmarks as follows:

$$e_{LN} = \frac{1}{n} \sum_{i=1}^{n} \|LN\|^2 .$$

Then, one defines S_{LN}^2 as the sample variance for the squared distance between landmarks L and N:

$$S_{LN}^2 = \frac{1}{n} \sum_{i=1}^{n} (\|LN\|^2 - e_{LN})^2 .$$

Likewise, for 2D objects one defines the quantity ω_{LN}^2 as

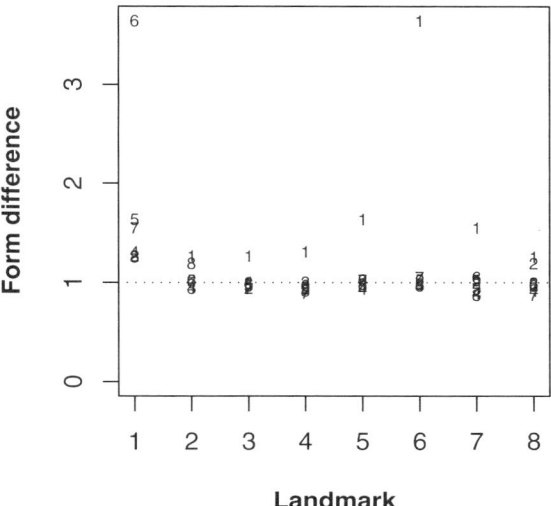

Fig. 4.19. A version of the graphical method for detecting influential landmarks using the EDMA approach. Only landmark 1 seems to really influence the difference between both configurations (this was expected since we have introduced an important change in this landmark for the generalized resistant-fit procedure)

$$\omega_{LN}^2 = (e_{LN}^2 - S_{LN}^2)^{\frac{1}{4}} \ .$$

For 3D objects, one calculates this quantity using the formula:

$$\omega_{LN}^2 = (e_{LN}^2 - \frac{3}{2}S_{LN}^2)^{\frac{1}{4}} \ .$$

The average form matrix $\overline{\omega}$ under its vector form corresponds to the series of ω for each interlandmark distance. The mEDMA function calculates this average from a set of configurations.

Function 4.35. mEDMA

> *Argument:*
>> A: $p \times k \times n$ *array containing configuration matrices.*
> *Value:*
>> $p \times p$ *average form matrix.*
> *Required function:* fm.

```
mEDMA<-function(A)
{n<-dim(A)[3];p<-dim(A)[1]; k<-dim(A)[2]
E<-matrix(NA,n,p*(p-1)/2)
```

Define a matrix of n rows of Euclidean interlandmark distances in their vectorized form (fm).

```
4   for (i in 1:n){E[i,]<-(fm(A[,,i]))^2}
```

Calculate each mean squared Euclidean distance.

```
5   Em<-apply(E,2,mean)
```

Compute the sample variance of each squared interlandmark distance.

```
6   S<-(apply(t((t(E)-Em)^2),2,sum))/n
```

Compute omega = ω for 2D or 3D data.

```
7    if (k==2){omega<-(Em^2-S)^0.25}
8    if (k==3){omega<-(Em^2-1.5*S)^0.25}
9    Om<-diag(0,p)
10   Om[row(Om)>col(Om)]<-omega; Om<-t(Om)
11   Om[row(Om)>col(Om)]<-omega
12   Om}
```

In general, the algorithm works well, but, in certain cases, problems may occur when the variance of the linear distances is larger than their mean. For instance, this occurred with the shortest distance concerning the first landmarks of the modified female gorilla dataset (gorf.dat with the introduction of the outlier).

```
>M<-mEDMA(gorf.dat)
>round(M[row(M)>col(M)],1)
 [1] 225.7 196.7 162.1   93.4   NaN 102.8 182.5   62.4   84.2
[10] 139.7 204.0 142.3   69.8  34.8 103.0 177.6 139.2 101.0
[19]  68.4 143.6 113.4   97.2  76.2  77.0 116.7  81.1 159.8
[28]  84.2
```

So far, this situation is rare for biological data (i.e., landmarks are well defined and not too close to each other in the configuration).

Lele [60] supplied the following algorithm to calculate the average form in terms of a $p \times k$ and $p \times p$ matrix. It relies on the squared average Euclidean distance $\overline{\omega}^2$ defined above. This is not less than an application of multidimensional scaling to the average form matrix. In order to compute the multidimensional scaling, one must first calculate the centered inner-product matrix of the mean form (called **B**). It is calculated by pre and post-multiplying the $\overline{\omega}$ matrix by a **H** matrix and by premultiplying by -0.5 as follows:

$$\mathbf{B} = -\frac{1}{2}\mathbf{H}\overline{\omega}\mathbf{H} ,$$

where **H**, the centering matrix, is a $p \times p$ matrix with diagonal elements equal to $1 - 1/p$ and off diagonal elements equal to $-1/k$.

Then the matrix is decomposed to appraise the $\lambda_1, \ldots, \lambda_p$ eigenvalues and the $u_1, \ldots, u_p$ eigenvectors. For 2D data one respectively calculates the x and y-coordinates by multiplying the square root of the first two eigenvalues by their corresponding eigenvectors. For 3D objects, one obtains the z-coordinates by multiplying the third eigenvector by the square root of the third eigenvalue. The new estimate of the mean Euclidean form $\overline{M}$ is calculated from the new configuration of x, y and eventually z-coordinates. It should be very close to the former estimator except it can be reflected (remember that reflections do not affect the average Euclidean form matrix).

For convenience, we write a MDS function to perform a multidimensional scaling on a matrix of Euclidean interlandmark distances. Remember that multidimensional scaling is concerned with constructing a configuration of p points and of k dimensions from a Euclidean distance matrix. The cmdscale function performs this task as well.

Function 4.36. MDS

Arguments:
      mat: p × p *matrix of Euclidean interlandmark distances.*
      k: *Number of desired dimensions.*
Value:
      Configuration matrix of p × k *dimensions.*

```
MDS<-function(mat, k)
{p<-dim(mat)[1]
C1<-diag(p)-1/p*matrix(1,p,p)
B<- -0.5*C1%*%mat^2%*%C1
eC<-eigen(B)
eve<-eC$vectors
eva<-eC$values
MD<-matrix(NA, p, k)
for (i in 1:k)
{MD[,i]<-sqrt(eva[i])*eve[,i]}
MD}
```

Then we use this function to estimate the mean form matrix. The whole computation is programmed in the mEDMA2 function. cimEDMA2 calculates the estimate of the mean form matrix in configuration matrix and in Euclidean interlandmark distance matrix format.

Function 4.37. mEDMA2

Argument:
      A: p × k × n *array of configuration matrices.*
Values:
      M: p × k *mean form matrix.*
      FM: *Mean form matrix (matrix of interlandmark Euclidean distances).*
Required functions: mEDMA, fm, MDS.

```
mEDMA2<-function(A)
{k<-dim(A)[2]
```

```
3 |  Eu<-mEDMA(A)
4 |  M<-MDS(Eu,k)
5 |  list("M"=M,  "FM"=FM(M))}
```

One can calculate the estimate of the variance-covariance matrix for a set of forms following the formula of Lele [60]. First, the configurations are centered on the origin, and the centered inner-product matrix ($\mathbf{Bs_i}$) is computed for each centered observation (Mi) following the equation:

$$\mathbf{Bs_i} = \mathbf{Bs_i}\mathbf{Bs_i}' \ .$$

One obtains the variance-covariance matrix $\mathbf{V_k}$ as

$$\mathbf{V_k} = \frac{1}{k}\left[\frac{1}{n}\sum_{i=1}^{n}\mathbf{Bs_i} - \overline{\omega\omega}'\right] \ ,$$

where $\overline{\omega}$ is formerly centered for each dimension.

Alternatively, one estimates $\mathbf{V_k}$ as:

$$\mathbf{V_k} = \frac{1}{k}\left[\frac{1}{n}\sum_{i=1}^{n}\mathbf{Bs_i} - \overline{MM}'\right] \ .$$

I used the second formula for writing the vEDMA function. This function computes the variance-covariance matrix from a set of configurations.

Function 4.38. vEDMA

Argument:
   A: $p \times k \times n$ *array of configuration matrices.*
Value:
   Variance-covariance form matrix
Required functions: mEDMA2, mEDMA, fm, MDS.

```
1  |  vEDMA<-function(A)
2  |  {p<-dim(A)[1]
3  |  k<-dim(A)[2]; n<-dim(A)[3]
4  |  Bs<-array(NA, dim=c(p,p,n))
5  |  for (i in 1:n){
6  |      Cc<-apply(A[,,i],2,mean)
7  |      Ac<-t(t(A[,,i])-Cc)
8  |      Bs[,,i]<-Ac%*%t(Ac)}
9  |  B<-apply(Bs, 1:2,mean)
10 |  M<-mEDMA2(A)$M
11 |  Ek<-(B-M%*%t(M))/k
12 |  Ek}
```

The variance-covariance matrix of EDMA usually has a rank equal to $p - 1$. However, diagonal elements may be negative when the variance of particular landmarks is very small, or when there are large covariances among some landmarks. Cole [21] suggests remedying this problem by defining the "positive semi-definite matrix that is most similar to the estimate" of possible negative elements in the diagonal of $\mathbf{V_k}$. In this respect, negative eigenvalues are set to 0, and the matrix is re-estimated. To perform this calculation, I excluded the three last eigenvalues in the following example:

```
>vcvgm<-vEDMA(gorm.dat)
>p<-dim(gorm.dat)[1]
>ei<-eigen(vcvgm)
>round(ei$vectors%*%diag(c(ei$values[1:(p-3)],
+      rep(0,3)))%*%t(ei$vectors),1)
        [,1]   [,2]   [,3]   [,4]  [,5]   [,6]   [,7]   [,8]
[1,]    45.7 -12.2 -15.6 -16.0  -0.7   36.0 -10.2 -27.0
[2,] -12.2  21.7    4.2    5.5  -3.6 -14.3   -4.2    2.9
...
```

Rao [87] and Mardia et al. [72] have presented methods that, like the Euclidean matrix analysis, work on interlandmark distances, except that they work on their logarithm. This allows a shape log-distance matrix to be defined from a form log-distance matrix as demonstrated below. Consider $\mathbf{G}(M)$ as being the form log-distance matrix. The $\mathbf{G_\star}(\mathbf{M})$ shape log-distance matrix is calculated using the equation

$$\mathbf{G_\star}(M) = \mathbf{G}(M) - \overline{g}\mathbf{1} \ ,$$

$\mathbf{1}$ being a square matrix of p 1, and $\overline{g}$ is the average of the log interlandmark distances.

The average $\mathbf{G}(\hat{u})$ of n form log interlandmark distances matrices $(\mathbf{G}(M)_i)$ is obtained as follows:

$$\mathbf{G}(\hat{u}) = \frac{1}{n} \sum_{i=1}^{n} \mathbf{G}(M)_i \ .$$

One can appraise the mean reflection size and shape by exponentiating and using multidimensional scaling as for the Euclidean distance matrix analysis approach. Likewise, the shape average is calculated by exponentiating $\mathbf{G}(\hat{u})$.

Rao [87] introduced two-sample tests procedures for mean shape and mean size. In working on the vectorized matrix d_i of $(p \times (p-1))/2$ log interlandmark distances, the sample mean $\hat{d}$ and the sample variance-covariance $\mathbf{S}$ are defined as:

$$\hat{d} = \frac{1}{n} \sum_{i=1}^{n} d_i; \qquad \mathbf{S} = \frac{1}{n-1}(d_1 d_1^t + \ldots + d_n d_n' - n\hat{d}\hat{d}^t) \ .$$

In R, we compute these means and variance-covariance using the log of our `fm` function applied to configuration matrices.

4.5 Angle-based Approaches for the Study of Shape Variation

Methods working on the matrices of all possible distances have the advantage of being invariant to rotation and translation. However, the shape space is not as simple as the Euclidean space described by $0.5 \times p \times p - 1$ dimensions. Indeed, the shape space occupies only a part of this Euclidean space, and its geometry is not a simple cartesian space. This intuitive idea is corroborated by the fact that the space used for Euclidean distance matrix analysis has many more dimensions than there are dimensions in the shape space defined from coordinates of landmarks.

Rao [88] has introduced a methodology based on angles that has the advantages of both describing objects based on the principle of invariance and objects that could have the same number of dimensions as the shape space for 2D objects. This method is based on a triangulation of landmarks in the whole configuration. The sum of the three angles of the triangle is constrained to π radians; thus, only two angles are necessary for describing a triangle. Since the number of triangles for describing a configuration in two dimensions is equal to $p - 2$, a minimum of $2p - 4$ angles can permit the description a 2D shape. Yet, there is a problem since there is no unique way to triangulate a configuration of landmarks; moreover, angles are circular data on which Euclidean statistics are not necessarily amenable. All the same, statistics have been developed for compositional and circular data, and some packages on R perform related statistical analyses applied to angles (CircStats, compositions). Here we will only write functions to compute the angles from the configuration and from a triangulation. Reconstructing configurations from angles is more difficult; indeed, problems can appear for reconstructing the shapes if some landmarks flip their position among configurations or if others are collinear. Unfortunately, more research in the statistics of angular data applied to shape analysis is needed.

Several options are possible with R to define a triangulation for a given configuration.

- We can first specify it by the matrix of $p - 2 \times 3$ triangle vertices that are chosen as triangulation for the configuration. Rows correspond to triangles and columns to vertices.
- One can use all possible triangles in the configuration. For this goal, one writes the alltri function.

Function 4.39. alltri
  Argument:
    p: *Number of landmarks.*
  Value:
    Matrix of triangle vertex combinations.

```
alltri<-function(p)
 {t1<-expand.grid(1:p,1:p,1:p)
 t2<-t1[t1[,1]<t1[,2],]
 t3<-t2[t2[,2]<t2[,3],]
 t3}
```

- We can also extract triangles produced from a Delaunay triangulation with the
  geometry package and the `delaunayn` function. If it returns a convenient set
  of triangles, the Delaunay triangulation usually generates more triangles than the
  minimum number of triangles necessary for describing a shape. We can eliminate
  rows that are not necessary. We display the Delaunay triangulation (see Fig. 4.20)
  by transforming the data in an object of `tri` class with the `tri.mesh` function
  and using the generic `plot` graphical function.

We can display the triangulation using a loop and the `polygon` function.

```
>gorm<-gorm.dat[,,4]
>plot(gorm, asp=1)
>library(geometry)
>tri<-delaunayn(gorm)
>n<-dim(tri)[1]
>for (i in 1:n){polygon(gorm[tri[i,],])}
```

After having defined the triangulation matrix, we calculate the angles at vertices
for each triangle. We use complex algebra to calculate these angles in a function that
we call `anglerao`.

Function 4.40. `anglerao`

Arguments:
> `M`: *Configuration matrix.*
> `triang`: *Triangulation matrix (giving the vertices position).*

Value:
> *Angle matrix depicting the configuration.*

```
1  anglerao<-function(M, triang)
2  {n<-dim(triang)[1]
3  triangl<-triang
4  for (i in 1:n){
5    t1<-triang[i,]
6    AB<-complex(real=M[t1[2],1]-M[t1[1],1],
7        imag=M[t1[2],2]-M[t1[1],2])
8    BC<-complex(real=M[t1[3],1]-M[t1[2],1],
9        imag=M[t1[3],2]-M[t1[2],2])
10   AC<-complex(real=M[t1[3],1]-M[t1[1],1],
11       imag=M[t1[3],2]-M[t1[1],2])
12   BAC<-Arg(complex(argument=-Arg(AB)+Arg(AC)))
13   CBA<-Arg(complex(argument=-Arg(BC)+Arg(-AB)))
14   ACB<-Arg(complex(argument=-Arg(-AC)+Arg(-BC)))
15   triangl[i,]<-c(BAC,CBA,ACB)}
16  triangl}
```

Computing the mean is a problem with angles close to $180°$. If one considers that
no landmarks flip in a distribution of shape, one avoids the problem by considering

the two smallest angles of each triangle for statistical inference as in the `meanrao` function programmed below. One can also use circular statistics and the inverse tangent. The `meanrao` function calculates the mean shape matrix from a configuration set and a triangulation matrix; in addition, it returns a matrix of 1 and 0 indicating, respectively, the largest angle and the two smallest angles for each triangle.

Function 4.41. `meanrao`

Arguments:
      `A`: *Array of configuration matrices.*
      `matt`: *Triangulation matrix.*
Values:
      `mean`: *Mean shape angular matrix.*
      `maxangle`: *Matrix indicating the position of the maximum angle for each triangle.*
Required function: `anglerao`.

```
meanrao<-function(A,triang)
{n<-dim(A)[3];p<-dim(A)[1];nt<-dim(triang)[1]
A1<-array(NA, dim=c(nt, 3, n))
A2<-matrix(NA, nt, 3)
A3<-matrix(0, nt, 3)
```

Compute angles for all configurations.

```
for(i in 1:n)
    {A1[,,i]<-anglerao(A[,,i], triang)}
```

Select each triangle from the configurations.

```
for(j in 1:nt)
    {Aj<-A1[j,,]
```

Compute the mean angles from the whole set of configurations.

```
    mAj<-apply(Aj,1,mean)
    MAj<-c(max(abs(Aj[1,])),max(abs(Aj[2,])),
        max(abs(Aj[3,])))
```

Find the angle of the triangle closest to 180°.

```
    MAji<-which(MAj==max(MAj))
    A3[j,MAji]<-1
```

Redefine the mean for the values that are closest to 180°.

```
    mAj[MAji]<-(pi-sum(abs(mAj[-MAji])))*
        sign(sum(mAj[-MAji]))
    A2[j,]<-mAj}
list(mean=A2, maxangle=A3)}
```

To compute the sample variance-covariance, we use only the smallest two angles (because π and $-\pi$ angles are equal). For interpreting this sample variance-covariance, the function should return the vertices of largest angles as in the previous function. These are stripped from the final variance-covariance. The `vcvrao` function computes the shape variance-covariance from a set of configurations.

Function 4.42. `vcvrao`

Arguments:
   A: *Array of configuration matrices.*
   `matt`: *Triangulation matrix.*
Values:
   VCV: *Angular variance covariance matrix.*
   `maxangle`: *Matrix indicating the position of the maximum angle for each triangle.*
   `strip`: *Indices of removed angles.*
Required functions: `meanrao, anglerao.`

```
1  vcvrao<-function(A,triang)
2  {n<-dim(A)[3];nt<-dim(triang)[1]
3  Aa<-t(matrix(A, nt*3,n))
4  VCV<-var(Aa)
5  maxangle<-meanrao(A,triang)$maxangle
6  strip<-which(maxangle==1)
7  VCV<-VCV[-strip, -strip]
8  list(VCV=VCV,maxangle=maxangle, strip=strip)}
```

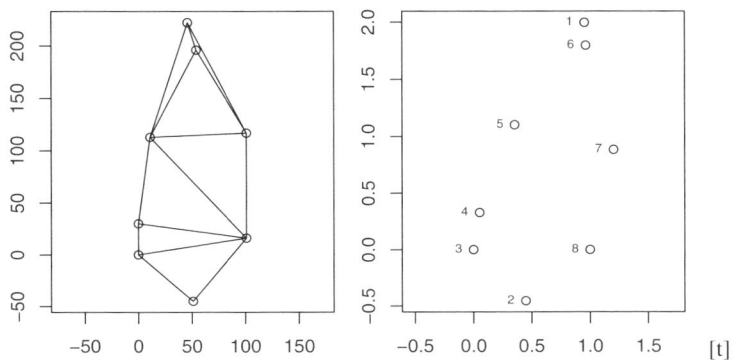

Fig. 4.20. Left: Delaunay triangulation for identifying a triangulation from the configuration of the sagittal section of the gorilla skull. The mean shape of the skull for the male gorilla dataset (`gorm.dat`) estimated from angles of the triangles of the configuration

It is useful to reconstruct the configuration from the angle matrix and the triangulation to visualize results or to make inferences in terms of shape difference.

We write the `raoinv` function for this goal. The total number of landmark positions is equal to the number of triangles of the minimal triangulation matrix $+2$. We set the first pair of landmarks in the triangulation to coordinates $(0, 0)$ and $(0, 1)$, and then the function calculates the coordinate of the third vertex using complex numbers. The next vertex belongs to the triangle that has at least two vertices in common with the first triangle of the triangulation. The program progresses until the number of landmarks is not equal to the number of vertices in the triangulation. Note that we have to take into account angle directions. To accomplish this, we check and correct the values of angles with conditional expressions.

Function 4.43. `raoinv`

Arguments:
    T: *Matrix of angles.*
    `triang`: *Triangulation matrix.*
Value:
    Configuration matrix.

```
1   raoinv<-function(T, triang)
2   {nt<-dim(triang)[1]
3   M<-NA
4   triang1<-triang
5   M1<-M[triang[1,1]]<-0+0i
6   M2<-M[triang[1,2]]<-1+0i
```

Calculate the coordinates of the third vertex of the first triangle.

```
7    M[triang[1,3]]<-complex(modulus=sin(T[1,2])*
8          Mod(M2-M1)/sin(T[1,3]),argument=
9          T[1,1]+Arg(M2-M1))+M[triang[1,1]]
10   itriang<-triang1[1,]
11   triang1<-triang1[-1,]
12   T1<-T[-1,]
13   i<-1
```

Until the number of vertices is not equal to the number of landmarks, calculate the coordinates of the next vertex, by checking the common base shared by formerly estimated triangles.

```
14   while (length(itriang)<length(unique(as.vector(triang))))
15     {while (length(which(itriang%in%triang1[i,]==TRUE))<2)
16        {i<-i+1}
17     iP<-which(itriang%in%triang1[i,]==TRUE)
18     P<-itriang[iP]
19     iN<-which(triang1[i,]%in%itriang==FALSE)
20     N<-triang1[i,iN]
21     P1<-P[1]; iP1<-which(triang1[i,]==P1)
22     P2<-P[2]; iP2<-which(triang1[i,]==P2)
23     if (iP1==1 & iP2==3)
24        {P3<-P1; P1<-P2; P2<-P3; iP1<-3; iP2<-1}
```

```
25      if (iP1==2 & iP2==1)
26          {P3<-P1; P1<-P2; P2<-P3; iP1<-1; iP2<-2}
27      if (iP1==3 & iP2==2)
28          {P3<-P1; P1<-P2; P2<-P3; iP1<-2; iP2<-3}
29      M1<-M[P1]
30      M2<-M[P2]
31      M[N]<-complex(modulus=sin(T1[i,iP2])*
32          Mod(M2-M1)/sin(T1[i,iN]),argument=
33          T1[i,iP1]+Arg(M2-M1))+M[P1]
34      T1<-T1[-i,]
35      itriang<-unique(c(itriang, triang1[i,]))
36      triang1<-triang1[-i,]
37      i<-1}
38  cbind(Re(M), Im(M))}
```

We can plot the mean shape of the sample, no matter whether the triangulation matrix is longer than the minimal one (see 4.20).

```
>msh<-meanrao(gorm.dat, tri)$mean
>plot(raoinv(msh, tri),asp=1)
>text(raoinv(ju, tri),labels=1:8, cex=0.7)
```

Methods based on angles are not commonly used. You can read [98] for comments concerning the geometry of the shape spaces described with those approaches. Actually, the choice of the triangulation affects the geometry of the shape space; results may differ depending on the triangulation that has been chosen. This tool, if more developed in the future, could provide exact estimates of dispersion and central tendency parameters.

Problems

4.1. Reconstructing configuration with the truss network strategy
Write a function that reconstructs locations of landmarks, after having defined an object of the truss class. The function must also receive a second argument for including the prototype.

4.2. Centroid size of a vectorized set of coordinates
Write a short function that returns the centroid size of a configuration written in its vectorized form.

4.3. Baseline size of a vectorized set of coordinates
Write a short function that returns the baseline size of a configuration written in its vectorized form.

4.4. Bookstein and Kendall coordinates
Check the relationships between Bookstein and Kendall coordinates calculated using the first configuration of the `gorf.dat` dataset of shapes.

4.5. Full and partial Procrustes superimposition
Write a comprehensive and unique function that computes both full and partial Procrustes superimpositions; add logical arguments for selecting among the types of superimpositions.

4.6. Weighted superimposition
Write a function that performs a weighted GPA for several groups of configurations. Each group may have unequal size, and the function should correct unbalance by assigning equal weigh to each group.

4.7. Resistant-fit superimposition
Write a unique function that can perform resistant-fit superimposition between two configuration matrices for either 2D or 3D data.

4.8. 3D TPS
Develop the function for 3D thin-plate splines. To achieve this, keep in mind that the interpolating function is no longer $U(r) = r^2 log(r^2)$ but $U(r) = \|r\|$. Do not forget to include the third dimension.

4.9. Estimating the mean shape
Compute the mean shape and scale it by the mean size using Procrustes methods, and compare it with the average form calculated with EDMA methods. Use the FDM approach for localizing eventual form difference.

4.10. EDMA and variance-covariance
Write a small function that estimates the variance-covariance matrix of a set of configurations. The function should consider the presence of negative eigenvalues and proceed by sorting eigenvalues for finding the first negative and initialize it and the following to 0.

4.11. Triangulation and mean shape
Compute the mean shape with triangulation and Procrustes methods and compare them using the interlandmark distances **FDM**. Repeat the operation with different triangulations. Compare your results with the ones obtained by Procrustes methods. Which method seems to return the more reliable estimate?

4.12. Procrustes test
Write a function to perform a Procrustes test (similarity of configurations in k dimensions). The test should authorize reflection for minimizing the Procrustes distance. See Section 3.4.4 for complementary information.

5

Statistical Analysis of Outlines

Rather than working on configurations of homologous landmarks, several morphometric methods have been applied to the study of outline data. Usually, morphometric methods applied to outline data are processed in four steps.

1. Sample landmarks along the outline. Users digitize a sample of points on the outlines according to what they think to be relevant or important. Those can be equally spaced points (for example, according to the curvilinear abscissa), or more densely digitized points when rapid changes in the curvature are observed. This sample yields a set of pseudolandmarks. Some methods need the number of pseudolandmarks to be the same among configurations if the goal is to analyze the variation within a set of outlines.

2. Align the configurations of pseudolandmarks. Basically, these alignments are based on superimposition methods (baseline alignment, Procrustes superimposition). The superimposition can use landmarks on the structure (on the outline or not) or peculiar geometric features of interest (e.g., major axis of the best-fitting ellipse on the outline) for removing rotation, size and position nuisances.

3. Fit a function for describing the outline as completely as possible. There is actually no homology between the landmarks sampled along the outline. In addition, although the sampling of pseudolandmarks can correspond very closely to the outline, the large number of coordinates and the redundancy within this data do not allow direct analysis. Prior to the analysis, the raw measurements (pseudolandmark coordinates) are compacted by fitting a function that describes parameters of the outline. It is not directly the relative positions of landmarks of the outline but the outline parameters themselves that yield the shape information. These parameters are appraised from the coefficients of the function adjusted to the sampled points. These coefficients are usually decreasing in order. According to their order, they correspond to more and more local shape features. Users helped by some protocol can choose to drop high-order coefficients to focus on fewer but more relevant parameters traducing shape variation.

4. Analyze the coefficients with multivariate statistics.

5.1 Open Outlines

Open outlines on a structure are delimited by a starting and an ending point. If one can arrange simple open curves after superimposition or rotation in order to express the y-coordinate as an injection of the x-coordinate, one can consider them to be simple open outlines. If for a given x, two or more y-coordinates correspond in two dimensions, one will speak about complex open outlines.

5.1.1 Polynomial Curves

One can exactly fit a polynomial equation of the form $y = b_0 + b_1 x + b_2 x^2 + \ldots + b_{p-1} x^{p-1}$ on a simple collection of p points on open outlines. In practice, because of redundancy in the data, dropping high-order coefficients should not significantly reduce the goodness-of-fit.

Let us consider the simple example in [108] with observations of coordinates $x = 1, 2, 4, 7, 9, 12$ and $y = 1, 5, 7, 8, 11, 12$. We will use the `booksteinM` function for aligning the outline (necessary if one wants to compare several similar outlines). We use the `poly` function of the **stats** package for specifying a polynomial regression in the formula (Fig. 5.1).

Align the starting and ending points of the curve onto the coordinates (-0.5,0); (0.5,0).

```
>M<-cbind(c(1,2,4,7,9,12), c(1,5,7,8,11,12))
>M1<-as.data.frame(booksteinM(M, 1,6))
>colnames(M1)<-c("x","y")
>plot(M1,asp=1,xlab="x",ylab="y")
```

Fit a polynomial regression to the data.

```
>fm1 <- lm(y ~ poly(x, 5, raw=T), data = M1)
```

Draw the fitted curve.

```
>d<-seq(-0.5, 0.5, length=200)
>lines(d, predict(fm1, data.frame(x=d)))
```

Return coefficients.

```
>coe<-coef{fm1}
>names(coe)<-c("Intercept","x^1","x^2","x^3","x^4","x^5")
>round(coe3)
Intercept        x^1        x^2        x^3        x^4        x^5
    0.088     -0.559      0.425      7.349     -3.101    -20.447
```

Since there are six points on the outline, the polynomial function passing through every point will be of degree five. The degree is specified as an argument of the `poly` *function. Note the use of* `predict` *for estimating the curve fitted on the outline.*

Decreasing the number of parameters by removing the higher-order coefficients will progressively smooth the curve. One can estimate the differences between the coordinates of sample points and estimated points. For this purpose, one can define a threshold according to measurement error.

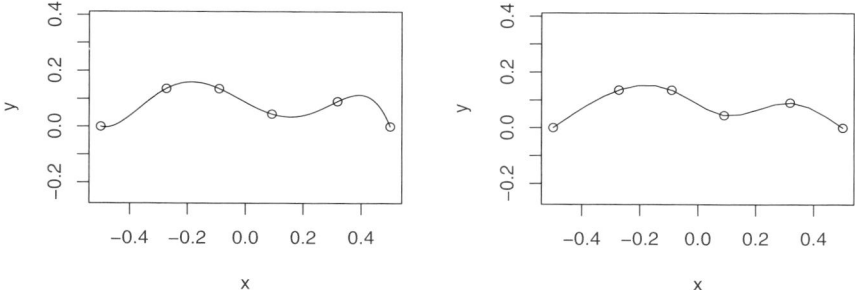

Fig. 5.1. Fitting curves to an open outline, the graph on the left corresponds to a polynomial regression, while the graph on the right corresponds to a natural cubic spline

5.1.2 Splines

Simple Cubic Splines

Technical information about estimating natural cubic spline parameters is supplied in [107]. If p points are collected, the cubic spline estimate fits a series of cubic polynomials to the data. The values of the second derivates of the interpolating polynomial functions correspond to the parameters of the curve. There are $p - 2$ variable parameters. One must resample fewer points on the curve to reduce the number of parameters, and then one must check whether the fit is still the same. The `splinefun` function of the **stats** package performs cubic spline interpolation on a given set of data points. Here I give an example of fit:

```
>fo<-splinefun(M1[,1], M1[,2], method="natural")
>ls(envir = environment(fo))
[1] "ties" "ux"    "z"
>splinecoef <- get("z", envir = environment(fo))
>splinecoef
$method
[1] 2

$n
[1] 6

$x
[1] -0.50000 -0.27272 -0.09090  0.09090  0.31818  0.50000

$y
[1] 0.00000 0.13636 0.13636 0.04545 0.09090 0.00000

$b
[1]  0.72095  0.35808 -0.42588 -0.15453 -0.04721 -0.72639
```

```
$c
[1]   0.0000 -1.5966 -2.7152   4.2076 -3.7354   0.0000

$d
[1]   -2.3416   -2.0508   12.6920 -11.6499    6.8483   0.0000
>plot(M1,asp=1, xlab="x",ylab="y")
>lines(spline(M1[,1], M1[,2], method="natural"))
```

Obtain the parameters of the curve.

```
>round(fo(M1[,1], deriv=2),4)
[1]   0.0000 -3.1932 -5.4305   8.4154 -7.4709   0.0000
```

Parametric Splines

One can express x and y-coordinates (or even z-coordinates) as parametric functions of the cumulative chordal distance for each point of the contour. This strategy produces more parameters but is useful for fitting complex open outlines (in which for a given x-value, several y-values can correspond), and even closed contour: the first point being the last one.

We first write a small `cumchord` function that calculates the cumulative chordal distance.

Function 5.1. `cumchord`

Argument:
   M: *Matrix of point coordinates.*
Value:
   Cumulative chordal distance vector.

```
1  cumchord<-function(M)
2  {cumsum(sqrt(apply((M-rbind(M[1,],
3         M[-(dim(M)[1]),]))^2,1,sum)))}
```

According to the number of dimensions, there are two or three cubic splines fitting between the cumulative chordal distance and each of the coordinates. Thus $2 \times p - 2$ coefficients contain shape information necessary for depicting the outline.

```
>M<-cbind(c(1,2,4,8,4,1,-3,-10),
+         c(1,5,7,-8,-11,-4,-3,-10))
>M1<-as.data.frame(booksteinM(M, 1,8))
>plot(M1,asp=1, xlab="x", ylab="y")
>z<-cumchord(M1)
```

Return coefficients.

```
>fo1<-splinefun(z,M1[,1],method="natural")
>fo2<-splinefun(z,M1[,2],method="natural")
>round(fo1(z, deriv=2),3)
```

```
[1]   0.0000   0.0052 -0.0059   0.0048 -0.0037   0.0000
>round(fo2(z, deriv=2),4)
[1]   0.0000 -0.0102 -0.0199   0.0300 -0.0263   0.0000
```

Draw the interpolated curve.

```
>lines(spline(z,M1[,1],method="natural",n=100)$y,
+      spline(z,M1[,2],method="natural",n=100)$y)
```

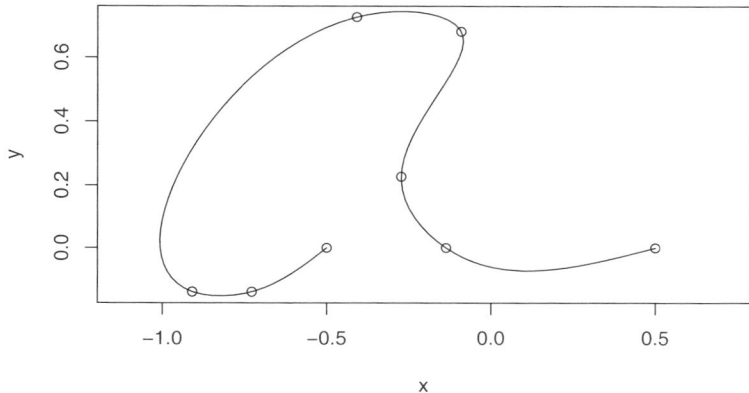

Fig. 5.2. Fitting curves with parametric splines

Actually one can replace the cumulative chordal distance by the curvilinear abscissa of unit regular step. However, this may have impacts on the values of coefficients and on the way shape variation is estimated; this mainly depends on the sampling procedure applied by users.

5.1.3 Bezier Polynomials

There are other ways to fit a curve to an open outline, such as Bezier polynomials. A Bezier polynomial of degree q is a function that passes to a sequence of $q+1$ points. The parameters of the function correspond to locations of points corresponding to vertices of a polygon around the observed points sampled on the curve. The first and last vertices are the same as the starting and ending points. One decomposes the matrix of sampled coordinates of points following the equation:

$$M = JB ,$$

where M corresponds to the location of sampled points (the configuration), J to the $q \times q + 1$ matrix of Bezier coefficients, and B to the matrix of coordinates for the Bezier vertices. One finds the coordinates of the Bezier vertices so that

$$B = (J'J)^{-1}J'M .$$

One computes the J matrix of coefficients for solving the former equation as

$$J_{iq} = \frac{q!}{(j-1)!(q-j+1)!} t_i^{j-1}(1-t_i)^{q-j+1} ,$$

where d is the distance between two sampled points, and t_i is the standardized chordal distance from the first data point to the ith points so that $t_1 = 0$ and $t_{q+1} = 1$.

We write a bezier function for extracting coefficients and Bezier vertices from a set of coordinates sampled on an outline. The function returns vertices and coefficients of order $n-1$. We set n by default to be equal to the number of sampled points.

Function 5.2. bezier

Arguments:
    M: *Matrix of point coordinates.*
    n: *Number of estimated Bezier vertices minus one.*
Values:
    J: *Matrix of Bezier coefficients.*
    B: *Coordinates of Bezier vertices.*
Required function: cumchord.

```
bezier<-function(M,n = dim(M)[1])
{p<-dim(M)[1]
if (n != p) {n<-n+1}
M1<-M/cumchord(M)[p]
t1<-1-cumchord(M1)
J<-matrix(NA, p, p)
for (i in 1:p){
for (j in 1:p){J[i, j]<-(factorial(p-1)/(factorial(j-1)*
    factorial(p-j)))*(((1-t1[i])^(j-1))*t1[i]^(p-j))}}
B<-ginv(t(J[,1:n])%*%J[,1:n])%*%(t(J[,1:n]))%*%M
M<-J[,1:n]%*%B
list(J=J, B=ginv(t(J[,1:n])%*%J[,1:n])%*%(t(J[,1:n]))%*%M)}
```

For drawing the curve, we compute the coordinates according to the Bezier polynomial given by the $B(t)$ function, so that

$$B(t) = \sum_{i=0}^{n} B_i b_{i,n}(t) ,$$

$t \in [0, 1]$ and $b_{i,n}(t)$ being the basis Bernstein polynomials calculated as

$$b_{i,n}(t) = \binom{n}{i} t^i (1-t)^{n-i}, \quad i = 0, \ldots n ,$$

with $\binom{n}{i}$ as the binomial coefficient.

These formulae are necessary for writing the `beziercurve` function that will return the p coordinates of equally spaced points on the Bezier curve estimated from the Bezier vertices B.

Function 5.3. `beziercurve`

Arguments:
    `B`: *Matrix of Bezier vertex coordinates.*
    `p`: *Number of points to be sampled on the curve.*
Value:
    Two-column matrix of interpolated coordinates.

```
 1  beziercurve<-function(B,p)
 2    {X<-Y<-numeric(p)
 3    n<-dim(B)[1]-1
 4    t1<-seq(0,1,length=p)
 5    coef<-choose(n, k=0:n)
 6    b1<-0:n; b2<-n:0
 7    for (j in 1:p)
 8      {vectx<-vecty<-NA
 9      for (i in 1:(n+1))
10          {vectx[i]<-B[i,1]*coef[i]*t1[j]^b1[i]*
11                    (1-t1[j])^b2[i]
12          vecty[i]<-B[i,2]*coef[i]*t1[j]^b1[i]*
13                    (1-t1[j])^b2[i]}
14      X[j]<-sum(vectx)
15      Y[j]<-sum(vecty)}
16    cbind(X, Y)}
```

Fitting Bezier polynomial curve has the advantage of being a good estimator for the curve even if one removes high order coefficients (i.e., considering only the first columns of J. One can estimate whether the fit is still reliable by removing one or more columns of J, and re-estimating the B matrix).

We try our functions for a collection of sampled landmarks `M`, and we draw the estimated curve (Fig. 5.3).

```
>M<-cbind(c(1,2,4,7,9,12), c(1,5,7,8,11,12))
>plot(M, pch=3, xlab="X",ylab="Y",ylim=c(0,13),asp=1)
>lines(beziercurve(bezier(M)$B,50))
```

In superimposing curves according to starting and ending landmarks, the vertices coordinates become the parameters of the curve. One can use them later as shape variables for statistical analyses.

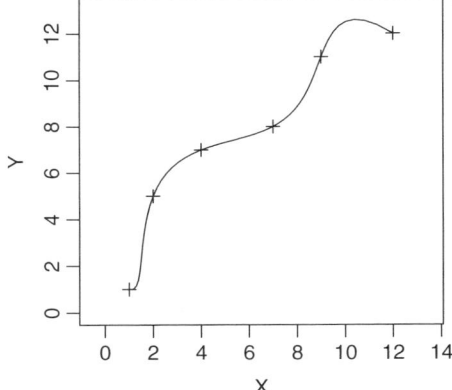

Fig. 5.3. Fitting curves with Bezier polynomials. The sampled points are drawn as crosses

5.2 Fourier Analysis

Closed outlines are more common in morphometric studies, and numerous studies have used Fourier series decomposition (by using the discrete Fourier transform) for fitting a periodic function to sampled pseudolandmarks along the outline. Fourier transforms use the Fourier series for decomposing and analyzing periodic signals (or functions) into a weighted sum of simpler sinusoidal component functions. The name Fourier is given in honor to Jean Baptiste Joseph Fourier, a French mathematician born in Burgundy during the 18th Century.

The general expression of the Fourier expansion for a periodic function $f(t)$ with $t \in \mathbb{R}$ and of period T, is defined as

$$f(t) = \frac{1}{2}a_0 + \sum_{n=1}^{\infty} [a_n \cos(\omega_n t) + b_n \sin(\omega_n t)] ,$$

where

$$\omega_n = n\frac{2\pi}{T}$$

is the nth harmonic of the function (in radians),
and where

$$a_n = \frac{2}{T} \int_{t_1}^{t_2} f(t) \cos(\omega_n t) \, dt$$

are the even Fourier coefficients,
and

$$b_n = \frac{2}{T} \int_{t_1}^{t_2} f(t) \sin(\omega_n t) \, dt$$

are the odd Fourier coefficients.

One can write $f(t)$ under its exponential form

$$f(t) = \sum_{n=-\infty}^{+\infty} c_n e^{i\omega_n t} ,$$

with

$$c_n = \frac{1}{T} \int_{t_1}^{t_2} f(t) e^{-i\omega_n t} \, dt .$$

For outlines, one cannot directly apply Fourier transforms since they are defined as a function of x and y-coordinates in two dimensions. There are at least two possibilities: The first is to express the outline as a function of one transformed variable. "Polar transform or Fourier analysis of equally spaced radii" and "Fourier analysis of the tangent angle to the outline" correspond to the first options. The third option is to separately decompose x an y-coordinates and to express them as functions of the curvilinear abscissa (elliptic Fourier decomposition).

5.2.1 Fourier Analysis Applied to Radii Variation of Closed Outlines

For illustrating the method, I will focus on a specific example that consists of the outline of the honeybee forewing. We first need to construct the outline. We obtain it from the the picture "wing.jpg" (see Fig. 5.4). For digitizing coordinates of pixels on the outline, we need the `Conte` function that was programmed in Section 2.2.4. The image is a gray-scale image and can be interpreted as a simple `imagematrix` object. For insuring that full rows and columns of white pixels border the `imagematrix`, we append one row of white pixels around the binarized image. We select the starting point at the wing insertion with the body.

```
>layout(matrix(c(1,2),1,2))
>library(rimage)
>wing<-read.jpeg("/home/juju/morph/wing.jpg")
>wing<-cbind(1, wing, 1)
>wing<-rbind(1,wing,  1)
>wing1<-wing<-imagematrix(wing, type="grey")
>wing1[which(wing1<0.95)]<-0
>plot(wing)
```

Digitize the first landmark of the outline.

```
>cont<-Conte(round(unlist(locator(1)),0),wing1)
>lines(cont$X, cont$Y,lwd=2)
```

We could smooth the outline to reduce noise introduced by the digitizing process. However, since we use landmarks spaced at equally spaced radii, this is not necessary. For extracting equally spaced radii, we need the `regularradius` function programmed in Chapter 2. In addition, we scale the data to retrieve raw coordinates, using the ruler on the bottom of the image. We can click on the two extreme graduations with `locator` to estimate the number of the pixels contained on 5 mm.

```
>scalecoord<-locator(2, type="p",pch=4,lwd=2)
>scalepixsize<-sqrt(diff(scalecoord$x)^2+
+               diff(scalecoord$y)^2)
>X<-cont$X*5/scalepixsize
>Y<-cont$Y*5/scalepixsize
>Xc <- mean(X)
>Yc <- mean(Y)
>wingr<-regularradius(X, Y, 64)
>plot(wingr$coord[,1], wingr$coord[,2],pch=16,
+        asp=1,xlab="X",ylab="Y")
>for (i in 1:64)
+       {segments(0,0,wingr$coord[,1],
+        wingr$coord[,2])}
```

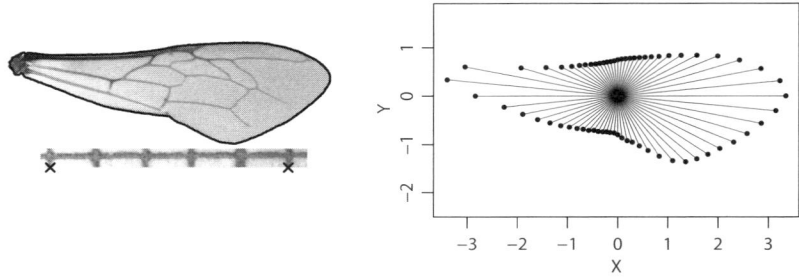

Fig. 5.4. Data acquisition for Fourier analysis of equally spaced radii. The two crosses on the left graph correspond to the landmarks digitized to obtain a scale on the millimeter paper, and to standardize coordinates by size

One can express the radius r as a periodic function of the angle θ. The relationship corresponds to the equation of the curve defined by the harmonics 0 to k:

$$r(\theta) = \frac{1}{2}a_0 + \sum_{n=1}^{k} [a_k \cos(\omega_k\theta) + b_k \sin(\omega_k\theta)] \ .$$

There cannot be more harmonics than half the number (p) of sampled points since there are two parameters for one harmonic and since the original data is univariate. One appraises the coefficients following the equations:

$$a_n = \frac{2}{p} \sum_{n=1}^{p} r_i \cos n\theta_i \ ,$$

$$b_n = \frac{2}{p} \sum_{n=1}^{p} r_i \sin n\theta_i \ ,$$

$$a_o = \sqrt{\frac{2}{p}} \sum_{n=1}^{p} r_i \ .$$

We write the `fourier1` function that calculates the Fourier coefficients from the original data. Here we develop it for the coordinates of the intersection between the equally spaced radii and the outline. The other argument is the number of harmonics that one wants to calculate. Using complex numbers facilitates the programming.

Function 5.4. `fourier1`

Arguments:
      M: *Matrix of sampled points.*
      n: *Number of harmonics.*
Values:
      ao: a_o *harmonic coefficient.*
      an: *Vector of* $a_{1 \to n}$ *harmonic coefficients.*
      bn: *Vector of* $b_{1 \to n}$ *harmonic coefficients.*

```
1   fourier1<-function(M, n)
2   {p<-dim(M)[1]
3   an<-numeric(n)
4   bn<-numeric(n)
5   Z<-complex(real=M[,1],imaginary=M[,2])
6   r<-Mod(Z)
7   angle<-Arg(Z)
8   ao<- 2* sum(r)/p
9   for (i in 1:n){
10      an[i]<-(2/p)*sum(r * cos(i*angle))
11      bn[i]<-(2/p)*sum(r * sin(i*angle))}
12  list(ao=ao, an=an, bn=bn )}
```

Using the function and entering correct arguments is straightforward:

```
>fourier1(wingr$coord, 12)
...
```

One can reconstruct the outline based on the k first given harmonics. The first harmonics are usually sufficient to describe the outline without wasting much of the information. For this task, we write the `ifourier1` function for reconstructing the outline from harmonics. We want to sample the estimated position of n points for drawing the outline from k harmonics functions.

Function 5.5. `ifourier1`

Arguments:

    `ao`: a_o *harmonic coefficient.*

    `an`: *Vector of* $a_{1 \to n}$ *harmonic coefficients.*

    `bn`: *Vector of* $b_{1 \to n}$ *harmonic coefficients.*

    `n`: *Number of interpolated points on the outline.*

    `k`: *Number of harmonics.*

Values:

    `angle`: *Radius angle to the reference.*

    `r`: *Vector of interpolated radii.*

    `X`: *Vector of* x-*interpolated coordinates.*

    `Y`: *Vector of* y-*interpolated coordinates.*

```
1  ifourier1<-function(ao, an, bn, n, k)
2  {theta<-seq(0,2*pi, length=n+1)[-(n+1)]
3  harm <- matrix (NA, k, n)
4  for (i in 1:k)
5     {harm[i,]<-an[i]*cos(i*theta)+ bn[i]*sin(i*theta)}
6  r<-(ao/2) + apply(harm, 2, sum)
7  Z<-complex(modulus=r, argument=theta)
8  list(angle = theta, r = r, X = Re(Z), Y = Im(Z))}
```

We apply the functions to our wing example.

```
>par(mar=c(2,2,2,2))
>f1<-fourier1(wingr$coord, 32)
>layout(matrix((1:9),3,3))
>for (i in 1:9)
+    {if1<-ifourier1(f1$ao, f1$an, f1$bn, 64, i)
+    plot(if1$X, if1$Y, asp=1, type="l", frame=F,
+    main=paste("Harmonics 0 to",i))}
```

We notice in Fig. 5.5 that the wing outline reconstruction is nearly correct at the seventh harmonic. Because of the constriction in the middle of the wing, the Fourier decomposition generates harmonics of a higher order than the order necessary to reliably describe the wing. One can decide to drop high order harmonics and to keep the first ones only, if the middle region of the wing is not considered as important for the study. Note that the starting point does not correspond to the one that we have first digitized.

The method reduces the number of parameters pretty well, and one can apply it to simple outlines. It will be more efficient on globular objects, where sample biases are lowest.

One must be careful when comparing several outlines: the angle of the starting radius must be the same among outlines. In this respect, the starting point should be a landmark. It is possible to rotate outlines so that the shapes are aligned on their first radius. To compare outlines, one must also always digitize the outline in the same direction (clockwise or counter clockwise). If one wants to compare left with right

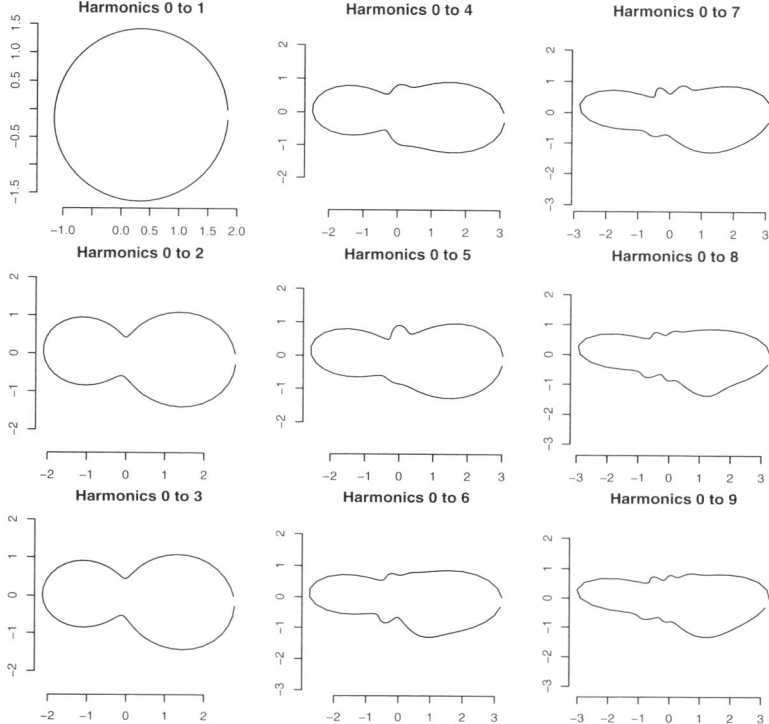

Fig. 5.5. Reconstruction of outlines by Fourier analysis applied to equally spaced radii

objects, one can digitize left and right outlines, respectively, clockwise and counter clockwise, and multiply one of the coordinate dimension times by -1 for a given side. If the digitization procedure always goes in the same direction, the raw indices of one side have to be reversed.

There can be analytical problems with equally spaced radii Fourier analysis: Indeed, the data itself can be flawed since the sampling strategy gathers more information in some parts of the outline and much less elsewhere. In addition, one cannot apply the method when a given radius twice intercepts the outline (it can arise when the outline presents pronounced convexities or concavities).

5.2.2 Fourier Analysis applied to the Tangent Angle

Zahn and Roskies [125] were the first to find a solution for applying Fourier analysis to outlines with important concavities that one cannot analyze with the method described above. Their method consists of describing the cumulative change in angle of a tangent vector to the outline ($\phi(t)$) as a function of the cumulative chordal distance t along the curve. For simplicity, the perimeter of the outline is first scaled to 2π. Then, one can express $\phi(t)$ as:

$$\phi(t) = \theta(t) - \theta(0) - t \; ,$$

where t is the distance along the perimeter, $\theta(t)$ is the angle of the tangent vector at the distance t, and $\theta(0)$ is the angle of the tangent vector for the starting point. $\theta(0)$ is removed for standardizing the data: this allows different individuals to be compared.

One can estimate the Fourier coefficients by multiple regression or by direct least-squares approximation.

$$a_n = \frac{2}{p} \sum_{n=1}^{p} \phi(t) \cos n\theta_i \; ,$$

$$b_n = \frac{2}{p} \sum_{n=1}^{p} \phi(t) \cos n\theta_i \; ,$$

and

$$a_o = \frac{2}{p} \sum_{n=1}^{p} \phi(t) \; .$$

One can estimate the angle of the tangent as the complex argument of the vector defined by two successive sampled points. This is useful for writing the corresponding function directly from the coordinates of equally spaced points along the outline. The `fourier2` function uses this strategy and returns the perimeter as well, computed as the sum of norms of the succession of unitary vectors making up the outline.

Function 5.6. `fourier2`

> *Arguments:*
>> M: *Matrix of sampled points on the outline.*
>> n: *Number of harmonics.*
> *Values:*
>> ao: *a_o harmonic coefficient.*
>> an: *Vector of $a_{1 \rightarrow n}$ harmonic coefficients.*
>> bn: *Vector of $b_{1 \rightarrow n}$ harmonic coefficients.*
>> phi: *Vector of variation of the tangent angle.*
>> t: *Vector of distance along the perimeter, expressed in radians.*
>> perimeter: *Perimeter of the outline.*
>> thetao: *First tangent angle.*

```
1  fourier2<-function(M,n)
2  {p<-dim(M)[1]
3  an<-numeric(n)
4  bn<-numeric(n)
5  tangvect<-M-rbind(M[p,],M[-p,])
6  perim<-sum(sqrt(apply((tangvect)^2, 1, sum)))
7  v0<-(M[1,]-M[p,])
8  tet1<-Arg(complex(real=tangvect[,1],
9          imaginary = tangvect [,2]))
```

```
10  tet0<-tet1[1]
11  t1<-(seq(0, 2*pi, length= (p+1)))[1:p]
12  phi<-(tet1-tet0-t1)%%(2*pi)
13  ao<- 2* sum(phi)/p
14  for (i in 1:n){
15      an[i]<- (2/p) * sum( phi * cos (i*t1))
16      bn[i]<- (2/p) * sum( phi * sin (i*t1))}
17  list(ao=ao, an=an, bn=bn, phi=phi, t=t1,
18  perimeter=perim, thetao=tet0)}
```

We program the reverse function to estimate ϕ and the corresponding coordinates from a given number of harmonic coefficients as in the ifourier2 function.

Function 5.7. ifourier2

Arguments:
> ao: a_o *harmonic coefficient.*
> an: *Vector of* $a_{1 \to n}$ *harmonic coefficients.*
> bn: *Vector of* $b_{1 \to n}$ *harmonic coefficients.*
> n: *Number of interpolated points on the outline.*
> k: *Number of harmonics.*
> thetao: *First tangent angle.*

Values:
> angle: *Position on the perimeter (in radians).*
> phi: *Vector of interpolated change of the tangent angle.*
> X: *Vector of* x-*interpolated coordinates.*
> Y: *Vector of* y-*interpolated coordinates.*

```
1   ifourier2<-function(ao,an,bn,n,k, thetao=0)
2   {theta<-seq(0,2*pi, length=n+1)[-(n+1)]
3   harm <- matrix (NA, k, n)
4   for (i in 1:k)
5     {harm[i,]<-an[i]*cos(i*theta)+bn[i]*sin(i*theta)}
6   phi<-(ao/2) + apply(harm, 2, sum)
7   vect<-matrix(NA,2,n)
8   Z<-complex(modulus=(2*pi)/n,argument=phi+theta+thetao)
9   Z1<-cumsum(Z)
10  list(angle=theta, phi=phi, X=Re(Z1), Y=Im(Z1))}
```

We are ready to apply this Fourier transformation to the outline of the honeybee forewing. Practically, the first step is to sample equally spaced landmarks on the outline. Prior to estimating these equally spaced landmarks, we smooth the outline to avoid imperfection due to automatic digitization. We have to determine the number i of smoothing iterations. Haines and Crampton [43] recommend having $ntot/p < \sqrt{i/2}$ where $ntot$ is the total number of pixels of the outline, and p is the number of points to be equally sampled on the outline.

```
>numb<-2*(length(X)/ 64)^2
>M<-smoothout(cbind(X,Y), numb+1)
```

The next step consists of selecting equally spaced points on the outline, which is made easy with the seq function.

```
>X64<-(M[,1][seq(1, length(X), length=65)])[-1]
>Y64<-(M[,2][seq(1, length(X), length=65)])[-1]
```

Finally, we use the function programmed above.

```
>f2<-fourier2(cbind(X64, Y64),32)
>if2<-ifourier2(f2$ao,f2$an,f2$bn,64,20,thetao=f2$thetao)
>layout(matrix(c(1,2),1,2))
>plot(if2$X, if2$Y, type="l", asp=1, xlab="X",
+        ylab="Y",main="harmonics 0 to 20")
```

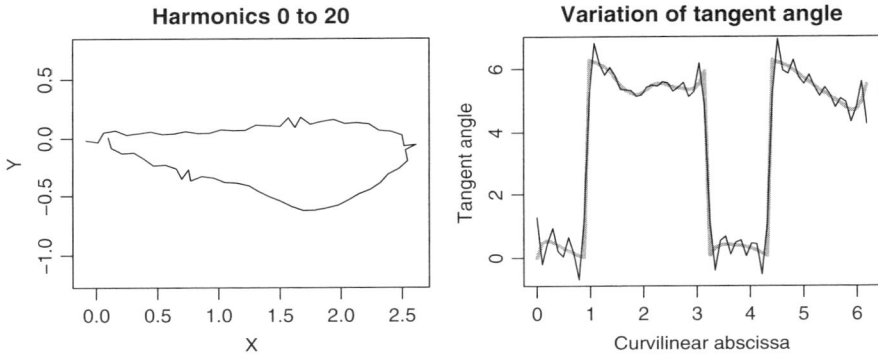

Fig. 5.6. Outline reconstruction from Fourier analysis applied to the tangent angle as a function of the perimeter: the graph on the left shows the reconstruction of the outline, while the graph on the right shows the fit of the signal by the Fourier series (black line) on the actual variation of the tangent angle (thick gray line)

The reconstruction of the outline is not always nice because estimating tangent angles can result in aberrant shapes (see Fig. 5.6). An alternative option consists of analyzing differences in reconstruction of variations of $\phi(t)$ according to t, the curvilinear abscissa (Fig. 5.6).

```
>plot(f2$t, f2$phi, type="l", lwd=2, col="grey70",
+      xlab="Curvilinear abscissa", ylab = "Tangent angle",
+      main="Variation of tangent angle")
>lines(if2$angle, if2$phi)
```

5.2.3 Elliptic Fourier Analysis

The elliptic Fourier analysis is another Fourier approach for fitting curves to complex closed outlines. Giardina and Kull [37] and Kuhl and Giardina [58] developed an algorithm for fitting Fourier series on x and y-coordinates as functions of the curvilinear abscissa. Crampton [23], and Rohlf and Archie [102] have provided a list of the advantages of the elliptic Fourier method over other Fourier applications to closed outlines: no need of equally spaced points, possible application to complex outlines, Fourier coefficients can be made independent of outline position, and can be normalized for size.

Let T be the perimeter of the outline, and this perimeter becomes the period of the signal. One sets $\omega = 2\pi/T$ to be the pulse. Then, the curvilinear abscissa, t varies from 0 to T. One can express $x(t)$ and $y(t)$ as:

$$x(t) = \frac{a_0}{2} + \sum_{n=1}^{+\infty} a_n \cos n\omega t + b_n \sin n\omega t ,$$

with

$$a_n = \frac{2}{T} \int_0^T x(t) \cos(n\omega t) dt ,$$

and

$$b_n = \frac{2}{T} \int_0^T x(t) \sin(n\omega t) dt ;$$

similarly,

$$y(t) = \frac{c_0}{2} + \sum_{n=1}^{+\infty} c_n \cos n\omega t + d_n \sin n\omega t ,$$

with

$$c_n = \frac{2}{T} \int_0^T y(t) \cos(n\omega t) dt ,$$

and

$$d_n = \frac{2}{T} \int_0^T y(t) \sin(n\omega t) dt .$$

The outline contains a k finite number of points. One can therefore calculate discrete estimators for every coefficient of the nth harmonics:

$$a_n = \frac{T}{2\pi^2 n^2} \sum_{p=1}^k \frac{\Delta x_p}{\Delta t_p} \left(\cos \frac{2\pi n t_p}{T} - \cos \frac{2\pi n t_{p-1}}{T} \right) ,$$

with

$$\Delta x_1 = x_1 - x_k ;$$

and

$$b_n = \frac{T}{2\pi^2 n^2} \sum_{p=1}^{k} \frac{\Delta x_p}{\Delta t_p} \left(\sin \frac{2\pi n t_p}{T} - \sin \frac{2\pi n t_{p-1}}{T} \right) .$$

c_n and d_n are calculated similarly.

a_o and c_o corresponds to the estimates of the coordinates of the centroid of the configuration. They are estimated by

$$a_o = \frac{2}{T} \sum_{i=1}^{p} x_i$$

and

$$c_o = \frac{2}{T} \sum_{i=1}^{p} y_i .$$

This methods inflates the number of coefficients by harmonic, but it may be expected that fewer harmonics are necessary to reliably describe the outline compared to the former methods. The parametric form makes transposition of Fourier analysis easy for 3D data: one simply needs to add $z(t)$. We transpose the formulae in a third function called efourier, that computes the Fourier coefficients $a_0, a_n, b_n, c_0, c_n, d_n$ from the M matrix of x, and y-coordinates of the sampled points. n the number of desired harmonics needed for reconstructing the outline is equal to half the number of sampled points. This number is set by default in the function that extracts harmonics from coordinates sampled on a given outline.

Function 5.8. efourier

Arguments:
    M: *Matrix of sampled points.*
    n: *Number of harmonics.*
Values:
    ao: a_o *harmonic coefficient.*
    co: c_o *harmonic coefficient.*
    an: *Vector of* $a_{1 \to n}$ *harmonic coefficients.*
    bn: *Vector of* $b_{1 \to n}$ *harmonic coefficients.*
    cn: *Vector of* $c_{1 \to n}$ *harmonic coefficients.*
    dn: *Vector of* $d_{1 \to n}$ *harmonic coefficients.*

```
1  efourier<-function(M, n=dim(M)[1]/2)
2  {p<-dim(M)[1]
3  Dx<-M[,1]-M[c(p,(1:p-1)),1]
4  Dy<-M[,2]-M[c(p,(1:p-1)),2]
5  Dt<-sqrt(Dx^2+Dy^2)
6  t1<-cumsum(Dt)
7  t1m1<-c(0, t1[-p])
8  T<-sum(Dt)
9  an<-bn<-cn<-dn<-numeric(n)
10 for (i in 1:n){
```

```
11    an[i]<- (T/(2*pi^2*i^2))*sum((Dx/Dt)*
12        (cos(2*i*pi*t1/T)-cos(2*pi*i*t1m1/T)))
13    bn[i]<- (T/(2*pi^2*i^2))*sum((Dx/Dt)*
14        (sin(2*i*pi*t1/T)-sin(2*pi*i*t1m1/T)))
15    cn[i]<- (T/(2*pi^2*i^2))*sum((Dy/Dt)*
16        (cos(2*i*pi*t1/T)-cos(2*pi*i*t1m1/T)))
17    dn[i]<- (T/(2*pi^2*i^2))*sum((Dy/Dt)*
18        (sin(2*i*pi*t1/T)-sin(2*pi*i*t1m1/T)))}
19  ao<-2*sum(M[,1]*Dt/T)
20  co<-2*sum(M[,2]*Dt/T)
21  list(ao=ao,co=co,an=an,bn=bn,cn=cn,dn=dn)}
```

For reconstructing outlines from a given set of harmonics, we write the reverse iefourier function. It estimates a curve from k harmonics, with n sampled points (passed as arguments). a_o and c_o determine the position of the configuration; we set their default value to 0 for convenience.

Function 5.9. iefourier

Arguments:
        an: *Vector of $a_{1 \rightarrow n}$ harmonic coefficients.*
        bn: *Vector of $b_{1 \rightarrow n}$ harmonic coefficients.*
        cn: *Vector of $c_{1 \rightarrow n}$ harmonic coefficients.*
        dn: *Vector of $d_{1 \rightarrow n}$ harmonic coefficients.*
        k: *Number of harmonics.*
        n: *Number of interpolated points on the outline.*
        ao: *a_o harmonic coefficient.*
        co: *c_o harmonic coefficient.*
Values:
        x: *Vector of x-interpolated coordinates.*
        y: *Vector of y-interpolated coordinates.*

```
1  iefourier<-function(an,bn,cn,dn,k,n,ao=0,co=0)
2  {theta<-seq(0,2*pi, length=n+1)[-(n+1)]
3   harmx <- matrix (NA, k, n)
4   harmy <- matrix (NA, k, n)
5   for (i in 1:k){
6      harmx[i,]<-an[i]*cos(i*theta)+bn[i]*sin(i*theta)
7      harmy[i,]<-cn[i]*cos(i*theta)+dn[i]*sin(i*theta)}
8   x<-(ao/2) + apply(harmx, 2, sum)
9   y<-(co/2) + apply(harmy, 2, sum)
10 list(x=x, y=y)}
```

One usually fewer less harmonics than half the number of original sampled points to fit a given outline with some reliability. One can appraise the goodness-of-fit using the sum of squared distances between the original data and reconstructed outline. Examining reconstructed outlines with the addition of successive contributing harmonics allows qualitative visualizations (Fig. 5.7).

We reconstruct the outlines on the honeybee wing outline.

```
>layout(matrix((1:9),3,3))
>par(mar=c(2,2,2,2))
>ef1<-efourier(M)
>for (i in 1:9)
+   {ief1<-iefourier(ef1$an,ef1$bn,ef1$cn,
+          ef1$dn,i,64,ef1$ao,ef1$co)
>plot(M,type="l",asp=1,frame=F,
+      main=paste("Harmonics 0 to",i),col="grey")
>polygon(M,col="grey",border=NA)
>lines(ief1$x,ief1$y,type="l")}
```

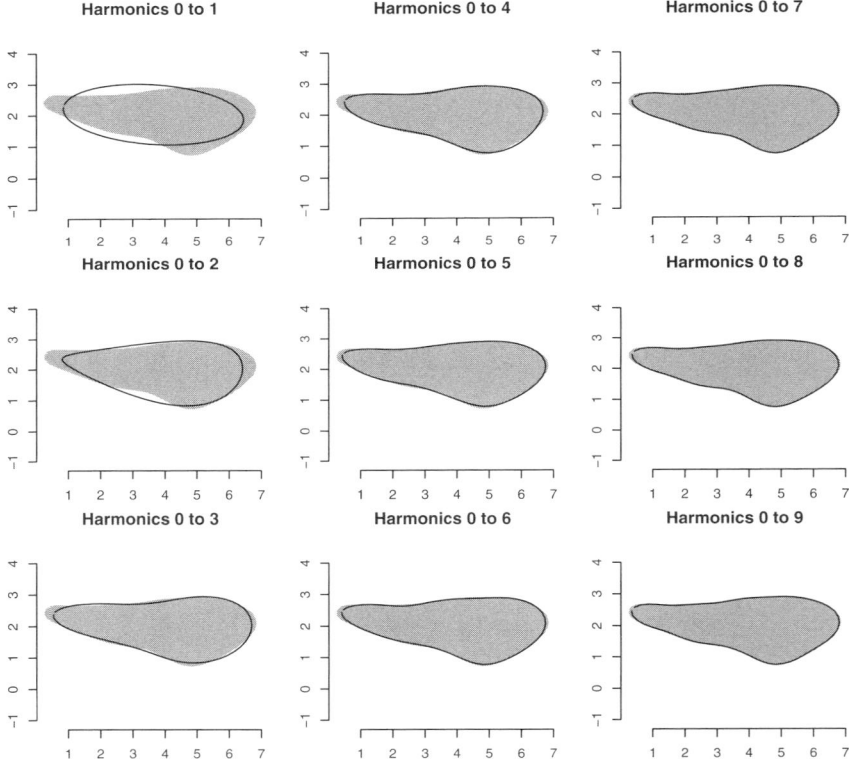

Fig. 5.7. Outline reconstruction from elliptic Fourier analysis applied to the outline coordinates of the honeybee wing. The original outline corresponds to the gray shape while reconstructed wings are thin black outlines: as early as the sixth harmonic, the outline is nearly perfectly reconstructed

For any harmonic, the $x(t)$ and $y(t)$ curves define an outline in the plane. Harmonics of higher order correspond to smaller ellipses, and actually the outline is approximated by the displacement of a point traveling around a series of superimposed ellipses (see [23, 58] for illustrations). The first harmonic defines an ellipse that best fits the outlines. One can use the parameters of the first harmonic to "normalize" the data so that they can be invariant to size, rotation, and starting position of the outline trace. This approach is referred to in the literature as the normalized elliptic Fourier [102]. It calculates a new set of Fourier coefficients A_n, B_n, C_n, D_n that one can use for further multivariate analyses.

One obtains the set of normalized coefficients following the equations

$$\begin{pmatrix} A_n & B_n \\ C_n & D_n \end{pmatrix} = \frac{1}{scale} \begin{pmatrix} \cos \psi & \sin \psi \\ -\sin \psi & \cos \psi \end{pmatrix} \cdot \begin{pmatrix} a_n & b_n \\ c_n & d_n \end{pmatrix} \cdot \begin{pmatrix} \cos n\theta & -\sin n\theta \\ \sin n\theta & \cos n\theta \end{pmatrix} ,$$

One estimates the scale as the magnitude of the semi-major axis of the ellipse defined by the first harmonic. The second right term corresponds to the orientation of the first ellipse (ψ being the rotation angle), the third to the original harmonic coefficients, and the last to the rotation of the starting point to the end of the ellipse (with a rotation angle of θ). Ferson et al. [31] supplied the following formulae to compute these parameters:

$$\psi = 0.5 \arctan \frac{2(a_1 b_1 + c_1 d_1)}{a_1^2 + c_1^2 - b_1^2 - d_1^2} \mod \pi ;$$

$$scale = \sqrt{a^{*2} + c^{*2}} ;$$

with $a^{*2} = a_1 \cos \psi + b_1 \sin \psi$ and $c^{*2} = c_1 \cos \psi + d_1 \sin \psi$;   and

$$\theta = \arctan(c^*/a^*) .$$

For writing the corresponding function, we write the code NEF. NEF calls the function we have written for computing ordinary elliptic Fourier coefficients. NEF can eventually perform the normalization of the starting point. This is passed through the third argument called start. We set the default value to be FALSE, meaning that the position of the starting point is not preserved according to the normalization. In other words, setting start = FALSE means that one does not consider the starting point as homologous.

Function 5.10. NEF

Arguments:

M: *Matrix of sampled points.*

n: *Number of harmonics.*

start: *Logical value telling whether the position of the starting point has to be preserved or not.*

Values:

A: *Vector of $A_{1 \to n}$ harmonic coefficients.*

B: *Vector of $B_{1 \to n}$ harmonic coefficients.*

C: *Vector of $C_{1 \to n}$ harmonic coefficients.*

D: *Vector of $D_{1 \to n}$ harmonic coefficients.*

size: *Magnitude of the semi-major axis of the first fitting ellipse.*

theta: *θ angle between the starting point and the semi-major axis of the first fitting ellipse.*

psi: *Orientation of the first fitting ellipse.*

ao: *a_o harmonic coefficient.*

co: *c_o harmonic coefficient.*

Required function: efourier.

```
 1  NEF<-function(M, n=dim(M)[1]/2,start=F)
 2    ef<-efourier(M,n)
 3    A1<-ef$an[1]; B1<-ef$bn[1]
 4    C1<-ef$cn[1]; D1<-ef$dn[1]
 5    theta<-0.5*atan(2*(A1*B1+C1*D1)/
 6          (A1^2+C1^2-B1^2-D1^2))
 7    Aa<-A1*cos(theta)+B1*sin(theta)
 8    Cc<-C1*cos(theta)+D1*sin(theta)
 9    scale<-sqrt(Aa^2+Cc^2)
10    psi<-atan(Cc/Aa)%%pi
11    size<-(1/scale)
12    rotation<-matrix(c(cos(psi),-sin(psi),
13                  sin(psi),cos(psi)),2,2)
14    A<-B<-C<-D<-numeric(n)
15    if (start){theta<-0}
16    for (i in 1:n){
17        mat<-size*rotation%*%matrix(c(
18          ef$an[i],ef$cn[i],ef$bn[i],ef$dn[i]),2,2)
19          %*%matrix(c(cos(i*theta),sin(i*theta),
20          -sin(i*theta),cos(i*theta)),2,2)
21    A[i]<-mat[1,1]
22    B[i]<-mat[1,2]
23    C[i]<-mat[2,1]
24    D[i]<-mat[2,2]}
25    list(A=A,B=B,C=C,D=D,size=scale,theta=theta,
26        psi=psi,ao=ef$ao,co=ef$co)}
```

These standardizations are useful because they allow comparisons between outlines to be achieved and the outline shape variation to be analyzed. One can apply traditional multivariate statistics to normalized elliptic Fourier coefficients. The first harmonic has three constant coefficients $A_1 = 1$, $B_1 = C_1 = 0$. The remaining term D_1 is associated with the harmonic excentricity (which corresponds more or less to the overall width-on-length ratio of the object). This first harmonic often contains a large part of the variation; however, the orientation of the object under the camera objective can largely influence this variation. One can inspect the percent of error that D_1 records to decide whether it should be conserved for further analyses. One can also include it in the analysis by normalizing all coefficients.

Crampton [23] discusses practical details of whether normalization should be performed with elliptical Fourier analysis. We should remember that coefficients are dependent on several factors:

1. Outline orientation
2. Position of the starting point
3. Outline size

Whether it is necessary to normalize for the starting point, outline orientation, and size should be decided prior to performing further analysis involving sets of Fourier coefficients representing several observations. The other remark is that the normalized elliptical Fourier provides only a single way to perform standardization. Rather than superimposing ellipses according to the first harmonics, one could choose, for example, to align configuration according to another method. Alternative options are available, if, in addition to the outline, one could have digitized some landmarks on the object. If these landmarks are available, one can perform a Procrustes alignment of outlines based on landmarks for normalizing the data. Once one has estimated rotation and scale parameters, one can use these new parameters to replace ψ and $scale$ in the former equation. Friess et al. [32] and Baylac et al. [6] have used a similar approach, superimposing outlines according to a set of defined landmarks for examining cranial shape variation in human populations.

The elliptic Fourier analysis is particularly efficient for reducing the number of variables of the original dataset (coordinates of sampled points along the outline). In our example, we know that the first six harmonics gathered nearly all the information necessary for reconstructing the outline of the honeybee wing. In addition, standardization for size and orientation makes three coefficients constant, so that there are only $4k - 3$ variables remaining for describing outline parameters (in the case of the honeybee wing this is summarized by 21 variables instead of the 2×64 original coordinates.

One evaluates the number of harmonics that one should retain according to different methods. The first way is qualitative: It consists of visualizing contour reconstructions produced by increasing the number of harmonics involved and comparing these reconstructions with the original one. This inspection usually shows that high-order harmonics record high frequency variation that one can assimilate in most cases to "noise variation." Crampton [23] proposes examining the average deviation from the original outline (that one reconstructs using $n/2$ harmonics) as a function of the number of harmonics used to reconstruct the outline. One can evaluate maximal deviation or 95% confidence interval from a sample of outlines (see Fig. 5.8). For a single outline, we can examine deviation between the reconstructed and original outlines as below:

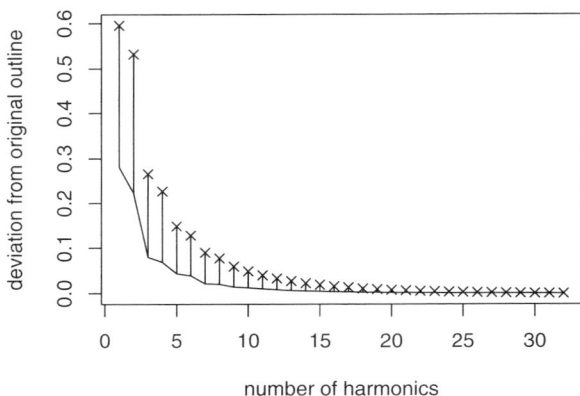

Fig. 5.8. Error resulting from outline reconstruction as a function of the number of harmonics involved. Averaged variation corresponds to the curve, while maximum deviations are symbolized by crosses

```
>ef<-efourier(M)
>n<-dim(M)[1]
>ief<-iefourier(ef$an,ef$bn,ef$cn,ef$dn,
+          32,64,ef$ao,ef$co)
>M1<-cbind(ief$x, ief$y)
>averagedev<-maxdev<-numeric(32)
>for (i in 1:32)
+    {ief<-iefourier(ef$an,ef$bn,ef$cn,ef$dn,
+          i,64,ef$ao,ef$co)
+     deviation<-sqrt(apply((M1-cbind(ief$x,
+          ief$y))^2,1,sum))
+     averagedev[i]<-mean(deviation)
+     maxdev[i]<-max(deviation)}
>plot(1:32, maxdev, pch=4, ylab=
+    "deviation from original outline",
+     xlab="number of harmonics")
>segments(1:32,maxdev, 1:32, averagedev)
>lines(1:32, averagedev)
```

We can also estimate the number of necessary harmonics after examining the spectrum of harmonic Fourier power. The power is proportional to the harmonic amplitude and can be considered as a measure of shape information. As the rank of harmonic increases, the power decreases and adds less and less information. We can evaluate the number of harmonics that we must select, so their cumulative power gathers 99% of the total cumulative power. The power of a given harmonic is calculated as

$$\text{Power}_n = \frac{A_n^2 + B_n^2 + C_n^2 + D_n^2}{2} .$$

We can observe the evolution of cumulative total power in our outline:

```
>ef1<-efourier(M)
>Power<-(ef1$an^2+ef1$bn^2+ef1$cn^2+ef1$dn^2)/2
>(cumsum(Power)/sum(Power))[1:6]
[1] 0.979027 0.987326 0.998174 0.998619 0.999406 0.999561
```

Most of the shape "information" is contained in the first harmonic. This is not surprising because this is the harmonic that best fits the outline, and the size of ellipses decreases as for explaining successive residual variation. However, one may think that the first ellipse does not contain relevant shape information, especially when differences one wants to investigate concern complex outlines. One can decide to remove the first harmonic for this computation.

```
>(cumsum(Power[-1])/sum(Power[-1]))[1:6]
[1] 0.395706 0.912975 0.934166 0.971708 0.979085 0.992371
```

In keeping the second to seventh harmonics, more than 99% of the information is remaining.

When working on a set of outlines, the strategy of harmonic selection must be adapted to the need of the analysis. Indeed, high rank-harmonics can contain information that may allow groups to be distinguished. In addition to our approaches, one can examine digitization error or shape variance explained as a criterion for harmonics number selection. For instance, Renaud et al. [90] coupled an analysis of cumulative power together with an estimation of measurement error (computed as the averaged coefficient of harmonic amplitude based on a sample of digitized outlines replicated a certain number of times). They showed that measurement error increased with harmonic order. Reducing the number of harmonics is thus a useful way not only to limit the number of variables but also to remove some part of measurement error.

5.3 Eigenshape Analysis and Other Methods

Other approaches for analyzing outlines have been proposed. If one superimposes several outlines according to an optimization procedure, one can directly use the superimposed coordinates in statistical analyses. This has sometimes received the name *analysis of coordinates*. One can then analyze shape variation using multivariate methods applied to superimposed coordinates as in the case of Procrustes superimposition. Whether one should consider it to be appropriate depends on the data structure and the purpose of the study. In addition, one must keep in mind that superimposed landmark (and especially pseudolandmark) coordinates are not independent observations, because the superimposition process removes degrees of freedom in the overall variation. Moreover, the sampling procedure for pseudolandmarks usually decreases further the number of shape space dimensions. The ranks of variation inferred from this kind of data are usually smaller than traditional Procrustes analysis. Bookstein [13] introduced a method for allowing pseudolandmarks to slip on curves (semi- or sliding landmark approach). It can be interesting to optimize the fit; some further 3D developments are provided in Gunz et al. [42].

Lohman [65] proposed a method for analyzing shape variation of the outlines of biological objects. Initially applied to microfossils, the method can be applied to other shape samples. Lohman and Schweitzer [66] gave an extensive description of the method. The main philosophy is to measure variation between shape functions. Shape functions as defined by the authors correspond to the Zahn and Roskies function of tangent angle change on equally spaced points sampled on the outline.

$$\phi^*(t) = \phi(t) - t \; ;$$

$\phi(t)$ corresponds to the direction of the tangent to the outline, and t corresponds to the position of a point moving around the outline scaled so that t ranges from 0 to 2π.

This representation is invariant to size or translation in the plane. To achieve normalization, one must find a way to rotate outlines according to a standard orientation. If one can define a starting point, one can use the first tangent angle for standardization, as was done in the second Fourier method presented in Section 5.2.2:

$$\phi^*(t) = \phi(t) - \phi(0) - t \; .$$

The previous section describes how to compute $\phi^*(t)$ from coordinate data.

One can decide to orient outlines according to another procedure (for example, a Procrustes superimposition, or by aligning outlines on the major axis of the best fitting ellipse defined by elliptic Fourier analysis).

One can omit variation in angularity (see [66, 95]), in standardizing the function to have a zero mean and a variance of 1. However, this can be undesirable since different shapes can become indistinguishable.

Given n objects with p sampled points, one can assemble the reoriented $\phi^*(t)$ functions in a Z matrix of n rows and p columns. These functions are transformed for Z^* that are centered original data (column means equal zero). The "covariation" or "correlation" $\mathbf{R_z}$ matrix is obtained by matrix multiplication: $Z^{*'}Z^*$. The resulting matrix is not exactly the variance-covariance or correlation matrix but it is proportional at the approximation of the number of degrees of freedom. This matrix is later decomposed according to a singular-value decomposition (the svd function of R performs this easily) so that

$$\mathbf{R_z} = \mathbf{VDV'} \; .$$

The elements $\mathbf{V}$ weighted by $\mathbf{D^{0.5}}$ are interpreted as the loadings of the objects onto the empirical shape function.

One calculates the scores $\mathbf{U}$ as

$$\mathbf{U} = \mathbf{Z}^* \mathbf{V} \mathbf{D}^{-0.5} .$$

Finally, one obtains

$$\mathbf{Z}^* = \mathbf{U} \mathbf{S}^{0.5} \mathbf{V}' .$$

The shape functions ($\mathbf{Z}$) are therefore decomposed into a set of empirical orthogonal shape functions, $\mathbf{U}$; each column of it being an eigenshape function (or a normalized eigenvector), itself associated with the ith column of $\mathbf{V} \mathbf{D}^{0.5}$ corresponding to the weights or projections of the object on the function. The eigenshape functions successively account for the maximum of variation represented by the original shape functions. One can consider the first eigenshape function as the mean of the shape original functions plus proportional variation in angle changes (angularity when working on the variance-covariance matrix). Other eigenshape functions describe additional differences among the shape functions along orthogonal directions.

One can appraise variation in the shape space by projecting the data onto $\mathbf{U}$. One can consider each original object as a weighted linear combination of the eigenshape functions. One can reduce the shape space dimensions to the first eigenshape functions. We need a function to extract $\mathbf{U}$, $\mathbf{V}$ and $\mathbf{V} \mathbf{D}^{0.5}$. This function works on the matrix of standardized shape functions Z, and we call it `eigenshape`.

Function 5.11. `eigenshape`

Argument:
    `Z`: *Matrix of standardized shape functions.*
Values:
    `eigenshape`: *Eigenshape functions.*
    `weigths`: *Eigenshape scores.*
    `eigenvalues`: *Eigenvalues.*

```
1  eigenshape<-function(Z)
2  {n<-dim(Z)[1]
3  Zz<-scale(Z, scale=F)
4  R<-t(Zz)%*%Zz
5  V<-svd(R)$u
6  S<-svd(R)$d
7  U<-Zz%*%V%*%(diag(1/sqrt(S)))
8  weight<-diag(sqrt(S))%*%t(V)
9  list(eigenshape=U, weights=weight[1:n,], eigenvalues=S)}
```

Rohlf [95] studied the relationship among eigenshape analysis, Fourier analysis and analysis of coordinates. If one retains all Fourier harmonics and eigenvectors, one can consider that both approaches are equivalent.

Applications to the study of outline variation are presented in the following chapter.

Problems

5.1. Three dimensional curve fitting
Write a function for fitting 3D curves. Several strategies are possible (parametric cubic splines, elliptic Fourier analysis.)

5.2. variation in digit3.dat
Using open curve fitting and Procrustes analysis, perform a principal component analysis of the `digit3.dat` dataset of the **shapes** package. Are observations ordinated in the same way using both methods?

5.3. Fourier analysis of equally spaced radii
Write a function that calculates the Fourier coefficients with arguments being the norm of equally spaced radii.

5.4. Estimating Fourier coefficients by multiple regression
Write a function using a regression model to estimate the Fourier coefficients estimated from cumulative change of the tangent angle.

5.5. Elliptic Fourier normalization
Write a function that standardizes classic Fourier coefficients to be independent for size and write one that rotates the first ellipse but that does not shift the starting point.

Statistical Analysis of Shape using Modern Morphometrics

While one can analyze morphometrics based on linear measurements with multivariate statistics and interpreted as any other set of independent variables, modern morphometrics has the advantage of describing and appraising shape variation among configurations. Since configurations themselves are multivariate sets of variables (i.e., coordinates, outline parameters, etc.), one analyzes their variation using multivariate techniques. However, since landmarks and pseudolandmarks are dependent on each other, one cannot directly apply these techniques to raw coordinates. One must transform these raw coordinates into shape variables or shape parameters for statistical issues (superimposed coordinates, Bookstein coordinates, Fourier coefficients, etc.). Most modern morphometrics offer ways to visualize shape change and variation in a qualitative way as depicted by transformation of configurations along shape variation components. These components can be either static – for a naive description of the shape space – or explicative – interpretation of shape variation. In addition, various tests are available for checking hypotheses. In biological sciences, where morphometrics is more commonly used, one can interpret the variation with regards to biological mechanisms or processes, taking into account the complexity of phenotype organization.

6.1 Explorative Analyses of the Shape Space

One can explore the shape space, without posing adhoc hypotheses, by analyzing principal components of variation. These components are estimated by principal component analyzes or related techniques. These techniques decompose shape variation into several components, decreasing in importance in terms of explained variation. Other techniques, which have been particularly developed to analyze Procrustes data, allow the shape variation to be decomposed in other distinct components: affine and nonaffine components or local and more general components of variation. We know that we can map thin-plate splines to configurations for interpreting morphological transformation in terms of deformation grids. Concepts behind deformation grids have made statistical shape analysis progress very rapidly during the last decade of the 20th Century and have been synthesized in two important publications [10, 27].

6.1.1 Landmark Data

The simplest way to describe variation and covariation within the configuration set is to perform a PCA of the superimposed coordinates of landmarks of the configurations. This analysis is possible on coordinates that have been projected onto the tangent space, as we have seen in Section 4.2.4.

A biplot with the variable contribution is not easy to interpret, especially, when the number of landmarks is important and when each landmark is 2D or 3D. However, since all deviations from centered data are expressed in the same metrics, it is possible to qualitatively visualize the shape change associated with a given principal component. Since principal components are independent, we can display the variation in configurations on a given axis by using the mean shape on which we add or substract the loadings of the corresponding unitary, eigen or singular, vector multiplied by a given score. Examining extreme shapes reconstructed along each principal component allows us to qualitatively visualize the signification of component in terms of extreme shape variation. Adding link segments between selected landmarks is helpful for understanding shape changes. Alternatively one can display deformation grids expressing shape changes between two selected scores on the PCs. Since shape changes can be subtle, it is often necessary to amplify shape variation by some amplification scalar. Finally, a look at eigenvalues (or singular values) provides information on how a shape-axis is important for explaining shape variation.

This approach is applied to the concatenated male and female gorilla datasets of shapes.

```
>library(shapes)
>data(gorf.dat); data(gorm.dat)
>gor<-array(c(gorf.dat, gorm.dat), dim=c(8,2,59))
```

The first step consists of superimposing configurations. Then we perform a PCA of the tangent shape space coordinates. For graphical purpose, we first rotate shapes on their principal axis with the aligne *function (see Section 4.2.3).*

```
>GOR<-aligne(gor)
>go<-pgpa(GOR)
>gos<-orp(go$rotated)
>m<-t(matrix(gos, 16, 59))
>pcs<-prcomp(m)
```

We plot the PCA scores, and we display the variance explained by each PC with a barplot.

```
>par(mar=c(4,4,1,1))
>layout(matrix(1:4,2,2))
>plot(pcs$x, pch=c(rep("f",30), rep("m", 29)))
>barplot(pcs$sdev^2/sum(pcs$sdev^2),ylab="% of variance")
>title(sub="PC Rank",mgp=c(0,0,0))
```

Note the use of the `mgp` *option in the* `title()` *command. It positions the subtitle within the graph and it is also available in other graphical functions. We estimate the mean configuration, and we record the maximal and minimal scores for the two first PCs. Extreme configurations are plotted, with segments between appropriate landmarks.*

```
>mesh<-as.vector(mshape(gos))
>max1<-matrix(mesh+max(pcs$x[,1])*pcs$rotation[,1],8,2)
>min1<-matrix(mesh+min(pcs$x[,1])*pcs$rotation[,1],8,2)
>joinline<-c(1,6:8,2:5,1,NA,7,4)
>plot(min1,axes=F,frame=F,asp=1,xlab="",ylab="",pch=22)
>points(max1,pch=17)
>title(sub="PC1",mgp=c(-4,0,0))
>lines(max1[joinline,],lty=1)
>lines(min1[joinline,],lty=2)
>max2<-matrix(mesh+max(pcs$x[,2])*pcs$rotation[,2],8,2)
>min2<-matrix(mesh+min(pcs$x[,2])*pcs$rotation[,2],8,2)
>plot(min2,axes=F,frame=F,asp=1,xlab="",ylab="",pch=22)
>points(max2,pch=17)
>title(sub="PC2",mgp=c(-4,0,0))
>lines(max2[joinline,],lty=1)
>lines(min2[joinline,],lty=2)
```

Males and females occupy different positions on the first two PCs of shape variation (Fig. 6.1). According to the position of individuals along PC axes, one sees that the first PC opposes skull outlines that are rather elongated antero-posteriorly (males) and more globular skulls (females). The second PC opposes forms having relatively larger neurocranium and reduced faces (females), and forms with relatively smaller neurocranium and more pronounced faces (males).

One can visualize shape variation along PC by using thin-plate splines, with the `tps` function defined in Section 4.3. The variation is easily displayed using the mean shape as references and extreme reconstructed shapes as targets. For producing a field of vector differences, we can opt for the same method as in Section 4.3.

```
>par(mar=c(0,1,2,2))
>msh<-mshape(gos)
>tps(msh, min1,12)
>points(min1, pch=21, bg="black")
>title("PC1: left extreme")
>tps(msh,max1,12)
>points(max1,pch=22, bg="black")
>title("PC1: right extreme")
```

The `shapepca` function of the **shapes** package directly plots PC scores and provides a graphical display for visualizing shape changes along axes. The function needs an object of the `procGPA` class. One can display shape variation using extreme configurations, vectors fields, tps, or even doing a small movie for depicting changes between both sides of the PC. Similarly, one can visualize PC scores and related variation with the **Rmorph** package and the `visu` function.

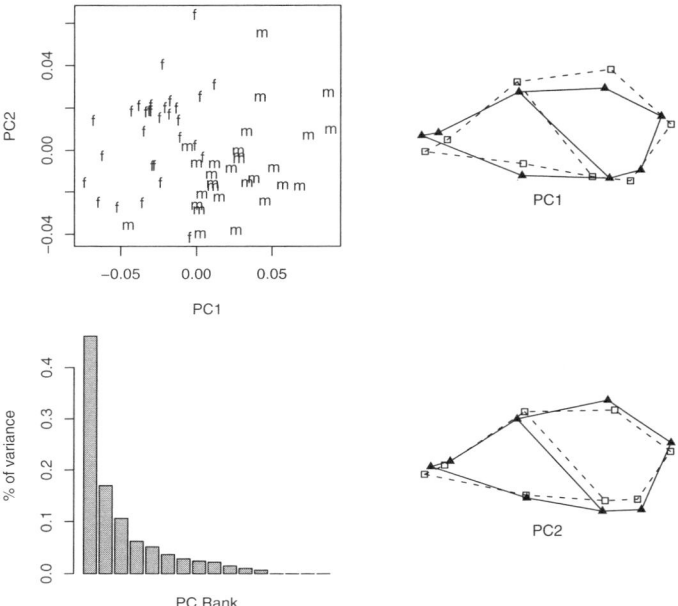

Fig. 6.1. Graphical display of shape variation depicted by the two first principal components of total shape variation of the concatenated `gorf.dat` and `gorm.dat` datasets; on the left: plot of PC scores and barplot of PC contributions; on the right: illustration of shape variation along each PC. Dotted links correspond to the minimal scores, while full links to maximal scores. The segment appearing inside the skull outline approximately corresponds to the boundary between the face and the neurocranium

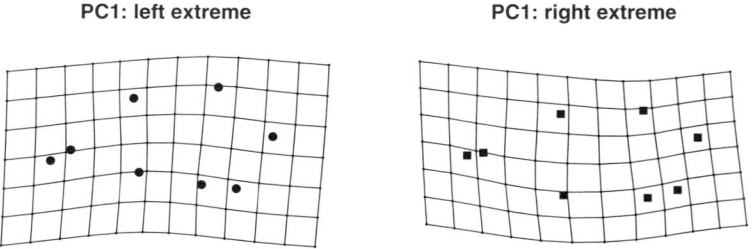

Fig. 6.2. Illustration of shape variation with deformation grids along the first axis of the PCA applied to configurations of the `gorf.dat` plus `gorfm.dat` datasets

One could also interpret shape change as resulting from an accumulation of growth gradients expressed from a global to a local scale. Shape variation can thus be decomposed into global to local components. Thin-plate splines and the bending energy needed for transforming one configuration to one other can help us to understand intuitively how to decompose shape variation into these components. Variation will be depicted into global deformation consisting of affine or nonaffine parts, and local deformation that necessarily consists of nonaffine deformation. It can be useful to distinguish between both components, because affine changes and variation deal with global flattening, shearing, or dilatation that may possibly receive a simple interpretation. Translated in terms of deformation grids, the affine components correspond to deformation that keeps parallelism between lines of the original grid (usually corresponding to a grid of squares). In other words, each square of the grid is transformed in a similar parallelogram throughout the configuration. This affine deformation is also called uniform deformation because it concerns the whole configuration equally.

Bookstein and other authors [10, 12, 27, 104] have provided several formulae for estimating uniform components of shape variation. The uniform and nonuniform components of shape variation rely on the bending energy matrix and are coined with the methods of principal, partial and relative warps.

For convenience, several authors prefer to work on the set of superimposed configurations, with the reference (the mean shape) aligned along its principal axes. This does not change anything in the superimposition; the dispersion of landmarks is just maximized on the x-axis. We produce this computation by combining and modifying the superimposition functions written in Section 4.2. The `procalign` function is a possible way to produce this alignment. Notice the use of `eigen` for avoiding problems with reflections. Alternatively we can obtain very similar results by superimposing configurations that have been previously rotated on their principal component of variation with the `aligne` function.

Function 6.1. `procalign`

Argument:
> `A`: *Array containing configuration matrices.*

Values:
> `rotated`: *Aligned configurations array. In addition to the superimposition, the configuration datasets have been projected onto the Euclidean Shape space according to a orthogonal projection (the mean shape being the tangent point between spaces).*
> `meansh`: *Mean shape used for the alignement, aligned along its principal axes.*

Required functions: `pgpa`, `trans1`, `centsiz`, `mshape`, `pPsup`.

```
1  procalign<-function(A)
2  {pA<-pgpa(A); n<-dim(A)[3]; k<-dim(A)[2]
3  A<-pA$rotated; msh<-pA$mshape
4  A1<-A
5  sv<-eigen(var(msh))
6  V<-sv$vectors;
7  rotmsh<-msh%*%V
8  for (i in 1:n)
9      {A1[,,i]<-pPsup(A[,,i],rotmsh)$Mp1}
10 list("rotated"=orp(A1), "meansh"=rotmsh)}
```

It is simple to compute the uniform shape vectors for a given specimen for 2D data. There are two uniform components of $k \times p$ elements. Rohlf [98] gives the following formulae:

$$U_1 = (\sqrt{\frac{\alpha}{\gamma}}y_i, \sqrt{\frac{\gamma}{\alpha}}x_i) \, ,$$

and

$$U_2 = (-\sqrt{\frac{\gamma}{\alpha}}x_i, \sqrt{\frac{\alpha}{\gamma}}y_i) \, ,$$

where $\alpha = \sum x^2$, and $\gamma = \sum y^2$, with x_i, y_i being the coordinate of the reference configuration (it is usually the mean shape).

One calculates the scores u_1 and u_2 by multiplying the V matrix of differences between the aligned specimens and the reference with the uniform components. The $\mathbf{V}$ matrix has $n \times kp$ dimensions, with the first p columns corresponding to the first dimensions, and the next p columns to the second dimension. U_1 represents a horizontal shear, while U_2 represents a vertical dilatation. In R environment, we can translate the formula as a function that returns both uniform components and scores with the uniform2D function.

Function 6.2. uniform2D

Argument:
    A: *Array containing 2D configuration matrices.*
Values:
    scores: *Scores of the uniform component for each observation. The two columns
        of this* matrix *object respectively correspond to the U1 and U2 scores.*
    uniform: *Two-column matrix of uniform component vectors.*
    meanshape: *Meanshape aligned along its principal axis.*
    rotated: *Superimposed and projected configurations.*
Required functions: procalign, pgpa, trans1, centsiz, mshape, pPsup.

```
1  uniform2D<-function(A)
2  {n<-dim(A)[3]
3  kp<-dim(A)[1]*dim(A)[2]
4  temp<-procalign(A)
5  msh<-temp$meansh
6  proc<-temp$rotated
7  X<-t(matrix(proc,kp,n))
8  V<-X-rep(1, n)%*%t(as.vector(msh))
9  alph<-sum(msh[,1]^2)
10 gam<-sum(msh[,2]^2)
11 U1<-c(sqrt(alph/gam)*msh[,2],sqrt(gam/alph)*msh[,1])
12 U2<-c(-sqrt(gam/alph)*msh[,1],sqrt(alph/gam)*msh[,2])
13 score<-V%*%cbind(U1,U2)
14 list("scores"=score,"uniform"=cbind(U1,U2),"meanshape"=msh,
15     "rotated"=proc)}
```

Associated deformation grids can illustrate the deformation related to the first and second uniform components as in the following example.

```
>gou<-uniform2D(gor)
>layout(matrix(c(1,1,1,1,2,4,3,5),2,4))
>msh<-gou$meanshape
>Un<-gou$uniform
>par(mar=c(5,4,4,2))
>plot(gou$scores,pch=c(rep("f",30),rep("m", 29)),asp=1)
>par(mar=c(4,1,4,1))
>tps(msh, matrix(as.vector(msh)+
+          Un[,1]*max(gou$scores[,1]),8,2),12)
>title("U1: left")
>tps(msh, matrix(as.vector(msh)+
+          Un[,2]*max(gou$scores[,2]),8,2),12)
>title("U2: top")
>tps(msh, matrix(as.vector(msh)+
+          Un[,1]*min(gou$scores[,1]),8,2),12)
>title("U1: right")
>tps(msh, matrix(as.vector(msh)+
+          Un[,2]*min(gou$scores[,2]),8,2),12)
>title("U2: bottom")
```

Rohlf and Bookstein [104] have provided strategies for computing the uniform (or affine) components of shape variation using different methods. They supply an approach permitting us to assess the uniform components using regression of the superimposed coordinates onto the coordinates of the reference shape. The resulting number of coefficients is redundant and higher than the number of uniform components (4 against 2 for 2D data, and 9 against 5 for 3D data). To resolve the problem, they use a singular-value decomposition calculation. The approach is convenient because it can be easily amenable to 3D data. In contrast to the former approach, every component contains a mixture of shear and dilatation, which makes them more

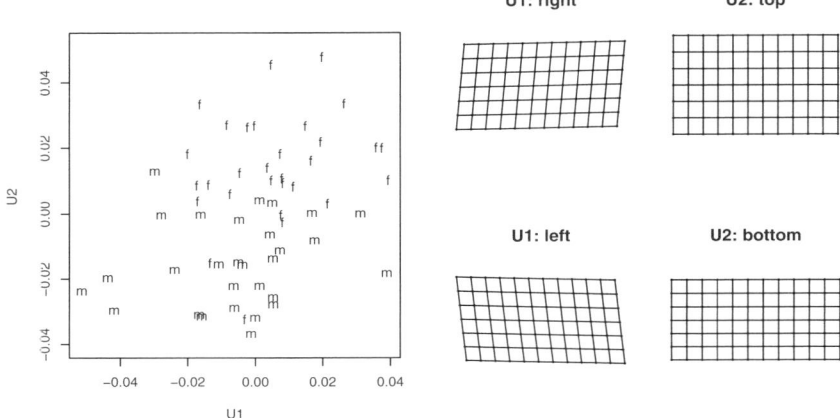

Fig. 6.3. Illustration of uniform shape variation within the female `gorf.dat` and male `gorfm.dat` dataset. Males correspond to "m" and females to "f". Males have relatively flatter and longer skull outlines than females on the U_2 component

difficult to interpret. However, these components correspond to a rotation of those described above. Taking into consideration that configurations are all centered on their centroid (no need to center the data), one can directly compute the regression coefficients for each coordinate as

$$B_x = (X'_r X_r)^{-1}(X'_r)\mathbf{X}'_{\mathbf{x}} \ ;$$

X_r corresponds to the reference configuration matrix (here the mean shape), $\mathbf{X}_{\mathbf{x}}$ to the $n \times p$ matrix of x-coordinates, and B_x is an $n \times k$ matrix of regression coefficients. B_y and eventually B_z are computed similarly. The coefficients are then combined in a single $\mathbf{B}$ matrix of n raws and k^2 columns so that

$$\mathbf{B} = [B'_x | B'_y | B'_z] \ .$$

To respect dimensionality and to obtain only a three-column matrix from $\mathbf{B}$, one performs a singular-value decomposition as follows:

$$\mathbf{LSR}' = \mathbf{B}(I_k \otimes X'_r) \ .$$

$\mathbf{L}$ is an n by n matrix of left singular vectors, $\mathbf{S}$ the diagonal of singular values and $\mathbf{R}$ is the kp by kp matrix of right singular vectors. The product $\mathbf{LS}$ equals the scores of the uniform components, while $\mathbf{R}$ are the coefficients of the uniform components. We write the `uniformG` function to compute uniform components and theirs scores of 2D or 3D data.

Function 6.3. `uniformG`

Argument:

  A: *Array containing configuration matrices.*

Values:

  `score`: *Scores of the uniform component for each observation.*

  `un`: *Matrix of uniform component vectors.*

Required functions: `procalign, pgpa, trans1, centsiz, mshape, pPsup.`

```
1  uniformG<-function(A)
2  {n<-dim(A)[3]; k<-dim(A)[2]; p<-dim(A)[1]
3  temp<-procalign(A)
4  msh<-temp$meansh
5  proc<-temp$rotated
6  Bn<-Xn<-list()
7  for (i in 1:k)
8      {Bn[[i]]<-solve(t(msh)%*%msh)%*%t(msh)%*%(proc[,i,])}
9  if (k==2)
10   {LSR<-svd(t(rbind(Bn[[1]],Bn[[2]]))%*%
11                    (diag(1,2)%x%t(msh)))
12    score<-LSR$u%*%diag(LSR$d)[,2:3]
13    Un<-LSR$v[,2:3]}
14  if (k==3)
15   {LSR<-svd(t(rbind(Bn[[1]],Bn[[2]],Bn[[3]]))%*%
16                    (diag(1,3)%x%t(msh)))
17    score<-LSR$u%*%diag(LSR$d)[,2:6]
18    Un<-LSR$v[,2:6]}
19  list("score"=score, "uniform"=Un)}
```

One can choose the strategy of relative warps presented in [10] to obtain the remaining components of variation; alternatively, one can simply project the shape coordinate residuals onto an orthogonal space that will contain nonaffine deformation only. One achieves this using the approach of Burnaby [16]. Indeed, one can consider any configuration as the result of the transformation of a given reference. One can decompose the transformation into a sum of affine and a sum of nonaffine transformations. One can thus define an affine subspace and a nonaffine subspace. The affine and nonaffine subspaces are orthogonal and their direct sum is equivalent to the shape space. Since it is possible to define the uniform portion in the shape space, one can calculate the nonaffine components by projecting the data in the complement space as we have proceeded with the Burnaby method (Section 3.4.3). We obtain them so that:

$$\mathbf{N_u} = I_{kp} - \mathbf{U}(\mathbf{U'U})^{-1}\mathbf{U'} ,$$

where I_{kp} is the identity matrix and $\mathbf{U}$ the uniform components of variation arranged in 2 or 5 columns. The matrix of Procrustes residuals $\mathbf{V}$ follows the relationship:

$$\mathbf{V} = \mathbf{X} - \mathbf{1}_n X_m ,$$

where $\mathbf{X}$ corresponds to the matrix of n by kp aligned coordinates, X_m to the mean shape. These Procrustes residuals are projected onto $\mathbf{V}$ by matrix multiplication, and a singular-value decomposition is performed on the result of the multiplication to appraise the left $\mathbf{L}$, right $\mathbf{R}$ singular vectors and singular values $\mathbf{S}$. We have

$$\mathbf{VN_u} = \mathbf{LSR'} .$$

The nonaffine components correspond to the right term $\mathbf{R}$. The product $\mathbf{LS}$ gives the scores on each nonaffine component. The method described here is not different in essence from the relative warp methodology (see [10, 98]). Deformation grids or displays of vector differences are helpful for understanding the changes occurring along components of nonaffine deformation.

For illustrating the analysis of the nonaffine deformation, we continue examining the `gorf.dat` and `gorm.dat` gorilla datasets, for which we have already computed uniform components.

```
>kp<-dim(gor)[2]*dim(gor)[1]
>n<-dim(gor)[3]
>X<-t(matrix(gou$rotated,kp,n))
>V<-X-t(t(rep(1,n)))%*%as.vector(msh)
>Ben<-diag(1,kp)-Un%*%solve(t(Un)%*%Un)%*%t(Un)
>LSR<-svd(V%*%Ben)
```

We compute the coefficients and the scores for each nonaffine component.

```
>score<-LSR$u%*%diag(LSR$d)
>NonUnif<-LSR$v
```

We plot the scores of individuals on nonaffine components on a graph; deformation grids associated with variation on axes are displayed as well.

```
>layout(matrix(c(1,1,1,1,2,3,4,5),2,4))
>plot(score[,1:2],pch=c(rep("f",30),rep("m",29)),xlab="RW1",
+    ylab="RW2",asp=1)
>tps(msh, matrix(as.vector(msh)+NonUnif[,1]*max(score[,1])
+    ,8,2),20)
>title("RW 1: right")
>tps(msh, matrix(as.vector(msh)+NonUnif[,1]*min(score[,1])
+    ,8,2),20)
>title("RW 1: left")
>tps(msh, matrix(as.vector(msh)+NonUnif[,2]*max(score[,2])
+    ,8,2),20)
>title("RW 2: top")
>tps(msh, matrix(as.vector(msh)+NonUnif[,2]*min(score[,2])
+    ,8,2),20)
>title("RW 2: bottom")
```

We can estimate the variance of shape explained by each affine and nonaffine component by computing the variance of scores. Only the first $k - 6$ nonaffine components contain variation, the last ones having 0 eigenvalues.

```
>paste("Nonaffine variation =",round(sum
+          (diag(var(score[,1:(kp-6)])))),5))
[1] "Nonaffine variation = 0.0023"
>paste("Affine variation =",round(sum
+          (diag(var(gou$score[,1:2])))),5))
[1] "Affine variation = 0.00081"
>paste("Total variation =", round(sum
+          (diag(var(X)))),5))
[1] "Total variation = 0.00311"
```

The total shape variation is indeed equal to the sum of the nonaffine and affine variations.

```
>diag(var(gou$score[,1:2]))/sum(diag(var(X)))
        U1          U2
0.1206839 0.1406331
>round(diag(var(score[,1:(kp-6)]))/sum(diag(var(X))),4)
[1] 0.3396 0.1238 0.0853 0.0495 0.0424 0.0337 0.0251
[8] 0.0189 0.0120 0.0083
```

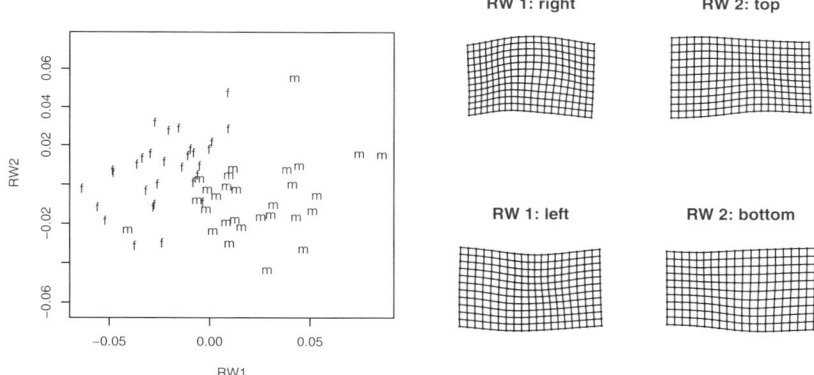

Fig. 6.4. Illustration of the first two nonuniform components of shape variation within the concatened `gorf.dat` and `gorm.dat` dataset. Males correspond to "m" and females to "f". Males and females are well characterized on the first component of nonaffine deformation. This component corresponds to a general bending of the configuration. The second component represents a differential of dilatation between the anterior and posterior parts of the skull cross section: differences between sexes are not characterized by this kind of morphological deformation

In our example, the variation explained by affine transformation summarizes 26% of the total variation. The second uniform component (dilatation) accounts for 14%. The nonaffine components summarize variation that decrease with their rank, the first explaining nearly 34% of the total shape variation of the sample.

6.1.2 Outlines

As for Procrustean data, it is possible to visualize shape changes along main components of shape variation using the variety of modern morphometric methods for analyzing outlines.

As an example, we will analyze outlines of fossil rodent molars belonging to two different species, *Megacricetodon tautavelensis* and *Megacricetodon aunayi*. For simplicity these species names are abbreviated in the text by *tauta* and *auna*. The original dataset contains 64 points sampled in a clockwise way on upper left molar of 29 and 31 individuals for each species. Since teeth have rather complex outlines (see Fig. 6.4), we select the elliptic Fourier methodology for defining shape parameters of the outlines. The datasets are stored in two data frames, of k columns, and $n \times p$ rows. For convenience, we transform them in two array objects, tauta and auna, of respective dimensions: 64, 2, 29 and 64, 2, 31.[1]

Plotting the sampled coordinates or the outline of a given individual is achieved with the familiar commands of R:

```
>tauta<-read.table("/home/juju/morph/tauta.R")
>auna<-read.table("/home/juju/morph/auna.R")
>taut<-array(NA,dim=c(64,2,29))
>taut[,1,]<-tauta[,1]; taut[,2,]<-tauta[,2]
>aun<-array(NA,dim=c(64,2,31))
>aun[,1,]<-auna[,1]; aun[,2,]<-auna[,2]
>plot(taut[,,1], asp=1, pch=3, xlab="X",ylab="Y")
>lines(taut[,,1]); points(t(taut[1,,1]),pch=16,cex=2)
```

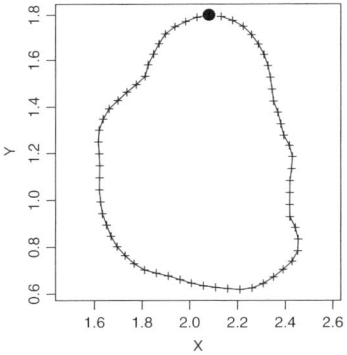

Fig. 6.5. Raw coordinates and outline of the first individual of the *Megacricetodon tautavelensis* dataset. The first digitized landmark is indicated by a circle

For outlines to be compared, one must superimpose outlines on their centroid and perform any necessary alignment, or one can use the normalized Fourier transform.

---

[1] The datasets are available in the onlline supplement.

Our choice is driven by the fact that there is no landmark available on the teeth; therefore the elongation of the first ellipse is used for aligning specimens and standardizing measurements. The same is true for the starting point. We perform the standardized elliptic Fourier analysis of both samples.

```
>coe<-matrix(NA,29,32*4)
>for (i in 1:29)
+       {N<-NEF(taut[,,i])
+        coe[i,]<-c(N$A,N$B,N$C,N$D)}
>coe1<-matrix(NA,31,32*4)
>for (i in 1:31)
+       {N<-NEF(aun[,,i])
+        coe1[i,]<-c(N$A,N$B,N$C,N$D)}
```

The high order harmonic coefficients should contain very little information concerning differentiation between groups, and are more likely to contain noise because of digitization error (see Chapter 5). When using the standardization, one can drop the first three coefficients of the first harmonic because they are constant. The last one is concerned with the relative elongation of the first ellipse. In some papers [89], the remaining variable coefficient of the first harmonic is not taken into account because it is said that it may not contain systematic information. Whether this coefficient must be included or analyzed separately depends on the way we think about harmonic signification. The variation contained in this coefficient corresponds to the relative width-on-length ratio. Prior to digitization, teeth were numerized under a stereographic camera. Depending on their orientation below the camera, teeth could actually look wider or longer. We choose therefore to forget the d_1 coefficient for limiting measurement error caused by orientation of teeth below the objective of the stereographic camera.

We need to estimate the number of Fourier harmonics to retain for future analyses. One can examine the total cumulative power and decide to select harmonics so that 99% of the total power is reached. Since we have stored harmonic coefficients into a `matrix` object, we can calculate this cumulative power with the `apply` function. We remove the first harmonic for this computation, since, besides being subject to error measurement, it summarizes most of the variance.

```
>coef<-rbind(coe, coe1)
>co<-coef^2
>tharm<-apply(co,2,sum)
>power<-apply(matrix(tharm, 32,4),1,sum)
>round(cumsum(power[-1])/sum(power[-1]),3)[1:9]
[1] 0.465 0.840 0.944 0.955 0.985 0.991 0.996 0.997 0.998
```

The first seven harmonics totalize more than 99% of the total power remaining after first removing variation contained in the relative length of the first ellipse. This option for selecting harmonics does not consider the actual variation in the sample. Instead, we can examine the cumulative variance explained by coefficients.

```
>vharm<-apply(coef,2,var)
>variation<-apply(matrix(vharm,32,4),1,sum)
```

```
>round(cumsum(variation[-1])/sum(variation[-1]),3)[1:18]
 [1] 0.357 0.530 0.686 0.763 0.854 0.890 0.928 0.945 0.959
[10] 0.969 0.975 0.980 0.984 0.987 0.990 0.992 0.994 0.995
```

The results are sensibly similar. Even in excluding the first harmonic, the first seven following ones totalize more than 92% of the total reconstructable variation. This means that by taking only seven harmonics, we are not losing too much information. It also means that most of the tooth variation is described by a few harmonic coefficients.

In addition, one can estimate the percent of error variation contained on each coefficient by replicating measurements on similar specimens, or by using the approach of Yezerinac et al. (Section 2.6, [124]). This other way consists of excluding harmonics that carry more relative error variation than explained variation.

To examine variation in the whole sample, we perform a PCA of the first 28 coefficients selected on basis of the cumulative total power and on the proportion of explained variation.

```
>pc<-princomp(coef[,c(2:8,32+2:8,64+2:8, 96+2:8)])
>layout(matrix(c(1,1,1,1,2,3),2,3))
>plot(pc$score, pch=c(rep("a",29), rep("t",31)), asp=1)
>(pc$sdev^2/sum(pc$sdev^2))[1:5]
     Comp.1      Comp.2      Comp.3      Comp.4      Comp.5
 0.37810540  0.17056417  0.14000524  0.07961182  0.06154700
```

The last line of the code calculates the percent of variance on each PC. The first two PCs summarize, respectively, 37.8% and 17.1% of the variance.

To reconstruct variation along axes, We must first estimate the Fourier coefficients of the mean shape by averaging the first harmonic coefficients, including the mean of the first harmonic coefficients, that we have ignored to perform the PCA. Once done, we use the scores and eigenvectors for reconstructing theoretical outlines at different positions along the principal axes.

```
>mshcoef<-apply(coef[,c(1:8,32+1:8,64+1:8, 96+1:8)],2,mean)
```

Although the first harmonic is not considered for explaining variation, the extreme scores are multiplied by the related eigenvectors and added to the corresponding coefficients of the meanshape to compute the coordinates of extreme shapes. This provides a means of describing extreme shape change explained by PC-axes.

```
>ev<-pc$loadings
>Mx1<-mshcoef+max(pc$score[,1])*c(ev[1:7,1],
+      0,ev[8:14,1],0,ev[15:21,1],0,ev[22:28,1],0)
>mx1<-mshcoef+min(pc$score[,1])*c(ev[1:7,1],
+      0,ev[8:14,1],0,ev[15:21,1],0,ev[22:28,1],0)
>Mx2<-mshcoef+max(pc$score[,2])*c(ev[1:7,2],
+      0,ev[8:14,2],0,ev[15:21,2],0,ev[22:28,2],0)
>mx2<-mshcoef+min(pc$score[,2])*c(ev[1:7,2],
+      0,ev[8:14,2],0,ev[15:21,2],0,ev[22:28,2],0)
```

We achieve the reconstruction of outlines by inverse Fourier transformation with the `iefourier` function.

```
>Mx1<-iefourier(Mx1[1:8],Mx1[9:16],Mx1[17:24],Mx1[25:32]
+      ,8,64)
>mx1<-iefourier(mx1[1:8],mx1[9:16],mx1[17:24],mx1[25:32]
+      ,8,64)
>Mx2<-iefourier(Mx2[1:8],Mx2[9:16],Mx2[17:24],Mx2[25:32]
+      ,8,64)
>mx2<-iefourier(mx2[1:8],mx2[9:16],mx2[17:24],mx2[25:32]
+      ,8,64)
>plot(Mx1,type="l",col="grey60",asp=1,frame=F,axes=F,
+      main="PC1",xlab="",ylab="",lwd=2)
>points(mx1, type="l",asp=1)
>plot(Mx2,type="l",col="grey60",asp=1,frame=F,axes=F,
+      main="PC2",xlab="",ylab="",lwd=2)
>points(mx2, type="l",asp=1)
```

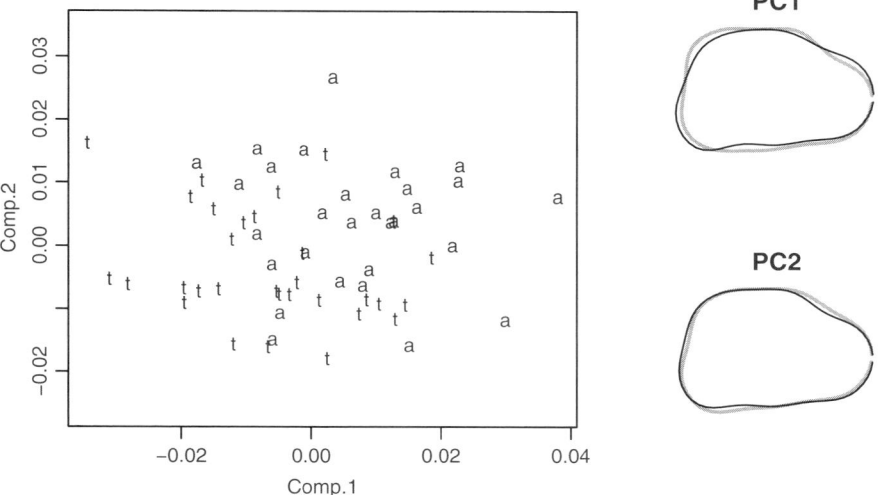

Fig. 6.6. PCA applied to the first $8 - 1$ harmonics defined from the outlines of the left first molar of fossil rodents of the genus *Megacricetodon*. "a" labels correspond to the "*auna*" species, "t" labels to the "*tauta*" species. Black and thin outlines correspond to minimal values along axes, and gray and thick outlines to maximal scores

We notice in Fig. 6.6 that the first two principal components oppose both species relatively well. The species "*tauta*" seems to have a more complex and angular outline than the species "*auna*."

6.2 Discriminant and Multivariate Analysis of Variance

When one compares several groups, one can choose either discriminant analysis for group discrimination and eventually inferring prediction concerning the category of a new observation, or multivariate analysis of variance for testing whether differences between groups are significantly greater than within-group variation.

6.2.1 Outlines

We will first examine differences between outlines using the outlines of fossil rodents as examples. We keep the number of coefficients previously selected for the principal component analysis of variance. These are the coefficients of the second to the eighth harmonics. The first group contains 29 individuals, while the second contains 31 individuals. To perform a one-way MANOVA, we simply code the explaining variable (species) as a factor.

```
>fact<-as.factor(c(rep(1,29), rep(2,31)))
>coeff<-coef[,c(2:8,32+2:8,64+2:8, 96+2:8)]
>summary(manova(coeff~fact), test="Hotelling")
          Df Hotelling-Lawley approx F
fact       1             3.8379    4.2491
Residuals 58

          num Df den Df    Pr(>F)
fact          28     31 7.601e-05 ***
Residuals
---
Signif. codes:  0 *** 0.001 ** 0.01 * 0.05 . 0.1    1
```

Both groups are significantly different.

For investigating the morphological effect on the linear discriminant axes, we nearly proceed as for the principal component analysis. However, since the within variance-covariance matrix has been used for standardizing the between-group variance, the linear discriminant loadings have to be premultiplied by the within-group matrix to be compatible with shape variation (see Fig. 6.7). There are six steps to follow:

1. Estimate the linear discriminant functions.
2. Project data onto the discriminant axes.
3. Compute the pooled within-group or error variance-covariance matrix.
4. premultiply linear discriminant axes by the within variance-covariance matrix, for removing effects of standardization.
5. Reconstruct outlines along axes: For this, one calculates the mean shape, on which one adds the rectified discriminant loadings multiplied by the desired scores. The scores we use are thus the mean group score, or extreme scores for individuals on axes, or even amplified scores for an easier visualization of shape differences – if those are small.

6. Use reverse Fourier functions for producing outlines.

We follow these six steps for depicting group discrimination between the two rodent species.

Step 1.

```
>library(MASS)
>mod1<-lda(coeff, fact)
>LD<-mod1$scaling
```

Step 2.

```
>ldascore<-predict(mod1)$x
```

Step 3.

```
>n<-dim(coeff)[1]
>mod3<-lm(coeff~fact)
>dfw<- n - length(levels(fact))
>SSw<- var(mod3$residuals) * (n-1)
>VCVw<-SSw/dfw
```

Step 4.

```
>LDs<-VCVw%*%LD
>layout(matrix(c(1,2),1,2))
>hist(ldascore[1:29],xlim=range(ldascore),main=
     "Score distribution", xlab="Discriminant score")
>hist(ldascore[30:60],col="grey",add=T)
```

Step 5: amp1 is the amplification coefficient – here extreme differences are exaggerated twice.

```
>mshcoef<-apply(coef[,c(1:8,32+1:8,64+1:8, 96+1:8)],2,mean)
>amp1<-2
>Mx1<-mshcoef+ amp1 * max(ldascore[,1])*c(LDs[1:7,1],
+           0,LDs[8:14,1],0,LDs[15:21,1],0,LDs[22:28,1],0)
>mx1<-mshcoef+ amp1 * min(ldascore[,1])*c(LDs[1:7,1],
+           0,LDs[8:14,1],0,LDs[15:21,1],0,LDs[22:28,1],0)
```

Step 6.

```
>Mx1<-iefourier(Mx1[1:8],Mx1[9:16],Mx1[17:24],Mx1[25:32]
+      ,8,64)
>mx1<-iefourier(mx1[1:8],mx1[9:16],mx1[17:24],mx1[25:32]
+      ,8,64)
>plot(Mx1,type="l",col="grey55",asp=1,frame=F,axes=F,lwd=2
+      ,main="Variation on LD1 x 2",xlab="",ylab="")
>points(mx1, type="l",asp=1)
```

The rodents differ in the more or less accentuated concavities and convexities appearing along the outline (Fig. 6.7).

Another possibility for observing differences between groups – when two groups only are to be differentiated – is to look at mean shape of each group. However, this

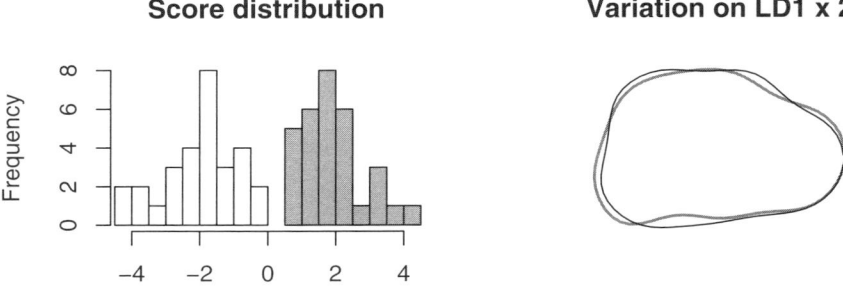

Fig. 6.7. Discriminant analysis based on the species category applied to the first $8-1$ harmonics defined from the outlines of left first molar of fossil rodents. Black outline and white bins in the histogram correspond to the species "*auna*", thick gray outline and gray bins in the histogram correspond to the species "*tauta*"

will not necessarily emphasize the significant shape components that differ between groups.

Compute averaged coefficients `mshcoef1` *and* `mshcoef2` *for each group.*

```
>mshtauta<-apply(coef[1:29,c(1:8,32+1:8,64+1:8,96+1:8)]
+        ,2,mean)
>mshauna<-apply(coef[30:60,c(1:8,32+1:8,64+1:8,96+1:8)]
+        ,2,mean)
```

For exaggerating differences between mean groups, we use the difference between each species mean shape and the overall mean shape.

```
>msh<-apply(rbind(mshtauta, mshauna), 2, mean)
>ampl<--4
>tauta2<-msh+(mshtauta-msh)*ampl
>auna2<-msh+(mshauna-msh)*ampl
>TAUT<-iefourier(tauta2[1:8], tauta2[9:16],
+       tauta2[17:24], tauta2[25:32],8,64)
>AUN<-iefourier(auna2[1:8], auna2[9:16],
+       auna2[17:24], auna2[25:32],8,64)
>plot(TAUT,type="l",asp=1,frame=F,axes=F,main=
+      "Differences between mean groups exagerated x 4",
+      xlab="",ylab="",col="grey55",lwd=2)
>points(AUN, type="l")
```

The differences between mean groups (Fig. 6.8) are nearly the same as the variation found on the linear discriminant axis.

Four–times exagerated

differences between group means

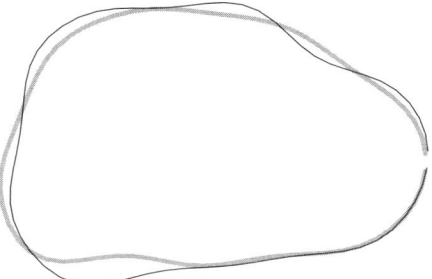

Fig. 6.8. Four-times amplified differences between mean shapes for each species of the rodent teeth dataset. The black and thin outline corresponds to the species "*auna*" while the thick and gray outline corresponds to the species "*tauta*"

6.2.2 Procrustes Data

For Procrustes data, things are becoming more difficult than classic multivariate statistics. When one works on superimposed coordinates, the number of dimensions of the shape space is not equal to the number of analyzed variables pk. Indeed, the inherent transformations (scaling, translation, rotation) accompanying the superimposition procedure reduces the rank number contained in original data. There are four dimensions lost for 2D data and seven for 3D data. This mismatch between the number of dimensions and the number of variables makes it difficult to use standard multivariate tests for Procrustes data. These tests need to be adapted.

One must slightly adapt multivariate tests to estimate the significance of shape difference between groups of configurations depicted by landmarks. Two parameters must be modified in the test:

1. The number of space dimensions should replace the number of variables in the transformation of the multivariate statistic to its F-approximation. We automatically obtain the number of dimensions by estimating the rank of the matrix of variance-covariance with the qr function. This rank substitutes the number of variables in the test.
2. Since between-group (or effect) variance is scaled by within-group (or error) variance by using matrix multiplication and matrix inverse, one encounters a computational problem with classic algorithms estimating matrix inverse. Indeed, one cannot invert a nonsingular matrix as simply as usual covariance matrices, and one must use the Moore-Penrose generalized inverse. Fortunately the ginv function of the **MASS** package calculates this inversion.

We modify the Hotelling-Lawley trace statistic and its F-approximation in a function called Hotellingsp that receives several arguments: sum of squares and cross-products of effect, sum of squares and cross-products of residual variation, and their respective degrees of freedom. We use the F-approximation following Section 3.4.3. For implementing the two possible approximations, the last argument of the function is set to `exact=FALSE`, or to `exact=TRUE`. In the latter case, the F-approximation follows the second moment estimate as suggested by McKeon [73].

Function 6.4. `Hotellingsp`

> *Arguments:*
>> `SSef`: *Sum of squares and cross-products of effect.*
>> `SSer`: *Sum of squares and cross-products of residual variation.*
>> `dfef`: *df for the effect term.*
>> `dfer`: *df for the error term.*
>> `exact`: *Logical value indicating whether one should use the estimate of McKeon [73].*
>
> *Value:*
>> *A summary table containing the degrees of freedom, the Hotelling-Lawley trace, the*
>> *F-approximation and relative degrees of freedom, and the p-value.*

```
1  Hotellingsp<-function(SSef, SSer, dfef, dfer, exact=F)
2    {library(MASS)
```

p *corresponds to the number of shape space dimensions.*

```
3   p <- qr(SSef+SSer)$rank
4   k<-dfef; w<-dfer
5   s<-min(k,p)
6   m<-(w-p-1)/2
7   t1<-(abs(p-k)-1)/2
8   Ht<-sum(diag(SSef%*%ginv(SSer)))
9   Fapprox<-Ht*(2 * (s*m+1))/(s^2*(2*t1+s+1))
10  ddfnum<-s*(2*t1+s+1)
11  ddfden<-2*(s*m+1)
12  pval= 1-pf(Fapprox, ddfnum, ddfden)
13
14   if (exact)
15      {b<-(p+2*m)*(k+2*m)/((2*m+1)*(2*m-2))
16       c1<-(2+(p*k+2)/(b-1))/(2*m)
17       Fapprox<-((4+(p*k+2)/(b-1))/(p*k))*(Ht/c1)
18       ddfnum<-p*k
19       ddfden<-4+(p*k+2)/(b-1) }
20
21  unlist(list("dfeffect"=dfef,"dferror"=dfer,"T2"=Ht,
22    "Approx_F"=Fapprox,"df1"=ddfnum,"df2"=ddfden,"p"=pval))}
```

Using the same strategy as before, we will examine differences between cross sections of ape skulls depicted by landmark data. We use the `gorf.dat`, `pan.dat`, and `pongo.dat` datasets of the **shapes** package. The datasets must be first reorganized because landmarks are not labeled in the same order for each set.

```
>library(shapes)
>gorf<-gorf.dat
>panf<-panf.dat[c(5,1,2:4,6:8),,]
>pongof<-pongof.dat[c(5,1,2:4,6:8),,]
>APE<-array(c(panf, gorf, pongof),dim=c(8,2,80))
>AP<-orp(pgpa(aligne(APE))$rotated)
>fact<-as.factor(c(rep("p",26),rep("g",30),rep("o",24)))
>m<-t(matrix(AP, 16, 80))
>n<-dim(m)[1]
```

When sets are assembled and superimposed, we apply a linear model to our data, the explaining factor being species. We use this model to compute the effect and error sum of squares and cross-products.

```
>mod1<-lm(m~as.factor(fact))
>dfef<- length(levels(fact))-1
>dfer<- n - length(levels(fact))
>SSef<-(n-1)*var(mod1$fitted.values)
>SSer<-(n-1)*var(mod1$residuals)
```

Finally, we test for differences between within-group and between-group variance-covariance matrices with the adapted Hotelling-Lawley multivariate test.

```
Hotellingsp(SSef, SSer, dfef, dfer)
 dfeffect    dferror        T2  Approx_F
  2.00000   77.00000  17.08507  46.27206

      df1         df2          p
 24.00000 130.00000    0.00000
```

The group means are significantly different.

We can use linear discriminant analysis and reconstruct configurations corresponding to extreme values along discriminant axes to determine differences between groups. Since in our case, there are three groups, there will be two discriminant axes. The linear discriminant axes are premultiplied by the within variance-covariance matrix and associated scores to reconstruct configurations depicted by scores in the discriminant space.

```
>mod1<-lda(m, fact)
Warning message:
variables are collinear in: lda.default(x, grouping, ...)
```

Plot individuals on the linear discriminant axes.

```
>score<-predict(mod1)$x
>layout(matrix(c(1,1,2,3),2,2))
>plot(score,pch=as.character(fact),asp=1)
```

Calculate averaged shapes for each group.

```
>LD<-mod1$scaling
>msh<-apply(m, 2, mean)
```

Estimate the within group variance-covariance.

```
>mod3<-lm(m~as.factor(fact))
>n<-dim(m)[1]
>dfw<- n - length(levels(fact))
>SSw<- var(mod3$residuals) * (n-1)
>VCVw<-SSw/dfw
>LDs<-VCVw%*%LD
```

Estimate configurations corresponding to changes onto linear discriminant axes.

```
>LD1M<-(matrix(msh+ max(score[,1])* LDs[,1],8,2))
>LD1m<-(matrix(msh+ min(score[,1])* LDs[,1],8,2))
>LD2M<-(matrix(msh+ max(score[,2])* LDs[,2],8,2))
>LD2m<-(matrix(msh+ min(score[,2])* LDs[,2],8,2))
>joinline<-c(1,6:8,2:5,1,NA,7,4)
>par(mar=c(5,1,2,1))
>plot(LD1M,axes=F,frame=F,asp=1,xlab="",ylab="",
+     pch=22,main="LD1")
>points(LD1m,pch=17)
>lines(LD1M[joinline,],lty=1)
>lines(LD1m[joinline,],lty=2)
>plot(LD2M,axes=F,frame=F,asp=1,xlab="",ylab="",
+     pch=22, main="LD2")
>points(LD2m,pch=17)
>lines(LD2M[joinline,],lty=1)
>lines(LD2m[joinline,],lty=2)
```

Note the warning message that appears due to the fact that matrices of variance-covariance are not singular. We could have avoided this by substituting original data by their projections on the $p - 4$ PC-axes: the result would have been strictly similar. On the projections of individuals, we notice that no individual has been misclassified. The first linear discriminant axis mainly shows differences between the Asian and the African species, while the second axis opposes the gorilla to the chimpanzee and orangutan. The orangutan has a more triangular skull cross section, while the chimpanzee and gorilla are more rectangular. On the second axis, the chimpanzee is opposed to both gorilla and orangutan: differences between species mainly concern the facial part of the skull – the base of the skull in chimpanzee is upturned anteriorly, while it is inflected downwards for both other species (Fig. 6.9).

For other kinds of shape variables, one can usually proceed as with independent multivariate datasets. When shape dimensions are lost during the transformation of original variables into shape variables, one can use the Hotelling-Lawley trace test developed before.

6.3 Clustering

As seen in Chapter 3, we can use clustering methods for investigating whether there is a grouping structure in the collection of observations. We will use a part of the

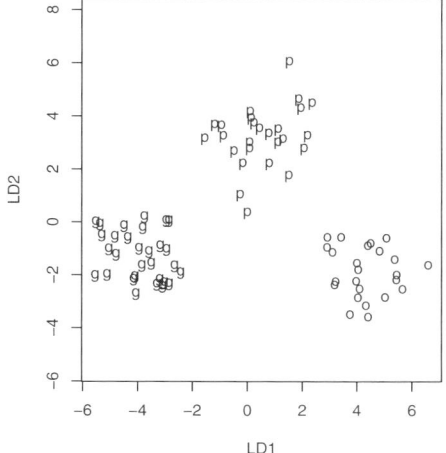
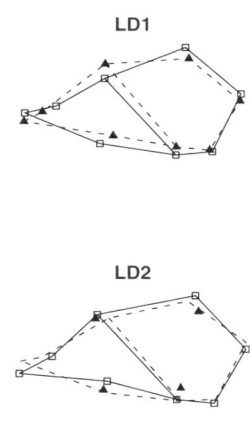

Fig. 6.9. Discriminant analysis among group of three different species of ape: "p" labels correspond to chimpanzee, "g" to gorilla, and "o" to orangutan. The morphological interpretations
of the linear discriminant axes are displayed on the right side of the graph: deformation corresponds to extreme individual scores on LD-axes. Square symbols and full lines correspond to
maximal values, triangle symbols and dotted lines to minimal values

configuration set for apes to investigate whether hierarchical and partitioning methods result in the same classification between female gorilla and chimpanzee skulls.
We first perform an UPGMA and a complete linkage on the Euclidean distance
in the tangent shape space. Once configurations are superimposed and projected,
we display the tree using a simple line of commands with the `hclust` and `dist`
functions.

```
>library(MASS)
>library(shapes)
>gorf<-gorf.dat
>panf<-panf.dat[c(5,1,2:4,6:8),,]
>pongof<-pongof.dat[c(5,1,2:4,6:8),,]
>APE<-array(c(panf, gorf, pongof),dim=c(8,2,80))
>APE<-array(c(panf, gorf),dim=c(8,2,56))
>AP<-orp(pgpa(APE)$rotated)
>m<-t(matrix(AP, 16, 56))
>par(mar=c(0.5,2,1,1))
>layout(matrix(c(1,2),2,1))
>plot(hclust(dist(m), method="average"),main="UPGMA"
+    ,labels=c(rep("P",26),rep("G",30)),cex=0.7)
>plot(hclust(dist(m), method="complete"),main="COMPLETE"
+    ,labels=c(rep("P",26),rep("G",30)),cex=0.7)
```

A single individual is misclassified in the UPGMA method, while the complete linkage method is able to identify correctly both groups (Fig. 6.10).

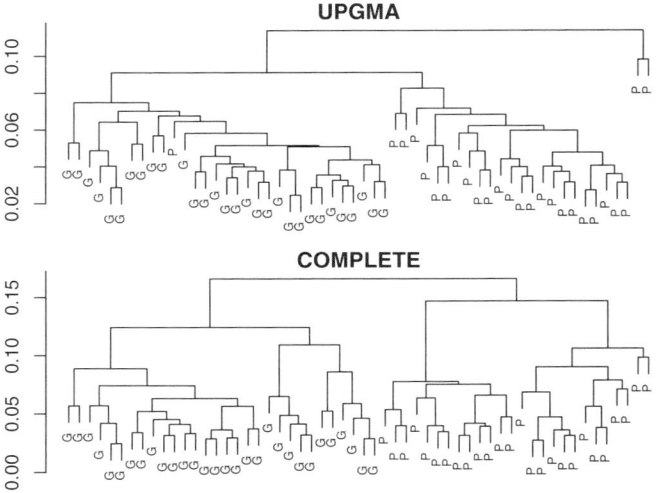

Fig. 6.10. UPGMA and complete linkage clustering applied to tangent shape space coordinates of the ape dataset containing the female chimpanzee and gorilla data. "G" labels are for gorillas and "P" for chimpanzees

Assuming that groups are not known in advance, one may want to estimate the number of groups within the data using a partitional clustering method and the elbow approach (see Section 3.4.3). In the following example, I used the clustering structure returned by the pam function and I computed the within- and between-group sum of squares using a linear model with response being the grouping factor.

```
>library(cluster)
>df<-dim(m)[1]-1
>SStot<-sum(diag(var(m)))*df
>expl<-0
>for (i in 2:20)
+    {mod<-pam(dist(m),i)
+     mod1<-lm(m~as.factor(mod$clustering))
+     expl[i]<-sum(diag(var(mod1$fitted.values)))*df/SStot}
>plot(1:20,expl,ylab="Explained variance"
+     ,xlab="Group number")
>lines(1:20,expl)
```

The elbow method used in Fig. 6.11 shows that the line breaks for two groups.

```
> pam(dist(m),2)$clustering
  [1] 1 1 1 1 1 1 1 1 1 1 1 1 1 1 1 1 1 1 1 1 1 1 1 1 1 1 1 1
```

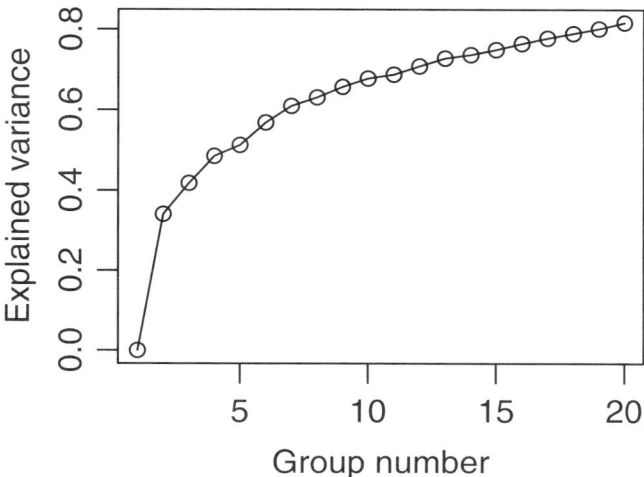

Fig. 6.11. Estimation of group numbers using partitional clustering with the `pam` function applied on the superimposed configurations of the female chimpanzee and gorilla datasets

```
[26] 2 2 2 2 2 2 2 2 2 2 2 2 2 2 2 2 2 2 2 2 2 2 2 2 2
[51] 2 2 2 2 2 2
> pam(dist(m),3)$clustering
 [1] 1 1 1 1 1 1 1 1 1 1 2 2 2 2 2 1 1 1 1 1 1 1 2 2 1 1
[26] 2 3 3 3 3 3 3 3 3 3 3 3 3 3 3 3 3 3 3 3 3 3 3 3 3
[51] 3 3 3 3 3 3
```

With an assumption of two groups, one individual of chimpanzee is misclassified in the gorilla group. Moreover, increasing the number of groups results in the nearly same partition, except that some clusters are truly monospecific.

Several other methods can perform clustering analysis of data. Users have to define the method and distance metrics; their choices are guided by the type of data. For example, one can eventually normalize elliptic Fourier coefficients since the first summarize more power and variance than the following ones. For clustering group means, it is recommended to use the Mahalanobis distance between groups since this will be scaled by the within-group variance-covariance.

6.4 Morphometrics and Phylogenies

There are many possible applications combining morphometric methods and phylogenies (for example, see Macleod and Forey [69]). Although using morphometric characters to estimate phylogenies is feasible, this is not always recommended (see [97, 100]). One can use morphometric exploration for identifying characters that could further be included in a morphological dataset to infer phylogenies by parsimony.

An interesting issue is to fit morphometric data onto a given phylogenetic tree. Several methods are available for estimating ancestral character states; nevertheless, in the case of Procrustes data [100] notes that one cannot use linear parsimony estimates. Actually, observations produced with Procrustean superimpositions are sensitive to the orientation of the reference shape (that one usually takes as being the mean shape). Squared parsimony or maximum likelihood methods for estimating ancestral character states are preferred over linear parsimony.

I used the mosquito dataset presented in Rohlf [96]. You can find the phylogeny and the set in data supplied by the free "tpstree" software developed by Rohlf [94].[2] We need the data concerning terminal tips of the tree, the phylogeny, and then to perform a Procrustes analysis of configurations defined by landmarks digitized on the wings of the eleven species of mosquitoes present in the set (Fig. 6.12). Their phylogenetic relationships are illustrated by a tree in Fig. 6.13.

Import the set of original configuration in R.

```
>dat<-scan("/home/juju/morph/mosqH.tps", what="char")
```

Arrange the set in an $p \times k \times n$ array object.

```
>taxa<-dat[c(1:11)*38]
>taxa<-sub("ID=", "", taxa)
>rem<-sort(c(c(1:11)*38, (c(1:11)*38)-37))
>dat<-as.numeric(dat[-rem])
>dat1<-matrix(dat,18*11,2, byrow=TRUE)
>dat2<-array(NA, dim=c(18,2,11))
>dat2[,1,]<-dat1[,1]
>dat2[,2,]<-dat1[,2]
```

Perform the superimposition and plot each species using different gray-scale levels.

```
>mosq<-orp(pgpa(dat2)$rotated)
>palette(gray(1:11/16))
>par(mar=c(1,1,3,1))
>plot(mosq[,1,],mosq[,2,],asp=T,cex=0,xlab="",axes=F,
+      frame=F,ylab="",main="Procrustes superimposition")
>joinline<-c(1:13,1,NA,10,18,9,NA,13,18,NA,8,17,7,NA,16,17,
+      NA,4,15,5, NA,2,14,15,NA,6,16,13,NA,3,14,13)
>for (i in 1:11)
+      {lines(mosq[joinline,,i],col=i)}
>points(mosq[,1,],mosq[,2,], asp=T, cex=0.7, pch=20)
```

We write the phylogeny, and transform the phylogram into a chronogram according the method of Sanderson [109]. We transform the phylogeny into a chronogram with the chronogram function of the **ape** package.

---

[2] http://life.bio.sunysb.edu/ee/rohlf/software.html

Procrustes superimposition

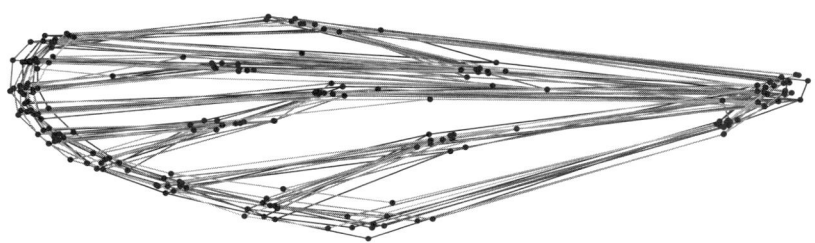

Fig. 6.12. Procrustes superimposition of the TPStree mosquito wings dataset

```
>library(ape)
>mosqp<-read.tree(text="(AN1:16,(cu101:2,((ma29:1,
+    (cl21:1,ae44:1):3):3,((ur17:1,(wy13:1,or28:6):4):2,
+    ((to12:1,de126:3):1,ps32:30):1):2):2):8);")
>mosqp$node.label<-c(12:21)
>layout(matrix(c(1,2),2,1))
>par(mar=c(3,2,3,2))
>mosqp<-root(mosqp, "AN1")
>plot(mosqp, show.node.label=T,main="Phylogram",cex=0.7)
>chrono<-chronogram(mosqp)
>plot(chrono,show.node.label=T,main="Chronogram",cex=0.7)
```

The combination of information from morphology and phylogeny permits us to address evolutionary questions regarding the relationship between divergence in time and in wing morphological evolution. One can determine the strength of the relationship using a Mantel test on the chronological distances and the Euclidean morphological distances. Before this, we must enter the order of superimposed configurations as they appear in the phylogeny. For this purpose, we use the match function.

```
>ordphyl<-match(mosqp$tip.label, taxa)
>mosq1<-mosq[,,ordphyl]
>procdata<-t(matrix(mosq1, 36, 11))
>chronodist<-cophenetic.phylo(chrono)
>euclidist<-as.matrix(dist(procdata))
>mantel.test(chronodist, euclidist)$p
[1] 0.427
```

Morphological disparity is not related to time divergence. We can address the question differently and analyze whether phylogenetic divergence is related to morphological variation.

```
>phyldist<-cophenetic.phylo(mosqp)
>mantel.test(phyldist, euclidist)$p
[1] 0.606
```

Phylogram

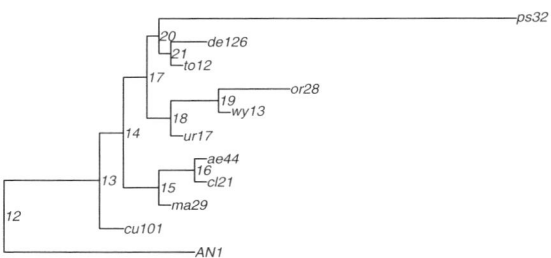

Chronogram

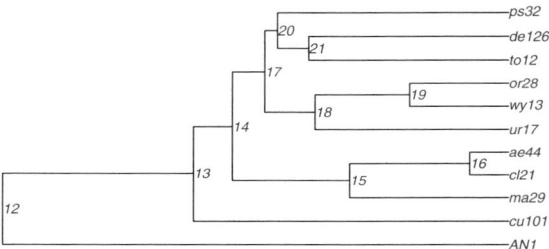

Fig. 6.13. Phylogeny and chronogram of the mosquito dataset. The nodes are labeled starting from the root with labels 12 to 21

The test does not tell that divergence in morphology is not related to time of divergence between species. The test only says that convergence may have occurred, and that rates of morphological evolution are probably not the same along every branch of the tree.

If one wants to determine whether there is a relationship between time of divergence and divergence in morphology, one must infer ancestral character states.

Although morphological divergence is not always proportional to time divergence, and this may preclude some reliable estimate of ancestral morphologies, we will proceed as if to show a strategy for estimating ancestral character states at tree nodes. We will follow the generalized linear model approach for reconstructing discrete and continuous characters as described in [24]. We have to determine the variance-covariance among terminal taxa $\mathbf{VAR}(\mathbf{Y})$. The `vcv.phylo` function receiving the `phylo` object as argument returns this variance-covariance matrix. The variance-covariance among terminal taxa corresponds to the shared history among taxa. We need to estimate the covariance between terminal taxa $\mathbf{Y}$ and ancestors

VAR(**Anc**, **Y**) as well; this corresponds to the shared history between terminal taxa and nodes. To obtain this matrix, we use the functions of ape. `Ntip` returns the number of terminal taxa, `dist.nodes` returns the distance between every node of the tree (terminal and internal), and `mrca` that returns the most common ancestor for each pair of nodes.

```
>n<-Ntip(chrono)
>dis <- dist.nodes(chrono)
>MRCA <- mrca(chrono, full = TRUE)
>M <- dis[n + 1, MRCA]
>dim(M) <- rep(sqrt(length(M)), 2)
```

M *corresponds to the vector of distances between the root and the most common ancestor of every pair of descendants. Directly using the* dim *function on* M *allows the coercion into a* matrix *object. The covariance between terminal taxa and nodes is obtained by logical indexing.*

```
>vcvay<-M[-(1:n), 1:n]
```

Ancestral character states are determined as the deviation of each node from the root value **Ro**. One must estimate the root. We choose the `ace` function with the `pic` method as implemented in the ape package to estimate the ancestral morphology of the root defined as a weighted average considering the topology of the tree (see Fig. 6.14 for an illustration of the estimated ancestral morphology).

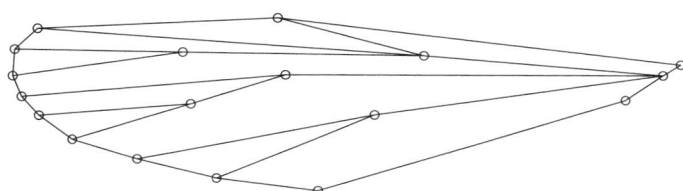

Fig. 6.14. The ancestral morphology of the mosquito wing estimated at the root of the tree. The estimated shape corresponds to a weighted average considering the phylogenetic variances and covariances along branches

```
>kp<-dim(procdata)[2]
>rootp<-numeric(kp)
>for(i in 1:kp)
+   {rootp[i]<-ace(procdata[,i],mosqp, method="pic")$ace[1]}
>ROOTP<-matrix(rootp, 18,2)
```

```
>par(mar=c(0,0,0,0))
>plot(ROOTP, asp=1, axes=F, frame=F)
>lines(ROOTP[joinline,])
```

In order to estimate ancestral character states **Anc**, one must perform a few matrix calculations such that

$$\mathbf{Anc} = \mathbf{VAR}(\mathbf{Anc}, \mathbf{Y})\mathbf{VAR}(\mathbf{Y})^{-1}(\mathbf{Y} - \mathbf{Ro}) + \mathbf{Ro} \ .$$

```
>n<-dim(procdata)[1]
>A<-vcvay[-1,]%*%ginv(vcv.phylo(chrono))%*%
+       (procdata-(rep(1,n)%*%t(rootp))) +
+       rep(1,(n-2))%*%t(rootp)
>ancestral<-array(t(rbind(rootp,A)),dim=c(18,2,10))
```

We have everything for examining which branches are concerned with rapid or slow morphological evolution (Fig. 6.15). We can estimate rates of morphological evolution from morphologies of descendants and ancestors. In our case, rates correspond to the amount of Euclidean distance scaled by branch length between two OTUs. Because of the randomness inherent in the course of phenotypic evolution, these rates are nevertheless underestimated.

```
>tipandnode<-rbind(procdata, rootp, A)
>DD<-as.matrix(dist(tipandnode))
>le<-dim(chrono$edge)[1]
>rate<-numeric(le)
>for (i in 1:le)
-       {rate[i]<-DD[chrono$edge[i,1],chrono$edge[i,2]]
-                    / chrono$edge.length[i]}
>par(mar=c(2,2,2,2))
>plot(chrono, edge.width=rate, label.offset=0.03)
```

In Figure 6.15, we notice an important acceleration in morphological divergence occurred at the split between the taxa abbreviated by "ae44" and "cl21."

One can address many other macro-evolutionary questions by combining a morphometric approach and a phylogenetic approach. I invite the reader to read Paradis [81] if he is interested in analysis of phylogenetics and evolution with R.

6.5 Comparing Covariation Patterns

Rather than examining whether two groups are different in shape, one can investigate for homogeneity or divergence in patterns of shape variance-covariance between groups. We have already developed several strategies (see Section 3.4.4). However, these strategies must be adapted for landmark data.

First one can perform a Mantel test between variance-covariance matrices. The Mantel test will simulate the distribution of correlation under the null hypothesis by randomly permuting rows and columns of one of the hemimatrices. For this purpose,

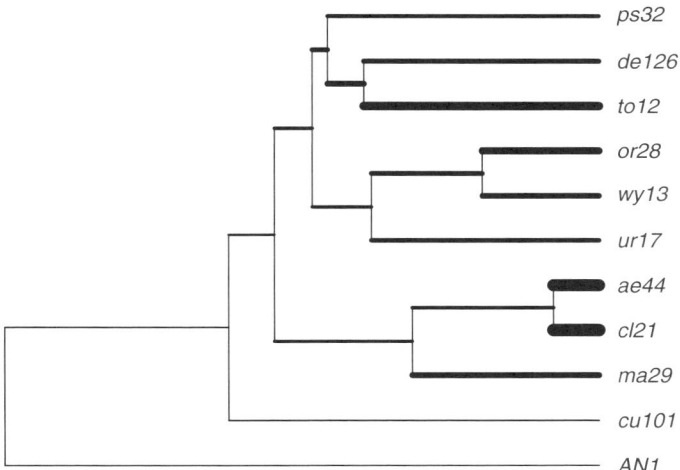

Fig. 6.15. Rates of morphological evolution estimated on branches of the chronogram of the mosquito dataset

one must include the diagonal of the variance-covariance within the test, because it contains the variance for each coordinate. In addition, since one landmark is defined by two or three coordinates, the rows and columns of the matrices corresponding to similar landmarks are not independent and are not interchangeable [56]. One should not randomly permute coordinates but pairs or triplets of coordinates associated with each landmark.

We have to write a function for performing the Mantel test on Procrustean data. For this goal, it is easier to use one previous function that already performs the Mantel test. We can learn how the `mantel.test` function of the ape package has been written. In typing `mantel.test` we can see that the function calls two other functions,(`mantz.stat` and `perm.rowscols`), which we will slightly modify. The `mantz.stat` function calls `lower.triang`. The latter returns the lower triangle of a square matrix. This latter function does not need to be modified because it specifies that the test considers the diagonal as well. We show below the required modifications and write a new Mantel test function that considers the geometry between landmarks as well.

We slightly modify the `mantz.stat` function into `mantrstat` to obtain the correlation between hemimatrices rather than the sum of the product between matrix elements; this will not change the test, but it provides some measure of correlation.

Function 6.5. `mantrstat`

Arguments:
> `m1`: *Square matrix.*
> `m2`: *Square matrix.*

Value:
> *Element-wise correlation coefficient between the lower triangles of each matrix.*

Required function: `lower.triang`.

```
mantrstat<-function (m1, m2)
{ cor(lower.triang(m1),lower.triang(m2))}
```

Then we modify the `perm.rowscols` internal function into`permrowscols` to perform the permutation between rows and columns. We add a third argument for specifying whether the configurations are organized either 1D, 2D, or 3D. The default value will be one, but one will be allowed to change this value in the argument of the general function that will perform the Mantel test. The `m1` matrix must be organized with the first p columns corresponding to the x-coordinates, and the next columns to the y, and eventually to z. The combination of the `sample` and `rep` functions allows us to sample rows and columns and helps us to handle the geometry contained in landmark data.

Function 6.6. `permrowscols`

Arguments:
> `m1`: *Square matrix.*
> `n`: *Number of landmarks* × *number of dimensions.*
> `coord`: *Number of dimensions*

Value:
> *Permuted matrix.*

```
permrowscols<-function (m1,n,coord)
{s <- sample(1:(n/coord))
 m1[rep(s,coord), rep(s,coord)]}
```

Finally, we must program a new Mantel test, `mantel.t2`, nearly copying and pasting the former one, changing the names of the two internal functions and adding the argument `coord`, which specifies the number of coordinates that have to be permuted together.

Function 6.7. `mantel.t2`

Arguments:

　　`m1`: *Square matrix.*
　　`m2`: *Square matrix.*
　　`coord`: *Number of dimensions.*
　　`nperm`: *Number of permutations.*
　　`graph`: *Logical value indicating whether a graph should be returned.*

Values:

　　`r.stat`: *Observed statistic (Element wise correlation coefficient between both hemi-*
　　　matrices).
　　`p`: *p-value.*

Required functions: `lower.triang, perm.rowscols, mantrstat.`

```
1  mantel.t2<-function(m1,m2,coord=1,nperm=1000,graph=FALSE,...)
2  {n<-nrow(m1)
3   realz<-mantrstat(m1, m2)
4   nullstats<-replicate(nperm,mantrstat(m1,
5                 perm.rowscols(m2,n,coord)))
6   pval <- sum(nullstats > realz)/nperm
7   if (graph) {
8       plot(density(nullstats), type = "l", ...)
9       abline(v = realz)    }
10  list(r.stat = realz, p = pval)}
```

We apply this test to the patterns of shape covariation of brain structure in control and schizophrenic patients. The question concerns whether these patterns are similar. We use the dataset of Bookstein [11] provided in the **shapes** package (Fig. 6.16). The dataset contains 13 landmarks on 14 control and 14 schizophrenic patients.

```
>library(shapes)
>schizo<-orp(pgpa(schizophrenia.dat)$rotated)
>control<-t(matrix(schizo[,,1:14],26,14))
>schizop<-t(matrix(schizo[,,15:28],26,14))
>plot(control[,1:13], control[,14:26],pch=3,
+            asp=1,xlab="", ylab="", frame="")
>points(schizop[,1:13], schizop[,14:26],pch=16, cex=0.7)
>mantel.t2(var(schizop), var(control), coord=2)
$r.stat
[1] 0.5150001

$p
[1] 2e-05
```

The test is significant, meaning that patterns in covariation in both groups are similar.

　　Using the `coli` function defined in Chapter 3, we can also test whether principal component-axes have the same direction. Here we only focus on the two first components, which totalize about 50% of the overall variation in both groups.

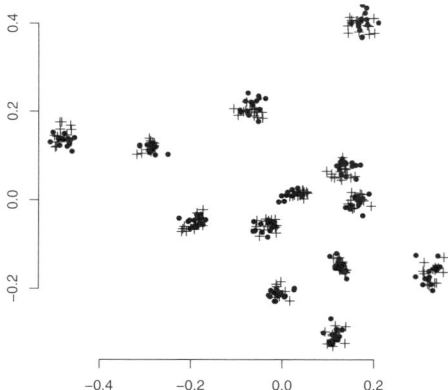

Fig. 6.16. Superimposition of landmarks digitized from MR images of the brains of control and schizophrenic patients. Crosses are controls, while circles are patients diagnosed with schizophrenia

```
>ei1<-eigen(var(control))
>ei2<-eigen(var(schizop))
>unlist(coli(ei2$vectors[,1], ei1$vectors[,1]))
    z.stat          p          angle
-0.0986313  0.6380000  1.6695882
>unlist(coli(ei2$vectors[,2], ei1$vectors[,2]))
    z.stat          p          angle
0.4320158 0.0180000 1.1240696
>unlist(coli(ei2$vectors[,1], ei1$vectors[,2]))
    z.stat          p          angle
-0.3038691  0.1420000  1.8795475
>unlist(coli(ei2$vectors[,2], ei1$vectors[,1]))
    z.stat          p          angle
 0.3999028  0.0340000  1.9822071
```

While the first eigenvectors do not share the same direction, the second eigenvectors have significantly similar direction. We notice that the second eigenvector of the schizophrenic patient is collinear to the first of the control group as well. We could have more qualitatively depicted the morphological meaning for each eigenvector by providing deformation grids or vectors associated with changes on each PC for each group. For this aim, we would have to estimate the mean shape, as well as scores of the PC.

The last possible question concerns whether geometry between variable ordinations is similar between both groups. One can answer using a Procrustes test. Since eigenvectors are of unit size, we scale them by the standard variation present on each axis (square root of the eigenvalue). We can remove the four last eigenvectors since the number of dimensions is reduced during the Procrustes superimposition.

```
conf1<-(ei1$vectors[,1:22])%*%(sqrt(diag(ei1$values[1:22])))
conf2<-(ei2$vectors[,1:22])%*%(sqrt(diag(ei2$values[1:22])))
protest(conf1, conf2)
Call:
protest(X = conf1, Y = conf2)

Correlation in a symmetric Procrustes rotation:  0.6493
Significance:  < 0.001
Based on 1000 permutations.
```

The relationships between variables are similar between both groups. We conclude that the schizophrenic and control patients do not differ in the covariation of internal brain structure described by the landmarks selected in this study.

Not only can we analyze the covariation patterns between groups, but we can also examine covariation between parts of a structure, or between one morphology and a set of predictor variables. For this latter task, we can follow the two-block partial least-squares approach defined in Section 3.4.4. The computation is similar with that for multivariate morphometrics and one can use our `pls` function. Moreover, as for PCA or discriminant analysis, it is possible to reconstruct shapes along the left and right singular vectors to interpret patterns of shape covariation. Since the method is quite trivial, this task appears in the exercise section.

6.6 Analyzing Developmental Patterns with Modern Morphometrics

6.6.1 Allometry

Analyzing growth patterns with modern morphometrics is not so different from traditional morphometrics. The advantage is that one can depict allometric changes in terms of configuration change more easily than with analyzing a collection of linear relationships between measurements. One must carefully work with degrees of freedom (or shape space dimension): observations as measured by modern morphometrics are often interdependent (because of superimposition or transformation of original coordinates), and one must choose an appropriate methodology. Monteiro [74] has proposed a multivariate regression model for searching causal factors in the analysis of shape variation using morphometrics. Shape is considered to be the dependent variable, while size or age (or any other vector) is considered to be the independent variable.

Since one can adopt traditional statistical methods for Fourier analysis and outline analysis, we will rather regard the method in detail for landmarks Procrustes analysis. The dimensions of the space covered by superimposed coordinates after translation, scaling and rotation are reduced by 4 degrees of freedom for 2D data and by 7 for 3D data. One can directly work on the first principal components defined on the variation of landmark displacements between individuals, or on Kendall coordinates that allows one to respect the geometry of the shape space; however, one can also work on superimposed coordinates modifying multivariate tests. Let $\mathbf{W}$ be the matrix of size n by kp containing superimposed and projected shape coordinates (projected onto the tangent Euclidean space). One can estimate the regression coefficient matrix ($\mathbf{B}$) by using the standard formula

$$\mathbf{B} = (X'X)^{-1}X'\mathbf{W} ,$$

where X corresponds to the centered matrix of independent variables.

The `lm` function can accept a multivariate dependent variable set, and can return coefficients, fitted values and residuals.

We will work on the famous rat calvarial dataset, which can be found on the Internet.[3] This dataset has been extensively worked on in the orange book [10]. It corresponds to 8 landmarks digitized in two dimensions on skull section of 21 rats, and these have been measured at regular ages of 7, 14, 21, 30, 40, 60, 90, and 150 days, yielding 168 observations. For some skulls, some landmarks were not recorded, so these observations will be excluded from the dataset. The final set corresponds to 164 observations. This dataset is ideal for estimating growth patterns since it contains the age; in addition, we can estimate the size as the centroid size from the landmark coordinates for each configuration. We first have to load the dataset on R. If the computer is directly connected to the web, we directly catch the data using `scan` or `read.table` by writing the web path to the file. Missing values have been recorded as 9999 in the original file. Each line contains the $x_1, y_1, x_2, y_2, x_3, y_3 \ldots$ coordinates. The first two columns correspond to the individual number and to the age of the rat in days. We must slightly adapt the format in R for our function to work correctly.

```
>rats<-read.table("http://life.bio.sunysb.edu/morph/data/
+        Book-VilmannRat.txt", skip = 2)
>rats<-as.matrix(rats)
>missing<-NA
>for (i in 1:168) {if (any(rats[i,]==9999)){missing[i]<-i}}
>missing<-as.numeric(na.omit(missing))
>rats<-rats[-missing,]
>RAT<-array(t(rats[,c((1:8*2+1),(1:8*2+2))]),dim=c(8,2,164))
```

Superimpose all configurations of the dataset.

```
>ratp<-pgpa(RAT)
>RAP<-ratp$rotated
>n<-dim(RAP)[3]; k<-dim(RAP)[2]; p<-dim(RAP)[1]
>rap<-t(matrix(RAP[,,],16,164))
```

Extract centroid size from the `ratp` *object.*

---

[3] `http://life.bio.sunysb.edu/morpho/data/Book-VilmannRat.txt`

```
>RAS<-ratp$cent.siz
>RAA<-rats[,2]
>RAI<-as.factor(rats[,1])
```

We have now a `rap` matrix containing shape variables, a `RAS` vector containing sizes, a `RAA` vector containing ages, and a `RAI` factor containing individual categories. First we will look whether size and shape are related. The `RAI` factor is a random variable, while size is a fixed independent variable. To determine whether allometry is present, we must test whether size is related to shape variables. Individual variation around the general allometric relationship corresponds to the interaction between the size and individual factors. Also, we could have first filtered by the effect individuals. We will only work on the raw relationships between shape and size, and without considering the problems linked to longitudinal data. Goodall [38] presented a statistical framework for analyzing Procrustes shape data, and he developed a possible F-test. We assess the significance of the regression model using the Goodall F-test which is a simple ratio of diagonal variances:

$$F_s = \frac{\sum d^2_{\bar{X}_i X_m}/q}{\sum d^2_{X_i \bar{X}_i}/(n-q-1)} \ .$$

This test is based on the Procrustes chord distance (see [38]), and should work depending on the assumption that variation is isotropic, and that variation is equal for each landmark. The numerator is the sum of squared Procrustes chord distances between estimated shapes and the mean shape of the sample, while the denominator is the sum of Procrustes distances between each X_i shape and its $\bar{X}_i$ estimate. These distances are the metric of the Kendall shape space, and since they are sum of squares, one can use them for measuring shape variance. In addition, these can be calculated as the trace of the sum of squares and cross-products of effects and residuals divided by the appropriate degrees of freedom. When variation is small, one can approximate mean-shapes with the averages of configurations coordinates, and one can directly use the trace of sum-of-squares and cross-products matrix. One can compare the F_s-value with an F-distribution with qm, and $(n - q - 1)m$ degrees of freedom, m being the number of shape space dimensions. One can use the ratio between variation explained by the model and total variation as a percentage of shape variation explained by a given factor (here the shape). One can visualize the effect of size with the estimated shapes for minimal and maximal values of size (see Fig. 6.17).

```
>mod1<-lm(rap~RAS)
>RAM<-apply(rap,2,mean)
```

Alternatively RAM<-as.vector(ratp$mshape).

```
>M<-rep(1,164)%*%t(RAM)
>num<-sum(apply((mod1$fitted.values-M)^2,1,sum))/1
>den<-sum(apply((rap-mod1$fitted.values)^2,1,sum))/
+        (n - 1 -1)
>Fs<-num/den
```

```
[1] 512.7921
>1-pf(Fs, 1 * k *p-4, (n-1-1)*(k*p-4))
[1] 0
>sum(diag(var(mod1$fitted.values)))/sum(diag(var(rap)))
[1] 0.7596167
```

Estimate configurations that are associated with minimal and maximal centroid sizes.

```
>fit.val<-mod1$fitted.values
>Ma<-matrix(fit.val[which(RAS==(max(RAS))),],p,k)
>mi<-matrix(fit.val[which(RAS==(min(RAS))),],p,k)
>par(mar=c(1,1,1,1))
>plot(rbind(Ma,mi),asp=1,xlab="",ylab="",axes=F,frame=F)
>polygon(Ma)
>polygon(mi, border="grey55",lwd=2)
```

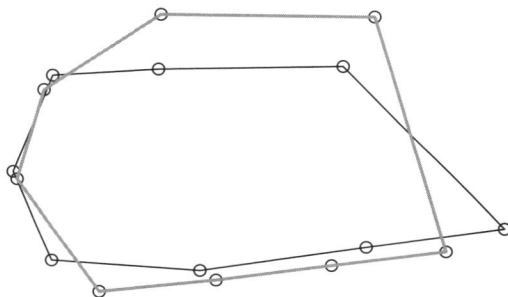

Fig. 6.17. Plot of the shape explained by size from the Villman calvarial growth dataset. The thick gray configuration corresponds to the estimated shape for the smaller individual, while the black corresponds to the estimated shape for the larger individual

The adjusted model is very significant, meaning that there is allometry in the dataset. However, we have not applied the appropriate model since the observations are not independent but partly longitudinal – the same individuals being measured several times during its growth. This model is also imperfect because the Goodall F-statistic assumes that variation is isotropic (each landmark having the same variance). If one wants to consider both the longitudinal nature of the data and the fact that variation is not isotropic, one should apply a multivariate test, such as the Hotelling-Lawley multivariate test. Since the independent variable is not a factor but a continuous variable, the degree of freedom is 1 for the effect, and is $n-2$ for the error term, n being the number of observations.

```
>mod<-lm(rap~RAS)
>Sef<-var(mod$fitted.values)*163
>Ser<-var(mod$residuals)*163
```

```
>ddef<-1
>dder<-162
>Hotellingsp(Sef, Ser, ddef, dder)
 dfeffect    dferror          T2   Approx_F
  1.00000 162.00000  17.98975 207.57405


       df1        df2          p
 13.00000 150.00000    0.00000
```

Notice that we can reduce the number of shape space dimensions by applying a principal component analysis to the original data. The studied space contains 13 dimensions; therefore, we take the first 13 columns of PC scores to perform the same test.

```
>anova(lm(prcomp(rap)$x[,1:13]~RAS))
Analysis of Variance Table

             Df    Pillai   approx F
(Intercept)   1 2.634e-29 3.039e-28
RAS           1     0.947    207.574
Residuals   162

num Df den Df Pr(>F)
    13    150      1
    13    150 <2e-16 ***
---
Signif. codes:  0 *** 0.001 ** 0.01 * 0.05 . 0.1   1
```

Since we have longitudinal replicated data, we should test the effect of size against the interaction between size and individual, rather than onto the residual variation. One can write the model of morphological growth as:

$$Y_{ij} = \mu_Y + \alpha_i + \beta(x_{ij} - \mu_{x_i}) + \gamma_i(x_{ij} - \mu_{x_i}) + \epsilon_{ij} ,$$

where Y_{ij} is the shape of one individual i at the time j; μ_Y the mean shape of the sample; α_i the shape difference between the mean shape of the sample and the mean shape for the ith individual; x_{ij} the size of one observation, x_i the averaged size for a given individual; β, the estimate of the mean growth slope; and γ_i, the estimate of the individual variation in growth. The morphology is thus equated with an individual effect, a size effect, an interaction effect (individual variation of growth pattern) and an error term. In the case of our example, we can compare the effect of size with the interaction between size and individual. The appropriate term for testing the significance of the allometric relationship should be the interaction between size and individual rather than the residual variation. Our design is not completely balanced, so effects will not be orthogonal. We can use type II mean squares estimation. In order to produce sum of squares and cross-products, the individual effect must first first estimated, while general growth must be placed as second term in the model.

```
>jj<-(lm(prcomp(rap)$x[,1:13]~RAI*RAS))
>SSef<-crossprod(jj$effects[jj$assign==2,])
>dfef<-length(which(jj$assign==2))
>SSre<-crossprod(jj$effects[jj$assign==3,])
>dfre<- length(which(jj$assign==3))
>Hotellingsp(SSef, SSre, dfef, dfre)
    dfeffect        dferror                T2      Approx_F
1.000000e+00 2.000000e+01 1.349225e+02 8.302924e+01
         df1            df2                 p
1.300000e+01 8.000000e+00 4.652088e-07
```

As in the previous tests, allometry significantly explains shape variation.

If we want to work on residual shape variation not explained by size (in other words, independently of growth), we can use the residual variation and eventually perform principal component analysis and other statistical tests on it.

6.6.2 Developmental Stability

Fluctuating asymmetry is often considered to be a measure of developmental instability in organisms [78]. One can measure fluctuating asymmetry from quantitative characters according to a variety of methods. The most commonly used method is the FA10 index based on a two-way ANOVA of continuous characters [78]. Measure of fluctuating asymmetry for complex traits evaluated by geometric morphometrics have been more recently developed in the papers of Auffray et al. [4], Mardia et al. [71], and Klingenberg et al. [55]. Mardia et al. [71] consider two kinds of bilateral symmetric structures: matching symmetry where two separate copies are mirror images of each other, and object symmetry where the structure itself is symmetric and contains the axes of symmetry.

One must reflect sides onto each other for them to be comparable in geometric terms. Strategy for transforming the datasets involves a preliminary reflection of one side for matching symmetry, while it introduces the generation of one reflected copy of the original object for object symmetry.

The reflection for matching symmetry is easily obtained by multiplying one of the coordinates (x or y or z) by -1. Once done, the set of *side1* and reflected *side2* is superimposed according to a Procrustes registration.

Producing a reflected copy in object symmetry involves two steps: First reverse the sign of one of the coordinates by -1, and second, relabel each paired landmark by its symmetric counterpart. Then the set of original objects and their reflected and re-labeled copy are superimposed according to a Procrustes registration.

In the model of Palmer [78], the significance of fluctuating asymmetry is tested against measurement error. One must measure the same structure several times in different sessions to estimate measurement error. Whether one works with object symmetry or matching symmetry has some influence since it affects degrees of freedom for each component of the two-way ANOVA.

The first available test for estimating the significance of fluctuating asymmetry and other components consists of what is called a "Procrustes two-way ANOVA" on the superimposed configurations (which coordinates have been projected onto the tangent shape space). The variance components included in this ANOVA correspond to the sum of the diagonal elements of the mean squares for each effect. This strategy ignores whether there are changes in the covariance structure between effects.

If variation is isotropic (the same for each landmark), the assumptions of the model are fullfilled and one can work with the Goodall F-statistic (see Section 6.6.1). The degrees of freedom are given in Table 6.1.

Table 6.1. Degrees of freedom for Procrustes symmetry and asymmetry studies as adapted from the two-way analysis of variance for fluctuating asymmetry [78]. n: number of individuals, p: number of paired landmarks, u: number of single landmarks (on the median symmetric plane), r: number of replicated measurements

2D		
Effects	Matching Symmetry	Object symmetry
Individuals	$(n-1) \times (2p-4)$	$(n-1) \times (2p+u-2)$
Sides	$2p-4$	$2p+u-2$
Individuals $\times$ sides	$(n-1) \times (2p-4)$	$(n-1) \times (2p+u-2)$
Error	$(r-1) \times 2n \times (2p-4)$	$(r-1) \times 2n \times (2p+u-2)$

3D		
Effects	Matching Symmetry	Object Symmetry
Individuals	$(n-1) \times (3p-7)$	$(n-1) \times (3p+2u-4)$
Sides	$3p-7$	$3p+u-3$
Individuals $\times$ sides	$(n-1) \times (3p-7)$	$(n-1) \times (3p+u-3)$
Error	$(r-1) \times 2n \times (3p-7)$	$(r-1) \times n \times (6p+3u-7)$

Under anisotropic conditions, one must use a more general multivariate test that can be an adaptation of the Hotelling-Lawley test for x groups with appropriate degrees of freedom (see Sections 6.2 and 3.4.3).

We will evaluate and test asymmetries in a dataset composed of 52 pairs of left and right jaws of *Mus musculus domesticus*.[4] The dataset is organized in a data frame: the first three columns respectively correspond to the individual label, side and session; the other columns are landmark location organized as $x_1, y_1, x_2, y_2 \ldots x_{16}, y_{16}$. We must first reflect the left side by multiplying one coordinate by -1, superimpose the configurations, and estimate variation components. We have digitized individuals twice to allow measurement error to be estimated. Superimposed configurations are illustrated in Fig. 6.18.

```
>DDO<-read.table("/home/juju/morph/Mouse.txt",header=T)
>Ddo<-as.matrix(DDO[,3+c(1:16*2-1,1:16*2 )],208,32)
>ddo<-array(t(Ddo),dim=c(16,2,208))
>ind<-as.factor(DDO[,1])
```

---

[4] The dataset is available in the online supplement.

```
>side<-as.factor(DDO[,2])
>for (i in 1:104) {ddo[,1,i*2-1]<- - ddo[,1,i*2-1]}
>ddosup<-orp(pgpa(ddo)$rotated)
>par(mar=c(2.5, 2.5, 1.5,1.5))
>plot(ddosup[,1,], ddosup[,2,], asp=1,xlab="X", ylab="Y",
+       axes=F, frame=F)
>DDOsup<-t(matrix(ddosup,32,208))
```

Fig. 6.18. Superimposition of the right and reflected left jaw configurations for the mouse dataset, replicated digitization included

```
>mod<-lm(DDOsup ~ ind * side)
>effects<-mod$effects
>SSind<-crossprod(effects[which(mod$assign==1),,drop=F])
>SSside<-crossprod(effects[which(mod$assign==2),,drop=F])
>SSinter<-crossprod(effects[which(mod$assign==3),,drop=F])
>SSres<-var(mod$residuals)*207
```

Note the special indexation with `drop=F`. This argument allows matrices of one row not to be coerced into a `vector` object.

The material has been digitized in 2D, and there is a total of 52 individuals, two replicates and two sides. We can consider jaws as belonging to the matching symmetry class. The degrees of freedom are then 1428 for individual, 28 for side, 1428 for interaction, and 2912 for error effects. We simply divide the trace of each sum-of-squares and cross-products matrices by their degrees of freedom to calculate the Procrustes mean squares.

```
>MSind<-sum(diag(SSind))/1428
>MSside<-sum(diag(SSside))/28
>MSinter<-sum(diag(SSinter))/1428
>MSres<-sum(diag(SSres))/2912
```

Individual and side effects are tested against interaction, and fluctuating asymmetry (interaction) is tested against error measurement.

```
>MSinter/MSres
[1] 2.620457
>MSind/MSinter
[1] 4.121681
>MSside/MSinter
[1] 10.74571
>1-pf(MSind/MSinter, 1428,1428)
[1] 0
>1-pf(MSside/MSinter, 28,1428)
[1] 0
>1-pf(MSinter/MSres, 1428,2912)
[1] 0
```

We can remove the residual variation from the fluctuating asymmetry term to obtain an index of fluctuating asymmetry. This is the Procrustean equivalent of the FA10 index used in fluctuating asymmetry studies. One can compare this level of shape asymmetry between different populations for further investigation.

One can use a more general multivariate model for comparing variance terms if one thinks that variation is anisotropic and of different levels for each landmark. For this task, we use the transformed Hotelling-Lawley trace test that we have developed for for Procrustes data (Section 6.2).

```
>Hotellingsp(SSinter, SSres, 51, 104)
    dfeffect        dferror              T2      Approx_F
   51.000000   104.000000    80.528171       4.233449
         df1            df2             p
 1428.000000   2102.000000    0.000000

>Hotellingsp(SSside, SSinter, 1, (52-1))
    dfeffect        dferror              T2      Approx_F
 1.000000e+00 5.100000e+01 1.248993e+01 1.070566e+01
         df1            df2             p
 2.800000e+01 2.400000e+01 5.712875e-08

>Hotellingsp(SSind, SSinter, 1, (52-1))
 dfeffect   dferror        T2 Approx_F
   1.0000   51.0000 215.4229 184.6482
        df1       df2         p
   28.0000   24.0000   0.0000
```

Since both tests (under isotropic and under anisotropic conditions) give similar results, we conclude that fluctuating asymmetry, directional asymmetry and individual variation are significant effects of bilateral variation in the mouse jaw sample.

The Procrustes test in the case of object symmetry is less obvious because shape dimensions of effects and error are not necessarily equal and orthogonal.

Klingenberg et al. [55] introduce a permutation approach for estimating the significance of the multivariate statistic.

6.6.3 Developmental Integration

In Chapter 3, we have explained how to analyze covariation patterns between two sets of variables. Klingenberg et al. [57] have used this approach for investigating the relationships between two sets of landmark measurements on configurations based on the fluctuating asymmetry and individual variance-covariance matrices. One can estimate the position of developmental modules (within which relationships between characters should be high, compared with covariation between modules) using different partitions of landmark sets. These sets should likewise have geometric continuity. In other words, a module should be recognizable as a continuous region of the complex structure under investigation. The rodent jaw has been hypothesized to be structured in two developmental and functional modules: the ascending ramus region and the alveolar region where the teeth are implanted. We can investigate different partitions of contiguous landmark sets to find which minimizes covariation between modules. As an example, we will use the previous mouse jaw dataset (Fig. 6.19). One can use the Rv coefficient as defined in Section 3.4.3 as a measure of association between different landmark sets. The Rv function works on the total variance-covariance with the indices of a first block of variable to calculate the Rv coefficient in a quicker way that the function pls.

Function 6.8. Rv

Arguments:
  VCV: Variance-covariance matrix.
  index1: Indices of variables included in the first block of the partition.
Value:
  The function returns the Rv coefficient.

```
1  Rv<-function(VCV, index1)
2  {VCV1<-VCV[index1,index1]
3   VCV2<-VCV[-index1,-index1]
4   VCV12<-VCV[index1,-index1]
5   VCV21<-VCV[-index1,index1]
6   sum(diag(VCV12%*%VCV21))/sqrt(sum(diag(VCV1%*%VCV1))
7       *sum(diag(VCV2%*%VCV2)))}
```

We will examine all contiguous partitions containing equal numbers of landmarks (8 and 8). Since our landmarks have been digitized along the mandible outline, the definition of partitions is not difficult using progressive indexing on matrix columns. First, we define all the different possible partitions using the landmark indices contained in the first submatrix.

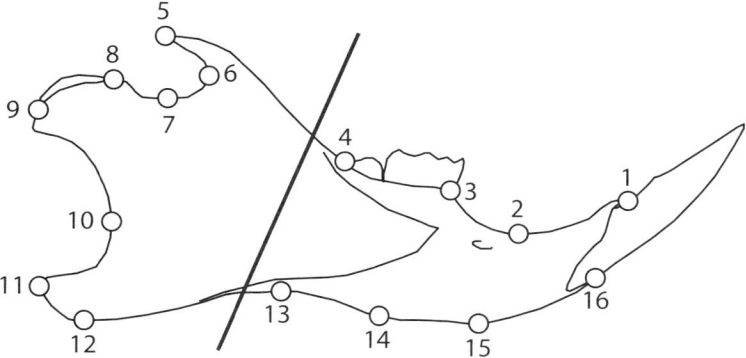

Fig. 6.19. Landmark labels for the mouse jaw dataset. The oblique black lines separate two partitions of landmarks that are assumed to correspond to different developmental and functional modules

```
>part1<-matrix(NA,16,8)
>for (i in 1:16)
+        {part1[i,]<-c(1:16,1:8)[i:(i+7)]}
```

The `part1` *matrix represents the 16 alternative partitions for which the Rv coefficient will be estimated for both interindividual and intra-individual covariance matrices. We can use the* `pls` *function (programmed in Section 3.4.2) to compute the Rv coefficient or the small function that we have written above. The Rv coefficients are stored in one vector later bound with the partition matrix.*

```
>Rvind<-Rvint<-numeric(16)
>for (i in 1:16)
+        {Rvind[i]<-Rv(SSind/51, part1[i,])
+         Rvint[i]<-Rv(SSinter/51, part1[i,])}
>data.frame(part1, Rvind, Rvint)
   X1 X2 X3 X4 X5 X6 X7 X8        Rvind         Rvint
1   1  2  3  4  5  6  7  8 0.4035437 0.2681725
2   2  3  4  5  6  7  8  9 0.3537373 0.2317912
3   3  4  5  6  7  8  9 10 0.3531442 0.2249909
4   4  5  6  7  8  9 10 11 0.3881753 0.2188373
5   5  6  7  8  9 10 11 12 0.3531832 0.1803118
6   6  7  8  9 10 11 12 13 0.3855864 0.2039900
7   7  8  9 10 11 12 13 14 0.3675017 0.2160983
8   8  9 10 11 12 13 14 15 0.3636016 0.1970535
9   9 10 11 12 13 14 15 16 0.4244084 0.3142359
10 10 11 12 13 14 15 16  1 0.4068938 0.3278042
11 11 12 13 14 15 16  1  2 0.3945449 0.3279134
12 12 13 14 15 16  1  2  3 0.4288943 0.3288309
13 13 14 15 16  1  2  3  4 0.3436903 0.2517959
14 14 15 16  1  2  3  4  5 0.3968735 0.3537576
15 15 16  1  2  3  4  5  6 0.3651967 0.3844970
16 16  1  2  3  4  5  6  7 0.4031166 0.3732866
```

The different Rv indices obtained for individual and fluctuating asymmetry variation indicate a lower association, respectively, for the third and the fifth partitions (Fig. 6.20). However, for individual variation the Rv coefficient obtained using the fifth partition is very close to the third. It seems that the module hypothesis as defined in Fig. 6.19 is the one that minimizes the covariation between sets of landmarks.

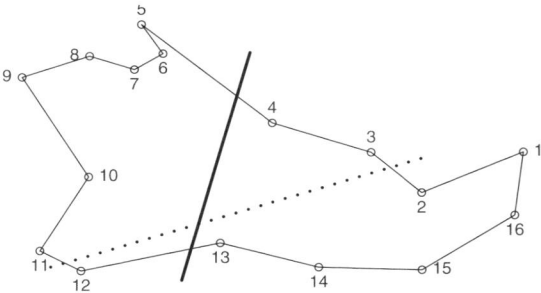

Fig. 6.20. First configuration of the mouse jaw dataset. The black line separates two set of landmarks between which the covariation is minimal considering the fluctuating asymmetry variance-covariance matrix. The dotted line identifies sets that minimize interindividual co-variation between parts. If we first filter allometry from shape variation, using both variance-covariance matrices, we find that the best partition for two modules corresponds to the black line

One can estimate the significance and the interval of confidence of the Rv coefficient by permuting rows of the matrix of original coordinates. In order to take into account the specificity of Procrustes data, Klingenberg et al. [57] have performed a resuperimposition of the data before recomputing simulated coefficients of association.

We can perform the same analysis by first filtering shape explained by growth (allometric variation). We filter allometric variation by multivariate regression, and then we work on residuals. The following calculations are the same as before.

```
>size<-pgpa(ddo)$cent.siz
>ddor<-lm(DDOsup~size)$residuals
>mod2<-lm(ddor ~ ind * side)
>SSind<-crossprod(mod2$effects[which(mod2$assign==1)
+        ,,drop=F])
>SSinter<-crossprod(mod2$effects[which(mod2$assign==3)
+        ,,drop=F])
>for (i in 1:16)
+    {Rvind[i]<-Rv(SSind/51, part1[i,])
+     Rvint[i]<-Rv(SSinter/51, part1[i,])}
>data.frame(part1, Rvind, Rvint)
```

	X1	X2	X3	X4	X5	X6	X7	X8	Rvind	Rvint
1	1	2	3	4	5	6	7	8	0.4192148	0.2730352
2	2	3	4	5	6	7	8	9	0.3607052	0.2376113
3	3	4	5	6	7	8	9	10	0.3573718	0.2333540
4	4	5	6	7	8	9	10	11	0.3570127	0.2264582
5	5	6	7	8	9	10	11	12	0.3465407	0.1862966
6	6	7	8	9	10	11	12	13	0.3766381	0.2045771
7	7	8	9	10	11	12	13	14	0.3832399	0.2196039
8	8	9	10	11	12	13	14	15	0.3826982	0.2037115
9	9	10	11	12	13	14	15	16	0.4844754	0.3204999
10	10	11	12	13	14	15	16	1	0.4629691	0.3325938
11	11	12	13	14	15	16	1	2	0.4508821	0.3349671
12	12	13	14	15	16	1	2	3	0.4399647	0.3351378
13	13	14	15	16	1	2	3	4	0.3355519	0.2601936
14	14	15	16	1	2	3	4	5	0.3714398	0.3581845
15	15	16	1	2	3	4	5	6	0.3824649	0.3895218
16	16	1	2	3	4	5	6	7	0.4162497	0.3808510

Here the fifth partition provides the lowest correlations between modules for both the fluctuating assymetry and the individual variations. This result is similar to that found in Klingenberg et al. [57].

Removing allometry has the advantage of eliminating the part of variation explained by common growth for both modules.

Problems

6.1. Significance of the Rv coefficient
Write a function that makes a permutation test for estimating the significance of the Rv coefficient.

6.2. Estimate PLS morphological meaning
Using the singular vectors of the partial two-block least-squares performed on mice jaws, provide an interpretative sketch of how the different modules covary using the partition that minimizes their covariation. For this goal, use the singular-value decomposition of the appropriate block.

7

Going Further with R

Chapter coauthored by: **Emmanuel Paradis and Julien Claude**

Rather than only using the available packages, and other sources of information, we can learn more, studying at the edge of shape statistics by observing the behaviour of tests. In addition, for gaining time in computation, we can interface R with other software and languages.

7.1 Simulations

Simulations are not commonly used in modern morphometrics, but they allow the behaviour of tests to be assessed. Moreover, one can use simulations for nonparametric testing in morphometrics.

One can perform a variety of simulations under R by using random sequences (Section 1.4.5) combined with loops and logical indexings. Since configurations are often depicted by several variables, we must generate random data following multivariate distributions. The mvtnorm package is well designed for the study of multinormal datasets. The `rmvt` function generates random multivariate distributions following a variance-covariance matrix passed as a `sigma` argument.

We can generate sets of configurations with raw coordinates, Bookstein coordinates, or Kendall coordinates drawn from a given multivariate distribution, and then look at the behaviour of Procrustes superimposition. We can define a mean shape of coordinates ($x_1 = 0$, $y_1 = 0$, $x_2 = 1$, $y_2 = 0$, $x_3 = 1$, $y_3 = 2$) around which there will be random deviation. As a first model, we simulate isotropic variation around these mean coordinates. By keeping two coordinates constant, and by generating variation onto the third coordinate, we simulate a variation that mimics the variation of Bookstein coordinates. For these simulations, we need to generate a matrix of random multinormal observations for each coordinate that we want to vary. For isotropic variation, we must specify the `sigma` argument of the `rmvt` function as a diagonal matrix with equal diagonal elements. The range of simulated variation depends on the value of the diagonal elements.

```
>layout(matrix(1:4, 2,2))
>par(mar=c(1,1,3,1))
```

Define the mean shape.

```
>Ms<-matrix(c(0,1,2,0,0,2),3,2)
```

Simulate random variation around each landmark of the mean shape.

```
>simu1<-rmvt(100,diag(0.05,6),df=99)[,c(1:3*2-1,1:3*2)]
>Simu1<-as.vector(Ms)+t(simu1)
>A1<-array(Simu1, dim=c(3,2,100))
```

Plot the simulated points.

```
>plot(A1[,1,],A1[,2,],asp=1,pch=1:3,axes=F,frame=F,
+   main="Simulation on 3 coordinates")
```

Simulate variation for the third landmark, mimicking multinormal covariation of Bookstein coordinates.

```
>simu<-rmvt(100,diag(0.05,2),df=99)
>simu<-cbind(0,0,simu[,1],0,0, simu[,2])
>Simu<-as.vector(Ms)+t(simu)
>A<-array(Simu, dim=c(3,2,100))
```

Plot the points of the second simulation.

```
>plot(A[,1,],A[,2,],asp=1,pch=1:3,axes=F,frame=F,
+   main="Simulation on 1 coordinate")
```

Observe the behaviour of Procrustes superimposition with both simulations of isotropic variation.

```
>pA1<-pgpa(A1)$rotated
>pA<-pgpa(A)$rotated
```

Plot the superimposed coordinates for each simulated dataset.

```
>plot(pA1[,1,],pA1[,2,],asp=1,pch=1:3,axes=F,frame=F,
+    main="Procrustes 3 sim coord")
>plot(pA[,1,],pA[,2,],asp=1,pch=1:3,axes=F,frame=F,
+    main="Procrustes 1 sim coord")
>mpA<-t(matrix(pA,6,100))
>mpA1<-t(matrix(pA1,6,100))
```

Observe the covariance matrices estimated from superimposed coordinates.

```
>var(mpA)
>var(mpA1)
```

We can observe that the superimposition procedure introduces biases that directly depend on the configuration. These biases will manifest themselves in the variance and covariance based on coordinates of superimposed configurations: the estimated variance-covariance will no more follow the initial assumptions of variation. Although we have simulated random isotropic variation, the effect of Procrustes superimposition reassigned different variances to the points according to their position

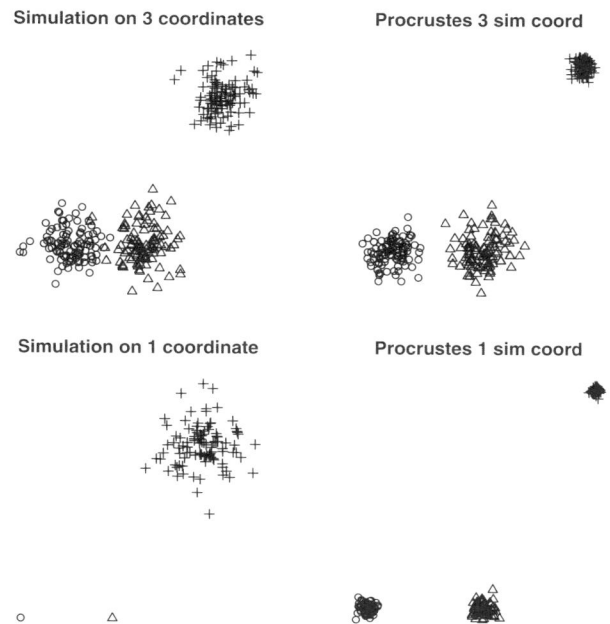

Fig. 7.1. Simulated isotropic variation of two kinds around a mean shape of coordinates ($x_1 = 0$, $y_1 = 0$, $x_2 = 1$, $y_2 = 1$, $x_3 = 2$, $y_3 = 1$). Up, variation is simulated around each landmark; while bottom, landmark variation is simulated only for one landmark mimicking behaviour of Bookstein coordinates: the two others are considered as the baseline

in the configuration. In this example, landmarks that are distant from the centroid concentrate less variation than landmarks that are close to the centroid. Landmarks close to the centroid absorb the simulated variation (Fig. 7.1).

One can simulate more complex variation by entering a symmetric matrix of variance-covariance for specifying the `sigma` argument of `rmvt`.

If Procrustes superimpositions introduce biases in estimating variance-covariance matrices and in multinormality in shape parameters, we must interpret standard multivariate tests with caution. Nonparametric distributions for the observed statistic may provide local solutions. Indeed, it is possible to obtain these distributions taking into account the effect of variable transformation (e.g., like Procrustes registration). Permutations of coordinates across observations can be a convenient way to simulate a null distribution of the statistic under interest. After the permutation, we should add the superimposition step to produce the new estimated and simulated statistics.

In Chapter 6, we used the Mantel test for investigating whether patterns of shape covariation between schizophrenic and control patients were similar. For running this test, we used the variance-covariance matrices estimated from superimposed coordinates in both groups. The permutation was concerned with the rows and columns of the variance-covariance matrix, but did take into account the superimposition procedure. We must develop an alternative test.

In the hypothesis of no common variation, one can permute landmarks among individuals in one of the groups to produce a covariance matrix whose structure is independent on the other covariance matrix. We can simulate the statistic considering the absence of similarity between patterns of covariation and compare it with an observed statistic (the linear correlation between lower triangle matrices elements). If the observed statistic is above the upper tail of the random distribution, one can conclude for similarity between matrices of covariance. The alternative hypothesis corresponds then to the similarity between matrices. After the permutation, we add a superimposition step, and generate the statistic that estimates the correlation between the variance-covariance matrix of one group and the simulated random variance-covariance matrix of the other group. We could write this new test in a new function; however, we will detail the procedure step-by-step so you can generate random variation for any other test. Notice that, like in the traditional Mantel test, only one set is resampled. Resampling the other set would not correspond to the hypothesis of independence, and it is even possible that most of the simulated statistics will be greater than the observed statistic (because Procrustes superimposition introduce covariation between new variables). We first write two functions to compute our statistic. The first extracts the lower triangle of the covariance matrix, while the second computes the linear correlation between elements of the lower triangle of both matrices.

```
>lcwertriang<-function (m)
+    {d<- dim(m)
+      m[col(m) <= row(m)]}
+ mstat<-function (m1, m2)
+ {cor(lower.triang(m1), lower.triang(m2))}
```

Then we compute the observed statistic.

```
>library(shapes)
>schizo<-orp(pgpa(schizophrenia.dat)$rotated)
>vcontrol<-var(t(matrix(schizo[,,1:14],26,14)))
>vschizo<-var(t(matrix(schizo[,,15:28],26,14)))
>stat.obs<-mstat(vcontrol,vschizo)
>stat.obs
[1] 0.5150001
```

We declare the stat.dis *vector to store simulated statistics under the null hypothesis. A loop starts for reiterating the permutation 500 times.*

```
>stat.dis<-numeric(500)
>schizosim<-controlsim<-array(NA, dim=c(13,2,14))
>for (j in 1:500){
```

In a second loop, landmarks are resampled among configurations of the first set.

```
+    for (i in 1:13)
+    {controlsim[i,,]<-schizo[i,,sample(1:14)]}
+    simu.dat<-array(c(schizo[,,15:28],controlsim),
```

```
+            dim=c(13,2,28))
+     simu<-orp( pgpa(simu.dat)$rotated )
+     vcontrols<-var( t(matrix(simu[,,1:14],26,14)) )
+     vschizos<-var( t(matrix(simu[,,15:28],26,14)) )
+     stat.dis[j]<-mstat(vcontrols,vschizos)}
```

After the permutation, we calculate the statistic between the new resampled set and the second variance-covariance matrix.

```
>sum(stat.dis>stat.obs)/500
[1] 0.042
>plot(1,type="n", xlim=c(0.2,0.7),ylim=c(0,14 ), main
    ="null density distributions", xlab="zstat")
>points(density(stat.dis), type="l", lwd=2)
>abline(v=stat.obs)
```

Since the observed value is in the upper tail of the distribution, we conclude for similarity between variance-covariance matrices in both the schizophrenic and control groups. We can visualize the position of the observed statistic and the density of the distribution under the null hypothesis with a simple graph (Fig. 7.2).

null density distributions

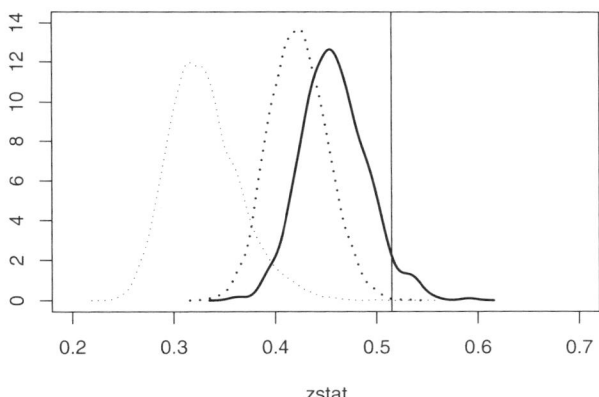

Fig. 7.2. Distribution density of the statistic measuring relationships between covariance matrices under the null hypothesis. The vertical line indicates the observed value. The thick full curve corresponds to the null distribution obtained by permuting landmarks across one set and resuperimposing configurations, the thick dotted curves to the null distribution obtained by permuting landmarks across one set but without resuperimposing configurations, and the thin dotted curve to the null density distribution of the Mantel test statistic as performed in Chapter 6

We can compare the null distribution that we have estimated with the one that has been estimated with the Mantel test as developed in Chapter 6. We can also compare these tests with a test that ignores the resuperimposition step.

We reiterate nearly the same permutation but without considering resuperimposition.

```
>stat.dis1<-numeric(5000)
>schizosim<-controlsim<-array(NA, dim=c(13,2,14))
>for (j in 1:5000){
+    for (i in 1:13){
+       controlsim[i,,]<-schizo[i,,sample(1:14)]}
+    simu.dat<-array(c(schizo[,,15:28], controlsim)
+              , dim=c(13,2,28))
+    simu<-simu.dat
+    vcontrols<-var( t(matrix(simu[,,1:14],26,14)) )
+    vschizos<-var( t(matrix(simu[,,15:28],26,14)) )
+    stat.dis[j]<-mstat(vcontrols,vschizos)}
>sum(stat.dis1>stat.obs)/5000
[1] 8e-04
>points(density(stat.dis1), type="l",lty=3, lw=2)
```

Our code for the permutation uses functions that were defined in Section 6.6. We change parts of the code in the Mantel test dealing with the null distribution. We name the new `mantel.t3` function.

Function 7.1. `mantel.t3`

Arguments:
> `m1`: *Square variance-covariance matrix.*
> `m2`: *Square variance-covariance matrix.*
> `coord`: *Number of dimensions.*
> `nperm`: *Number of permutations.*
> `graph`: *Logical value indicating whether a graph should be returned.*

Values:
> `r.stat`: *Observed statistic (Element wise correlation coefficient between both hemi-matrices).*
> `null`: *Simulated data for the null distribution.*

Required functions: `lower.triang`, `perm.rowscols`, `mantrstat`.

```
1  mantel.t3<-function(m1,m2,coord=1,nperm=5000,graph=FALSE)
2  {n <- nrow(m1)
3  realz <- mantrstat(m1, m2)
4  nullstats<-replicate(nperm,mantrstat(m1,
5             perm.rowscols2(m2, n, coord)))
6  list(r.stat = realz, null=nullstats)}
```

```
>jj<-mantel.t3(var(schizop),var(control),coord=2)
>sum(jj$null>stat.obs)/5000
[1] 2e-04
>points(density(jj$null),type="l",lty=3)
```

We notice that when we permute landmarks and resuperimpose the configurations, the test does not reject the null hypothesis in more cases than the former tests. This is true as well, but in a lesser extent if we ignore the resuperimposition procedure (Fig. 7.2). It is possible that the last two tests are inflating the type I error rate. It seems clear that the Mantel test for Procrustes data, as it has been firstly formulated [56], is accepting the alternative hypothesis more easily than the two other tests.

One can also run simulations to estimate the power and specificity of tests applied to morphometric data. Since shape statistics is a rather new discipline, and since the geometry of shape spaces are not always Euclidean, this is probably an important field that needs to be developed in modern morphometrics. Simulations of random or nonrandom shapes is necessary for estimating the false positive rate (proportion of negative instances that were erroneously reported as being positive: this fraction is usually reported as α) and the false negative rate of a given test (proportion of false negatives/number of positive instances: this fraction is reported as β and is equal to 1 - the power of the test). For examining type I or II error rates, one can combine the approach of simulated variation in shape (randomized or not) with a nonparametric test as illustrated before.

7.2 Writing Functions and Implementing Methods

Because R is an interpreted language, all commands are in fact *expressions* of the language, and there are many examples of these in this book. As such, R stands apart from the other languages because all users are confronted with it. Even if a user prefers one of the several GUIs available for R, almost all of these provide a scripting output of the commands clicked on the menus (e.g., the Rcmdr package, the Cocoa-GUI for Apple OS, or the RKWard package for KDE). Consequently, programming in R may be approached from the computer scientist's side, as choosing this language among others, but also from the user's side, as extending the commands used in data analysis.

Writing functions in R is the canonical way to generalize R commands. We have previously seen that a series of R commands can be saved in a script file, and then called from R, so we will concentrate in this section on writing functions. We will first see a few interesting concepts about writing computer functions, and the choices that have to be made to complete the task. Then, we will detail a worked example on contour acquisition.

7.2.1 Generalities and Strategies

Computer functions in general, and in R in particular, have two main features: they are created and used for a repetitive task, and some elements may have different values for each of these repetitions. These elements are called *variables* in the computing jargon, but this is a very different concept than the *statistical* variables. In most computer languages, including R, one can create functions without defining variables:

```
> specialMessage <- function()
+             cat("\nYou did something wrong.\n")
> specialMessage()

You did something wrong.
```

However, in practice, even in this simple example, it is often useful to define a variable controlling the output:

```
> specialMessage <- function(out)
+ {
+     if (out) cat("\nYou did something wrong.\n")
+     else cat("\nCongratulations!\n")
+ }
> specialMessage(0)

Congratulations!
> specialMessage(1)

You did something wrong.
```

Here, out is a *formal argument* of the function: every occurrence of this variable in the body of the function will be replaced by the value given when the function is called (here 0 or 1).

What happens if an object name is used inside the function but this is not a formal argument? R uses a mechanism called *lexical scoping* to solve this problem: if the object has not been created inside the function, then it looks for it the environment above the function call.

```
> specialMessage <- function(out)
  {   if (out) cat(badMessage)
+        else cat(goodMessage) }
> badMessage <- "\nYou did something wrong.\n"
> goodMessage <- "\nCongratulations!\n"
> specialMessage(0)

Congratulations!
> specialMessage(1)

You did something wrong.
```

Every object created inside a function is local to the function, and deleted when its execution is finished. There are two ways for an R function to return a result: either R implicitly returns the last expression of the function (the preferred way), or the return function is explicitly used anywhere in the function, in which case its argument is returned and the execution of the function is halted.

The above considerations give a (very) brief outline of some guidelines in programming functions in R (you can find more details in the manuals installed with

R). Apart from these, all considerations regarding the use of R functions equally apply to R programming. A motto that you should follow is simplicity. In R, simple expressions are much faster and efficient than more complicated ones. In particular, one can often avoid using `for` and/or `if` statements by using vectorization and logical indexing. For instance, if we want to replace all negative values in a x vector by 0, we could use one of the two commands:

```
>x[x < 0] <- 0
>for (i in 1:length(x)) if (x[i] < 0) x[i] <- 0
```

Note that the first one is both simpler and much more efficient. Functions useful to achieve simplicity include `match`, and `lapply`....

In spite of the flexibility of R, it is not always possible to achieve simplicity and efficiency in R programs. For some repeated computations based on complex criteria, it is almost impossible to avoid the `for` and `if` statements, which may result in long execution times. If this is the case, and one wishes to write general functions that are likely to be used by other users, then it is possible to write some of the codes in a compiled language (C, C++, or FORTRAN). This solution is particularly well-suited when one is writing functions that are distributed in a package through CRAN. In this situation, the *R Core Development Team* handles the compilation of the package, which depends on the operating system.

The degree to which R codes are transferred to a compiled language may vary greatly. In the extreme form, all computations, including object handling, are written in compiled languages that require you to use special functions to set object attributes (e.g., mode and length). Here we will focus on a more practical approach that uses C codes only for some parts of the computation. The advantage of the C language is that simple functions may be written. The rationale of writing as few codes in C as possible is that you cannot modify them without recompiling.

A general programming strategy in R may be defined as follows.

1. Write all the code in R paying some attention on simplicity.
2. If you can not avoid complicated use of `for` and/or `if` statements, try to avoid parts of the codes that take some time, possibly using R profiling tools (see `?Rprof`).
3. Once you have identified computational bottlenecks, and you think these are are bug-free, write a C function and replace the corresponding R codes by a call to `.C`.

This strategy helps to make a function lasting and evolving in a positive way because the way end-users call it is stable through time while its efficiency improves.

7.2.2 A Worked Example in R+C Programming: Contour Acquisition Revisited

We have seen a function in Chapter 2 that aims to extract the contour of an object from a digital image read with the **pixmap** package. The present section illustrates how to perform the same task calling a C program. Because it is a worked example, comments around the R code are in normal font here, while the ones around C code are in slanted font.

A strength of the interface between C and R is that is very simple, and C programs can be written with a minimum of overhead work. This makes this approach accessible with little background on the C language. We will see step-by-step through this example how the program is developed. You can find some introductions about the C language and summaries of its syntax on the Internet.

```
1  #include <R.h>
```

The file R.h *tells the C compiler where to find all the information it needs to create the executable library (will be a* *.dll *file under Windows,* *.so *file for the other operating systems).*

```
2  #define THRESHOLD 0.1
```

It is possible to define constants or blocks of commands (i.e., macros). This is clearly useful if one uses the same value at several places in the program.

```
3  int Modulo8(int x)
4  {      while (x < 0) x += 8;
5      while (x >= 8) x -= 8;
6      return x;}
```

We can also define any C function that we may need. Here this function takes as argument an integer and returns its value modulo 8. It obviously returns an integer.

```
7  void contour (double*x,int*nrow,int*start,int*cont,int*l)
```

Now we can start writing the function that will receive the data from R. Note that all data are passed from R to C as pointers. An important detail is that matrices in R are actually vectors, so they are passed to C as one-dimensional (and not 2D) arrays, and their elements are accessed with the familiar syntax x[1] *(and not* x[1][1]*). This implies a slight arithmetic manipulation in C to access the contiguous elements of a matrix. We also need to know that the elements of a matrix in R are filled column-wise, so that* x[5, 3] *and* x[2*nrow(x) + 5] *are the same. Another detail to keep in mind is that indexing in C starts at 0, so* x[1] *returns the second element of* x.

```
8   {int i, a=0, sel, o[8], left, center, right, loc=*start,
9       init = *start - *n, n = *nrow;
10      double diff, ndiff;
```

The body of the function starts with a curly brace similarly to R (the R syntax is superficially similar to C's). In comparison to R, all variables used in a C function must be declared. The distinction between integers and reals is explicit in C.

```
11      o[0] = -n;  o[1] = -n - 1;
12      o[2] = -1;  o[3] = n - 1;
13      o[4] = n;   o[5] = n + 1;
14      o[6] = 1;   o[7] = 1 - n;
```

We use the integer o *to find the indices of the eight neighbors around a given pixel. Here we see the use of the one-dimensional indexing logic explained above. Later, we will store the index of the central pixel in* loc, *so the index of the pixel on its left will be given by* loc + o[0], *the index of the pixel on the upper left-hand side will be* loc + o[1], *and so on in a clockwise way.*

```
15    while (abs(x[loc] - x[init]) < THRESHOLD) {
16      init -= n;
17      loc -= n; }
18    init = loc;
19    sel = 2;
```

This loop finds the actual starting point of the contour using the same criterion as in Conte. *Note that the index is moved to the left of the matrix, so it is decreased by the number of rows.* sel *stores the direction where the next step will be processed.*

```
20    while (loc != init || a < 3) {
```

The acquisition of the contour starts with an iteration that is ended as long as the focus pixel has not come back to its starting point (we add a further condition because loc *and* init *initially store the same value).*

```
21        left = loc + o[Modulo8(sel - 1)];
22        center = loc + o[Modulo8(sel)];
23        right = loc + o[Modulo8(sel + 1)];
```

*We find the indices of the three pixels in the selected direction. The three pixels are those in the same direction (*sel*), on the left-hand side (*sel - 1*), and on the right-hand side (*sel + 1*). We use the* Modulo8 *function defined above in case* sel *has gone beyond the permitted interval (see below).*

```
24        if (abs(x[loc] - x[left]) > THRESHOLD) {
25            if (abs(x[loc] - x[center]) < THRESHOLD){
26                if (abs(x[loc] - x[right]) < THRESHOLD) {
27                    loc = cont[a] = center;
28                        a++; }}
29                else sel++; }
30            else sel--; }
```

If the "center" and "right" pixels are within the object, and the "left" one is not, the algorithm progresses forward and the index of the "center" pixel is stored as within the contour.
*If the "center" pixel is outside the object, the positions of the three pixels is shifted clockwise (*else sel++*) and the evaluation is repeated.*
*If both the "center" and "left" pixels belong to the object, then the positions of the three pixels is shifted counter clockwise (*else sel-*) and the evaluation is repeated.*

```
31    *l = a; }
```

Once the acquisition is finished, the number of pixels found during the process is stored in *l *which is returned by the function, and a right curly brace marks the end of the function.*

The C program will be called from R with the .C function, which sends to the compiled library the objects matching the arguments defined in the C function. It is convenient to prepare a little information in R beforehand, We assume that we have the same data as for `Conte`.

```
>N <- nrow(imagematrix)
>n <- length(imagematrix)
>start <- x[1] + N*(x[2] - 1) - 1
```

The result of the call is stored in the `ans` object.

```
>ans <- .C("contour", as.double(imagematrix), as.integer(n),
+          as.integer(start), integer(n), as.integer(1),
+          PACKAGE = "imaginos")
```

The arguments of `.C` are the name of the C function, the variables passed to C, and the package where to find the C function (there are other options; see `?.C`). The call `.C` returns, in a `list` object, the variables possibly modified by the C program. Note the fourth argument, which is simply a vector of n integers (initialized to 0's), and is destined to store the indices of the contour (`cont` in C). This illustrates a common problem when interfacing C with R: the size of the objects passed to C must be *a priori* determined, but here we do not know how many pixels will be in the contour (Some memory allocation can be done in C with the standard `malloc` functions, but this cannot be returned to R). A simple and safe solution, that is used here is to create a vector sufficiently large and to keep track of how many elements are used to finally delete the useless 0's once the calculation has been done.[1] Because n is the number of pixels of the image, the contour cannot include more than n points. This may require a significant amount of computer memory for very large images (e.g., for a 10 megapixel image, 40 Mb will be allocated since an integer is stored on 4 bytes), but the unneeded memory can be released by the following command:

```
>cont <- ans[[4]][1:ans[[5]]]
>j <- cont %/% N
>i <- cont %% N
```

The two last commands convert the vector-style indices stored in `cont` in matrix-style indices.

It will be more convenient to include all these R commands within a function, such as:

```
Conte2 <- function(imagematrix, x)
{   N <- nrow(imagematrix)
....
    list(X = i, = j)}
```

It is easy to further develop such an R+C program. For instance, subsequent uses of the program may make it necessary to use different values of the THRESHOLD constant: it is straightforward to define this variable in R and pass it to C.

---

[1] Another solution is using the `.Call` and `.External` interfaces to C, but these require using specific functions to manipulate R objects in C.

7.3 Interfacing and Hybridizing R

R is efficient in many applications, but not in all where computers are useful. Indeed, in practice, data analyses use a combination of computer programs chosen according to their respective features and merits. There is a difference between using several programs with no logical or practical connections, and using tools that are integrated and communicate in a logical way. The latter is possible with open source software such as R and other tools that may be more appropriate for some tasks. The integration of computer programs needs to be developed in its own right like any other computer tool.

Why should we bother with such integration? There are at least two good reasons. First, a developer may want to integrate a given application in his/her programs. If he develops an R function for this task, this work will be available to any user familiar with R. Additionally, standard R programming would be used. Second, integrating other applications with R allows one to use the best of both (or several) programs. Other programs may be better than R, for instance for image analyses, but R has appropriate statistical and graphical functions that are likely to be better for analyses of the results of an image analysis.

7.3.1 Example 1: Creating an Animation with R and ImageMagick

Here we build an `animate` function that serially displays a series of images given either as a vector of `character` mode, or as a directory where all JPEG files are read. The `delay` argument is the time in seconds between each image. By default, a 400×400 display is used. This requires ImageMagick[2] to be installed. Under Windows, you will have to install an X server to run the X11-based programs like "animate", "display", and "import" that exist in Imagemagick.

Function 7.2. `animate`

Arguments:
    `files`: *Vector containing the name of the image files.*
    `dir`: *Alternatively, the name of the directory containing the image files.*
    `delay`: *Time in seconds between each image.*
    `width`: *Width of the display.*
    `height`: *Height of the display.*
Value:
    The animation produced from the inputed image files.

```
1  animate <- function(files, dir = NULL, delay = 2,
2                      width = 400, height = 400)
3  {if (!is.null(dir)) {
4    pwd <-getwd()
5      if (pwd != dir) {
```

---

[2] www.imagemagick.org

```
 6        setwd(dir)
 7        on.exit(setwd(pwd))}
 8      files<-base::dir()
 9     files<-files[grep("//.jpe?g$",files,ignore.case=TRUE)]}
10   cmd<-paste("␣-delay␣", 100*delay, "␣-size␣", width, "x",
11             height, "␣", files, sep = "", collapse = "␣")
12   system(paste("animate", cmd))}
```

Producing an animation is a pedagogical way to visualize shape changes along axes. We use it for deforming a template along the first axis of variation using the thin-plate splines strategy.

We first have to define a template; this template is a collection of points defining an object. We will draw it from the rodent jaw example formerly used in the analysis of asymmetry. The first digitization jjM corresponds to the 16 landmarks used in the mouse dataset, the following digitizations corresponds to pseudolandmarks strategically sampled on different curves composing the drawing (see Fig. 7.3).

```
>library(pixmap)
>jaw<-read.jpeg("/home/juju/morph/jaw2.jpg")
>par(mar=c(0,0,0,0))
>plot(jaw)
>jjM<-locator(16, type="p", pch=8, col="white")
>jj2<-locator(type="l", pch=8, col="white")
>jj3<-locator(type="l", pch=8, col="white")
>jj4<-locator(type="l", pch=8, col="white")
>jj5<-locator(type="l", pch=8, col="white")
>jj6<-locator(type="l", pch=8, col="white")
>jj7<-locator(type="l", pch=8, col="white")
>jj8<-locator(type="l", pch=8, col="white")
>X<-c(jjM$x,NA,jj2$x,NA,jj3$x,NA,jj4$x,NA,jj5$x,
+    NA,jj6$x,NA,jj7$x,NA,jj8$x)
>Y<-c(jjM$y,NA,jj2$y,NA,jj3$y,NA,jj4$y,NA,jj5$y,
+    NA,jj6$y,NA,jj7$y,NA,jj8$y)
```

One can display the template using conventional graphical functions. Note that NA is used for separating the different lines making the outlines of structures of the jaw (Fig. 7.4).

```
>par(mar=c(0,0,0,0))
>mat<-cbind(X,Y)
>plot(1, type="n", xlim=c(0,1030),ylim=c(0, 530),
+     frame=F, axes=F, xlab="", ylab="",asp=1)
>lines(mat[-(1:17),])
>points(mat[1:16,], cex=1.3, pch=19)
```

Now we have to calculate the meanshape and the superimposed coordinates from the dataset.

```
>DDO<-read.table("/home/juju/morph/Mouse.txt",header=T)
```

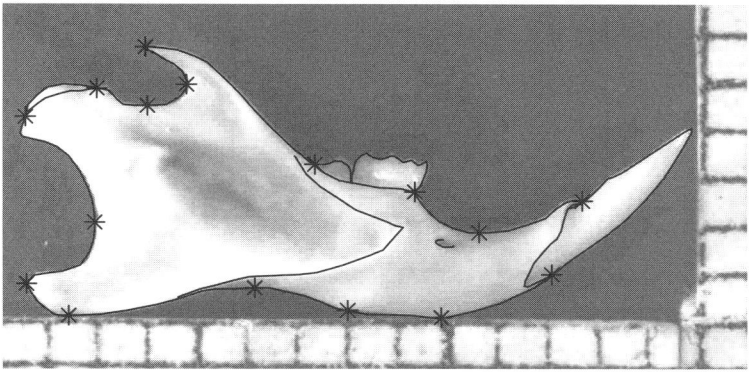

Fig. 7.3. Digitization of landmarks, curves, and outlines for producing a template of a mouse jaw. Stars correspond to landmark data, while lines correspond to curves: seven different curves have been used for producing the rodent jaw template

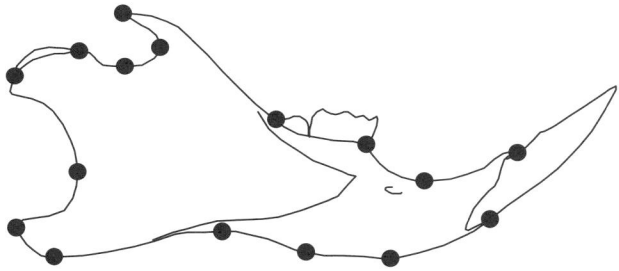

Fig. 7.4. Illustration of the template digitized on the mouse jaw

```
>Ddo<-as.matrix(DDO[,3+c(1:16*2-1,1:16*2 )],208,32)
>ddo<-array(t(Ddo),dim=c(16,2,208))
>ind<-as.factor(DDO[,1])
>side<-as.factor(DDO[,2])
>for (i in 1:104) {ddo[,1,i*2-1]<- - ddo[,1,i*2-1]}
>super<-pgpa(ddo)
>ddosup<-super$rotated
>DDOsup<-t(matrix(ddosup,32,208))
>meansh<-super$msh
```

We use the `tps2d` function to perform a thin-plate splines warping of the template using the meanshape as a target. The reference corresponds to the landmark digitized on the template. We compute the PCA of coordinates in the tangent shape space for extracting the first axis of variation.

```
>meansho<-tps2d(mat[-(1:16),], mat[1:16,], meansh)
>PCA<-princomp(DDOsup)
```

We produce a series of 30 configurations that correspond to changes along the first principal axis and by amplifying the deformation twice. This is done using the `seq` function, and the `M` and `outM` objects of the `list` class. Each configuration corresponds respectively to the estimate of landmark position when the shape evolve along the first axis. The template shape is evaluated from these configurations. We display each deformed configuration on a different graph. For each iteration of the loop, we store the produced graphs in one empty subdirectory. The name of files is given by the `Name` vector.

```
>setwd("/home/juju/morph/temp")
>M<-list()
>outM<-list()
>Name<-paste("image",1:30, sep="")
>Name<-paste(Name, "jpeg", sep=".")
>par(mar=c(0,0,0,0))
>seq1<-seq(min(PCA$score[,1]),max(PCA$score[,1]),
+          length=30)*2
>for (i in 1:30){
+    plot(meansh, type="n",asp=1, xlab="", ylab="", frame=F,
+       axes=F, xlim=c(-0.5,0.7))
+    M[[i]]<-matrix(seq1[i]*PCA$loadings[,1]+
+         as.vector(meansh),16,2)
+    outM[[i]]<-tps2d(meansho, meansh, M[[i]])
+    points(M[[i]], cex=1.3, pch=21, bg="black")
+    lines(outM[[i]])
+    jpeg(Name[i])}
```

We can display extreme deformation as a graph as well.

```
>plot(meansh, type="n", asp=1, xlab="", ylab="",
+     frame=F, axes=F, xlim=c(-0.35,0.55))
>lines(outM[[1]], col="grey50")
>lines(outM[[30]])
```

Fig. 7.5 shows deformation of the template of the jaw on the principal axis. We have amplified this deformation three times. We can notice the important shape changes occurring in the angular apophysis. Although thin-plate splines are interpolations, these interpolations fit quite well with the reality of mouse jaw variation.

The animation is simply displayed using the `animate` function that was created just before.

```
>animate(Name)
```

7.3.2 Example 2: Using ImageMagick to Display High Resolution Images

Because pixmap may be very slow to read and plot very large images and because the plot is static (no zoom), we may want an alternative solution eventually using

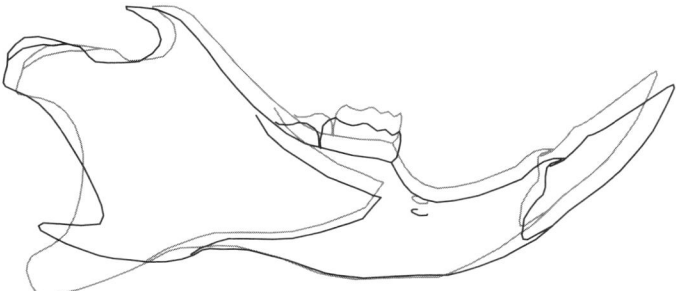

Fig. 7.5. Deformation on the first PC-axis for the mouse dataset, the dotted outline corresponds to the deformation associated with minimal extreme, while the full line corresponds to the deformation associated with the maximal extreme. Variation has been amplified twice for rendering visualization easier

an external application. ImageMagick includes a program called `display` that displays an image in a specific window. We directly call it from R with the function `display` written below. The `color` argument specifies whether to display the gray-scale image or the image in its original colors (the default).

Function 7.3. `display`

> *Arguments:*
> > `file`: *Name of the image file.*
> > `color`: *Logical telling whether gray-scale or color image should be displayed.*
> > `width`: *Width of the display.*
> > `height`: *Height of the display.*
> *Value:*
> > *Display of the image on the screen.*

```
1 display <- function(file,color=TRUE,width=400,height=400)
2 {cmd <- "display"
3  if (!color) cmd <- paste(cmd, "-monochrome")
4  cmd <- paste(cmd," -size ",width,"x",height,sep ="")
5  system(paste(cmd, file))}
```

7.4 Conclusion

Modern morphometrics is becoming an important area of statistics. This is not only a tool but also a scientific discipline in expansion. One can analyze variation in complex morphologies via a large number of strategies and methods that have been published during the last three decades; and that will be published in the near future. The purpose of this book is not to explain and summarize all of them, but to illustrate how to implement them in R. Most algorithms are present in the literature, and more are to come. Any user or developer having some basic knowledge in the R environment and language can develop his/her own toolbox and produce standard or customized graphs, helped with the increasing number of packages available on R.

We have seen several good reasons for using R in this domain. There are improvements in morphometrics for which R can become a favorite companion:

- Developing new statistical methods for the analysis of shapes and estimating their efficiency
- Appraising the power and specificity of tests applied to shape variables

For tasks expensive to compute, interfacing R with some other freewares can become the appropriate strategy. Another one would be to develop other libraries or packages, especially concerning the management and interaction with image processing and 3D objects. This is certainly a good perspective open for morphometricians and R users. This will depend, however, on the investments of morphometricians for using R more systematically, rather than homemade, sometimes obscure software, that is also not always free.

Problems

7.1. Nonparametric MANOVA
Develop a function to perform a nonparametric, one-way multivariate analysis of variance. This can be achieved by permuting landmarks between observations and resuperimposing newly simulated configurations.

Appendix A: Functions Developed in this Text

Function	Short Description
myfun	Sum of squares of a given vector
angle2d	Angle between two vectors in 2D
ild	Distance between two landmarks
angle	Unoriented angle between two vectors
angle3	Oriented angle between two vectors
regularradius	Extract coordinates equally spaced radii on an outline
smoothout	Outline smoothing
landmark.addition	Add pseudolandmarks on an outline
eigenrotation	Rotation of the major axis of a configuration along the x-axis
Conte	Outline extraction
ELLI	2D confidence ellipse coordinates
coli	Collinearity test between two vectors
isojoli	Multivariate test of isometry
pls	Two blocks partial least-squares
Angle	Cosinus angle between two vectors
centsiz	Scale a configuration to unit centroid size
centcoord	Coordinates of the centroid of a configuration
basesiz	Baseline size of a configuration
booksteinM	Baseline registration of a configuration
booksteinA	Baseline registration for a set of configurations
mbshape	Mean shape using Bookstein coordinates
tranb	Translation of a configuration according baseline coordinates
transb	Translation and scaling of a configuration according baseline coordinates
bookstein3d	Bookstein registration for 3D configurations
transl	Translate the configuration onto the origin
trans1	Translate the configuration onto the origin
helmert	Helmert matrix
helmertM	Helmertize a configuration
kendall2d	Kendall coordinates for a configuration

Function	Short Description
fPsup	Full Procrustes superimposition
ild2	Interlandmark distances between two configurations
pPsup	Partial Procrustes superimposition
mshape	Averaged shape from a set of configurations
aligne	Align configurations on their princiapl axes
fgpa	Generalized full Procrustes analysis, first algorithm
fgpa2	Generalized full Procrustes analysis, second algorithm
stp	Stereographic projection onto the tangent space
orp	Orthogonal projection onto the tangent space
medsize	Median size of a configuration
argallvec	Angles between homologous vectors of configurations
oPsup	Orthogonal resistant fit
rrot	Resistant fitting rotation in 3D
r3sup	3D orthogonal resistant fit
grf2	Generalized resistant fit for 2D configurations
grf3	Generalized resistant fit for 3D configurations
tps2d	2D tps interpolation
tps	Deformation grids from thin-plate splines
FM	Form matrix (EDMA)
fm	Form vectorized matrix (EDMA)
mEDMA	Averaged mean form (EDMA)
MDS	Multidimensional scaling
mEDMA2	mean form matrix (EDMA)
vEDMA	Variance covariance matrix (EDMA)
alltri	All possible triangulations in a configuration
anglerao	Rao angles in a configuration
meanrao	Mean shape (based on Rao angles)
vcvrao	Variance covariance matrix (based on Rao angles)
raoinv	Configuration coordinates from Rao angles
cumchord	Cumulative chordal distance
bezier	Bezier coefficients and vertices
beziercurve	Bezier curve fitting
fourier1	Radial Fourier analysis
ifourier1	Reverse Radial fourier outline fitting
fourier2	Tangent angle Fourier analysis
ifourier2	Reverse tangent angle Fourier outline fitting
efourier	Elliptic Fourier analysis
iefourier	Reverse elliptic fourier outline fitting
NEF	Normalized elliptic fourier analysis
eigenshape	Extract eigenshape from outlines
procalign	partial GPA with mean shape aligned on the major axis
uniform2d	2D uniform components and scores for shape variation
uniformG	Uniform components and scores for shape variation
Hottelingsp	Multivariate T^2_{HL} for shape data
vcvay	Covariance phylogenetic matrix between descendant and ancestors
mantel.t2	Mantel test for Procrustes data
Rv	Rv correlation coefficient of Escoufier
mantel.t3	Alternative Mantel test for Procrustes data
specialMessage	Tutorial function
Conte2	Outline extraction in C
animate	Movie from a sequence of images
display	Display high resolution images

Appendix B: Packages Used in this Text

Package	Short Description
graphics	R functions for base graphics
rgl	3D visualization device (OpenGL)
base	Base R functions
utils	R utility functions
stats	R statistical functions
datasets	Base R datasets
grDevices	Graphics devices and support for base and grid graphics
MASS	Functions and datasets to support Venables and Ripley, "Modern Applied Statistics with S" (4th edition)
shapes	Functions and datasets for statistical shape analysis
ade4	Multivariate data analysis and graphical display
Rcmdr	The R commander GUI for R
R.matlab	Read and write of MAT files together with R-to-Matlab connectivity
pixmap	Functions for handling .ppm images ("Pixel Maps")
rimage	Image Processing Module for R
rtiff	A .tiff reader for R
scatterplot3d	3D scatter plot
dynamicGraph	Interactive graphical tool for manipulating graphs
sp	Classes and methods for spatial data
splancs	Spatial and space-time point pattern analysis
car	Companion to Applied Regression
nlme	Linear and nonlinear mixed effects models
lmtest	Testing linear regression models
lme4	Linear and generalized linear mixed-effects models
HH	Statistical analysis and data display
gee	Generalized Estimating Equation solver

Package	Short Description
VGAM	Vector Generalized linear and Additive Models
ape	Analyses of Phylogenetics and Evolution
PHYLOGR	Phylogenetically based statistical analyses
smatr	(Standardized) Major axis estimation and testing routines
Flury	Datasets from Flury, 1997
Rmorph	Morphometric analyses with R
cluster	Cluster analysis extended
vegan	Community ecology package
mclust	Model-based clustering and normal mixture modeling including Bayesian regularization
CCA	Canonical correlation analysis
Hmisc	Miscellaneous functions from F. E. Harrell Jr.
CircStats	Circular Statistics
compositions	Compositional data analysis
geometry	Mesh generation and surface tesselation
mvtnorm	Multivariate normal and t-distributions

References

[1] Adams D. C., Rohlf F. J. and Slice D. E. 2004. Geometric morphometrics: Ten years of progress after the "revolution". *Italian Journal of Zoology* **71**: 5–16.

[2] Atchley W. R. and Anderson D. 1978. Ratios and the statistical analysis of biological data. *Systematic Zoology* **27**: 71–78.

[3] Atchley W. R., Gaskins C. T. and Anderson D. 1976. Statistical properties of ratios. I. empirical results. *Systematic Zoology* **25**: 137–148.

[4] Auffray J.-C., Alibert P., Renaud S., Orth A. and Bonhomme F. 1996. Fluctuating asymmetry in *mus musculus* subspecific hybridization: traditional and Procrustes comparative approach. In: *Advances in Morphometrics*, Marcus L. F., Corti M., Loy A., Naylor G. J. P. and Slice D. E., editors, pages 275–283. Plenum Press, New York.

[5] Bailey R. C. and Byrnes J. 1990. A new, old method for assessing measurement error in both univariate and multivariate morphometric studies. *Systematic Zoology* **39**: 124–130.

[6] Baylac M. and Friess M. 2005. Fourier descriptors, procrutes superimposition, and data dimensionality: an example of cranial shape analysis in modern human populations. In: *Modern Morphometrics in Physical Anthropology*, Slice D. E., editor, pages 145–165. Kluwer Academic / Plenum, New York.

[7] Bookstein F. L. 1984. A statistical method for biological shape comparisons. *Journal of Theoretical Biology* **107**: 475–520.

[8] Bookstein F. L. 1986. Size and shape spaces for landmark data in two dimensions (with discussion). *Statistical Science* **1**: 181–242.

[9] Bookstein F. L. 1989. Principal warps: thin-plate splines and the decomposition of deformations. *IEEE Transactions on Pattern Analysis and Machine Intelligence* **11**: 567–585.

[10] Bookstein F. L. 1991. *Morphometric tools for landmark data: Geometry and Biology*. Cambridge University Press, Cambridge.

[11] Bookstein F. L. 1996. Biometrics, biomathematics and the morphometric synthesis. *Bulletin of mathematical biology* **58**: 313–365.

[12] Bookstein F. L. 1996. A standard formula for the uniform shape component in landmark data. In: *Advances in Morphometrics*, Marcus L. F., Corti M., Loy A., Naylor G. J. P. and Slice D. E., editors, pages 153–168. Plenum Press, New York.

[13] Bookstein F. L. 1997. Landmark methods for forms without landmarks: Localizing group differences in outline shape. *Med. Image. Anal.* pages 225–243.

[14] Bookstein F. L. 2005. After landmarks. In: *Modern Morphometrics in Physical Anthropology*, Slice D. E., editor, pages 49–71. Kluwer Academic/Plenum, New York.

[15] Bookstein F. L., Chernoff B. L., Elder R. L., Humphries J. M., Smith G. and Strauss R. E. 1985. Morphometrics in evolutionary biology. *Academy of Natural Sciences of Philadelphia Special Publ.* **15**: 1–277.

[16] Burnaby T. P. 1966. Growth-invariant discriminant functions and generalized distances. *Biometrics* **22**: 96–110.

[17] Campbell N. A. and Mahon R. J. 1974. A multivariate study of variation in two species of rock crab of the genus *leptograpsus*. *Australian Journal of Zoology* **22**: 417–425.

[18] Carpenter K. E., Sommer J. I. and Marcus L. F. 1996. Converting truss interlandmark distances to cartesian coordinates. In: *Advances in Morphometrics*, Marcus L. F., Corti M., Loy A., Naylor G. J. P. and Slice D. E., editors, pages 103–111. Plenum Press, New York.

[19] Chambers J. M., Cleveland W. S., Kleiner B. and Tukey P. A. 1983. *Graphical Methods for Data Analysis*. Wadsworth, Brooks and Cole,.

[20] Cheverud J. M., Dow M. M. and Leutenegger W. 1985. The quantitative assessment of phylogenetic constraints in comparative analyses: sexual dimorphism in body weight among primates. *Evolution* **39**: 1335–1351.

[21] Cole III T. 2002. *WinEDMA User's Guide Version 1.0.1 beta*,.

[22] Cole III T. and Richtsmeier J. T. 1998. A simple method for visualization of influential landmarks when using euclidean distance matrix analysis. *American Journal of Physicla Anthropology* **107**: 273–283.

[23] Crampton J. S. 1995. Elliptic fourier shape analysis of fossil bivalves: some practical considerations. *Lethaia* **28**: 179–186.

[24] Cunningham C. W., Omland K. E. and Oakley T. H. 1998. Reconstructing ancestral character states: A critical reappraisal. *Trends in Ecology and Evolution* **13**: 361–366.

[25] Dalgaard P. 2002. *Introductory Statistics with R*. Springer Verlag, New York.

[26] Diniz-Filho J. A. F., Sant'Ana C. and Bini L. 1998. An eigenvector method for estimating phylogenetic inertia. *Evolution* **52**: 1247–1262.

[27] Dryden I. E. and Mardia K. V. 1998. *Statistical Shape Analysis*. Wiley, Chichester.

[28] Dryden I. E. and Walker G. 1999. Highly resistant regression and object matching. *Biometrics* **55**: 820–825.

[29] Escoufier Y. 1973. La dépendance de deux aléas vectoriels critères et visualisation. *Revue de statistique appliquée* **21**: 5–16.

[30] Felsenstein J. 1985. Phylogenies and the comparative method. *The American Naturalist* **125**: 1–15.

[31] Ferson S., Rohlf F. J. and Koehn R. K. 1985. Measuring shape variation of two-dimensional outlines. *Systematic Zoology* **34**: 59–66.

[32] Friess M. and Baylac M. 2003. Exploring artificial cranial deformation using elliptic Fourier analysis of Procrustes aligned outlines. *American Journal of Physical Anthropology* **122**: 11–22.

[33] Galton F. 1907. Classification of portraits. *Nature* **76**: 617–619.

[34] Garland Jr. T. and Ives A. 2000. Using the past to predict the present: confidence intervals for regression equations in phylogenetic comparative methods. *The American Naturalist* **155**: 346–364.

[35] Garland Jr. T. and Janis C. 1985. Does metatarsal/femur ratio predict maximal running speed in cursorial mammals. *Journal of Zoology* **125**: 1–15.

[36] Gayon J. 2000. History of the concept of allometry. *The American Zoologist* **40**: 748–758.

[37] Giardina C. R. and Kuhl F. P. 1977. Accuracy of curve approximation by harmonically related vectors with elliptical loci. *Computer graphics and image processings* **6**: 236–258.

[38] Goodall C. R. 1991. Procrustes methods in the statistical analysis of shape (with discussion). *Journal of the Royal Statistical Society, Series B* **53**: 285–339.

[39] Gower J. C. 1975. Generalized Procrustes analysis. *Psychometrika* **40**: 33–50.

[40] Grafen A. 1989. The phylogenetic regression. *Philosophical Transactions of the Royal Society London B* **326**: 119–156.

[41] Green B. F. 1952. The orthogonal approximation of an oblique structure in factor analysis. *Psychometrika* **17**: 429–440.

[42] Gunz P., Mitteroecker P. and Bookstein F. L. 2005. Semi landmarks in three dimensions. In: *Modern Morphometrics in Physical Anthropology*, Slice D. E., editor, pages 73–98. Kluwer Academic / Plenum, New York.

[43] Haines A. J. and Crampton J. S. 2000. Improvements to the method of fourier shape analysis as applied in morphometric studies. *Palaeontology* **43**: 765–783.

[44] Humphries J., Bookstein F. L., Chernoff B., Smith G. R., Elder R. L. and Poss S. G. 1981. Mutivariate discrimination by shape in relation to size. *Systematic Zoology* **30**: 291–308.

[45] Hurley J. R. and Cattel R. B. 1962. The Procrustes program: producing direct rotation to test an hypothesized factor structure. *Behavioural Science* **7**: 258–262.

[46] Huxley J. S. 1924. Constant differential growth-ratios and their significance. *Nature* **114**: 895–896.

[47] Huxley J. S. 1932. *Problems of Relative Growth*. Methuen, London 1st edition.

[48] Huxley J. S. and Teissier G. 1924. Terminology of relative growth. *Nature* **137**: 780–781.

[49] Jolicoeur P. 1963. The degree of generality of robustness in martes americana. *Growth* **27**: 1–28.

[50] Jolicoeur P. and Mossimann J. 1960. Size allometry: size and shape variables with characterizations of the lognormal and generalized gamma distribution. *Journal of the American Statistical Association* **330**: 930–945.

[51] Jolicoeur P., Pirlot P., Baron G. and Stephan H. 1984. Brain structure and correlation patterns in insectivora, chiroptera, and primates. *Systematic Biology* **33**: 14–29.

[52] Jolliffe I. 2002. *Principal Component Analysis*. Springer, New York 2nd edition.

[53] Kendall D. G. 1983. The shape of Poisson-Delaunay triangles. In: *Studies in Probabilities and related topics*, Demetrescu M. C. and Iosifescu M., editors, pages 321–330. Nagard, Montreal.

[54] Kendall D. G. 1984. Shape manifolds, Procrustean metrics and complex projective spaces. *Bulletin of the London Mathematical Society* **16**: 81–121.

[55] Klingenberg C. P., Burluenga M. and Meyer A. 2002. Shape analysis of symmetric structures: quantifying variation among individuals and asymmetry. *Evolution* **56**: 1909–1920.

[56] Klingenberg C. P. and McIntyre G. S. 1998. Geometric morphometrics of developmental instability: analyzing patterns of fluctuating asymmetry with Procrustes methods. *Evolution* **54**: 1363–1375.

[57] Klingenberg C. P., Mebus K. and Auffray J.-C. 2003. Developmental integration in a complex morphological structure: how distinct are the modules in the mouse mandibule. *Evolution and Development* **5**: 522–531.

[58] Kuhl F. P. and Giardina C. R. 1982. Elliptic Fourier features of a closed contour. *Computer graphics and image processings* **18**: 236–258.

[59] Lele S. 1991. Some comments on coordinate free and scale invariant methods in morphometrics. *American Journal of Physical Anthropology* **85**: 407–415.

[60] Lele S. 1993. Euclidean distance matrix analysis (edma): estimation of mean form difference. *Mathematical Geology* **25**: 573–602.

[61] Lele S. and Cole III T. M. 1996. A new test for shape differences when variance-covariance matrices are unequal. *Journal of Human Evolution* **31**: 193–212.

[62] Lele S. and Richtsmeier J. T. 1990. Statistical models in morphometrics: are they realistic? *Systematic Zoology* **39**: 60–69.

[63] Lele S. and Richtsmeier J. T. 1991. Euclidean distance matrix analysis: a coordinate free approach to comparing biological shapes using landmark data. *American Journal of Physical Anthropology* **86**: 415–428.

[64] Lele S. and Richtsmeier J. T. 1992. On comparing biological shapes: detection of influential landmarks. *American Journal of Physical Anthropology* **87**: 49–65.

[65] Lohman G. P. 1983. Eigenshape analysis of microfossils: a general morphometric procedure for describing changes in shape. *Mathematical Geology* **15**: 659–672.

[66] Lohman G. P. and Schweitzer P. N. 1990. On eigenshape analysis. In: *Proceedings of the Michigan Morphometrics*, Rohlf F. J. and Bookstein F. L.,

editors, pages 147–165. The University of Michigan Museum of Zoology, Ann Arbor.

[67] Lynch M. 1991. Methods for the analysis of comparative data in evolutionary biology. *Evolution* **45**: 1065–1080.

[68] Macleod N. 1990. Digital images and automated image analysis systems. In: *Proceedings of the Michigan Morphometrics*, Rohlf F. J. and Bookstein F. L., editors, pages 21–35. The University of Michigan Museum of Zoology, Ann Arbor.

[69] Macleod N. and Forey P. L. 2002. *Morphology, shape and phylogeny*. Taylor and Francis, London.

[70] Mantel N. 1967. The detection of disease clustering and a generalized regression approach. *Cancer Research* **27**: 209–220.

[71] Mardia K. V., Bookstein F. L. and Moreton I. 2000. Statistical assessment of bilateral symmetry of shapes. *Biometrika* **87**: 285–300.

[72] Mardia K. V., Coombes A., Kirkbride J., Linney A. and Bowie J. L. 1996. On statistical problems with face identification from photographs. *Journal of Applied Statistitics* **23**: 655–675.

[73] McKeon J. J. 1974. *F*-approximations to the distribution of hotelling's *T* square. *Biometrika* **61**: 381–383.

[74] Monteiro L. R. 1999. Multivariate regression models and geometric morphometrics: the search for causal factors in the analysis of shape. *Systematic Biology* **48**: 192–199.

[75] Mosimann J. E. 1970. Size allometry: size and shape variables with characterizations of the lognormal and generalized gamma distributions. *Journal of the American Statistical Association* **65**: 930–948.

[76] Neter J., Kutner M. H. and Nachtstheim C. J. 1996. *Applied Linear Regression Models*. McGraw-Hill/Irwin, Cambridge 3rd edition.

[77] Neter J., Kutner M. H., Nachtstheim C. J. and Wasserman W. 1996. *Applied linear statistical models*. Irwin, Chicago, Bogota, Boston 4th edition.

[78] Palmer A. R. 1994. Fluctuating asymmetry: a primer. In: *Developmental instability: its origins and implications*, Markow T., editor, pages 335–364. Kluwer, Dordrecht.

[79] Palmer A. R. and Strobeck C. 1986. Fluctuating asymmetry: Measurement, analysis, patterns. *Annual Review of Ecology and Systematics* **17**: 391–421.

[80] Paradis E. 2002. *R for beginners*,.

[81] Paradis E. 2006. *Analyses of phylogenetics and evolution with R*. Springer, New York 1st edition.

[82] Paradis E. and Claude J. 2002. Analysis of comparative data using generalized estimating equations. *Journal of Theoretical Biology* **218**: 175–285.

[83] Paradis E., Claude J. and Strimmer K. 2004. Analyses of phylogenetics and evolution in R language. *Bioinformatics* **20**: 289–290.

[84] Peres-Neto P. and Jackson D. 2001. How well do multivariate data sets match? the advantages of a Procrustean superimposition approach over the Mantel test. *Oecologia* **129**: 169–178.

[85] Pinheiro J. C. and Bates D. M. 2000. *Mixed-Effects Models in S and S-PLUS*. Springer, New York 1st edition.

[86] Qannari E. M., Vigneau E. and Courgoux P. 1998. Une nouvelle distance entre variables. apllication en classification. *Revue de statistique appliquée* **46**: 21–32.

[87] Rao C. R. and Suryawanshi S. 1996. Statistical analysis of shape of objects based on landmark data. *Proceedings of the National Academy of Science USA* **93**: 12132–12136.

[88] Rao C. R. and Suryawanshi S. 1998. Statistical analysis of shape through triangulation of landmarks: a study of sexual dimorphism in hominids. *Proceedings of the National Academy of Science USA* **95**: 4121–4125.

[89] Renaud S., Michaux J., Jaeger J.-J. and Auffray J.-C. 1996. Fourier analysis applied to *Stephanomys* (Rodentia, Muridae) molars: non-progressive evolutionary pattern in a gradual lineage. *Paleobiology* **22**: 251–261.

[90] Renaud S., Michaux J., Mein P., Aguilar J. P. and Auffray J. 1999. Patterns of size and shape differentiation during the evolutionary radiation of the european miocene murine rodents. *Lethaia* **32**: 61–71.

[91] Reyment R. A. 1991. *Multidimensional palaeobiology*. Pergamon, Oxford, New York, Seoul, Tokyo 1st edition.

[92] Richtsmeier J. T., Cheverud J. M. and Lele S. 1992. Advance in anthropoligical morphometrics. *Annual Review of Anthropology* **21**: 283–305.

[93] Richtsmeier J. T., Deleon V. B. and Lele S. R. 2002. The promise of geometric morphometrics. *Yearbook of Physical Anthropology* **45**: 63–91.

[94] Rohlf J. F. Tpstree.

[95] Rohlf J. F. 1986. Relationships among eigenshape analysis, Fourier analysis, and analysis of coordinates. *Mathematical Geology* **18**: 845–654.

[96] Rohlf J. F. 1993. Relative warp analysis and an example of its application to mosquito wings. In: *Pp. 131–159 in Contributions to Morphometrics*, Marcus L. F., E. B. and Garcia-Valdecasas A., editors, pages 131–159. Museo Nacional de Ciencias Naturales, Vol. 8, Madrid.

[97] Rohlf J. F. 1998. On applications of geometric morphometrics to studies of ontogeny and phylogeny. *Systematic Biology* **47**: 147–158.

[98] Rohlf J. F. 1999. Shape statistics: Procrustes superimposition and tangent spaces. *Journal of Classification* **16**: 197–223.

[99] Rohlf J. F. 2000. Statistical power comparisons among alternative morphometric methods. *American Journal of Physical Anthropology* **111**: 463–478.

[100] Rohlf J. F. 2002. Geometric morphometrics and phylogeny. In: *Morphology, Shape and Phylogeny*, MacLeod N. and Forey P. L., editors, pages 175–193. London.

[101] Rohlf J. F. 2003. Bias and error in estimates of mean shape in morphometrics. *Journal of Human Evolution* **44**: 665–683.

[102] Rohlf J. F. and Archie A. W. 1984. A comparison of Fourier methods for the description of wing shape in mosquitoes (Diptera: Cuculidae). *Systematic Zoology* **33**: 302–317.

[103] Rohlf J. F. and Bookstein F. L. 1987. A comment on shearing as a method for size correction. *Systematic Zoology* **36**: 356–367.

[104] Rohlf J. F. and Bookstein F. L. 2003. Computing the uniform component of shape variation. *Systematic Biology* **52**: 66–69.

[105] Rohlf J. F. and Corti M. 2000. Use of two-block partial least-squares to study covariation in shape. *Systematic Biology* **49**: 740–753.

[106] Rohlf J. F. and Marcus L. F. 1993. A revolution in morphometrics. *Trends in Ecology and Evolution* **8**: 129–132.

[107] Rohlf J. F. and Slice D. E. 1990. Extension of the Procrustes method for the optimal superimposition of landmarks. *Systematic Biology* **39**: 40–59.

[108] Rohlf J. 1990. Fitting curves to outlines. In: *Proceedings of the Michigan Morphometrics*, Rohlf F. J. and Bookstein F. L., editors, pages 167–177. The University of Michigan Museum of Zoology, Ann Arbor.

[109] Sanderson M. J. 1997. A nonparametric approach to estimating divergence times in the absence of rate constancy. *Molecular Biology and Evolution* **14**: 1218Ű–1231.

[110] Siegel A. F. and Benson R. H. 1982. A robust comparison of biological shapes. *Biometrics* **38**: 341–350.

[111] Siegel A. F. and Pinkerton J. R. 1982. Robust comparison of three dimesional shapes with an application to protein molecule configurations. *Technical Report, department of statistics, Pinceton University* **224**.

[112] Slice D. E. 1996. Three-dimensional, generalized resitant fitting and the comparison of least-squares and resistant fit residuals. In: *Advances in Morphometrics*, Marcus L. F., Corti M., Loy A., Naylor G. J. P. and Slice D. E., editors, pages 179–199. Plenum Press, New York.

[113] Slice D. E. 2001. Landmark coordinates aligned by Procrustes analysis do not lie in Kendall's shape space. *Systematic Biology* **50**: 141–149.

[114] Small C. G. 1996. *The statistical theory of shape*. Springer, New York.

[115] Sneath P. H. A. 1967. Trend-surface analysis of transformation grids. *Journal of Zoology* **151**: 65–122.

[116] Strauss R. E. and Bookstein F. L. 1982. The truss: body form reconstruction in morphometrics. *Systematic Zoology* **31**: 113–135.

[117] Thompson D. A. 1917. *On growth and form*. Cambridge University Press, Cambridge.

[118] Valeri C. J., Cole III T., Lele S. and Richtsmeier J. T. 1998. Capturing data from three-dimensional surfaces using fuzzy landmarks. *American Journal of Physical Anthropology* **107**: 113–124.

[119] Venables W. N. and Ripley B. D. 1997. *Modern Applied Statistics with S-plus*. Springer-Verlag, 2nd edition.

[120] Venables W. N., Smith D. M. and the R Development Core Team. 2005. *An introduction to R,*.

[121] Verbeke G. and Molenberghs G. 2000. *Linear Mixed Models for Longitudinal Data*. Springer, New York 2nd edition.

[122] Walker J. 2000. Ability of geometric morphometric methods to estimate a known covariance matrix. *Systematic Biology* **49**: 686–696.

[123] Warton D., Wright I., Falster D. and Westoby M. 2006. Bivariate line-fitting methods for allometry. *Biological Reviews* **81**: 259–291.

[124] Yezerinac S. M., Loogheed S. C. and Handford P. 1992. Measurement error and morphometric studies: statistical power and observer experience. *Systematic Biology* **41**: 471–482.

[125] Zahn C. T. and Rosckies R. Z. 1972. Fourier descriptors for plane closed curves. *IEEE Transcation on computer* **21**: 269–281.

Index

Data Manipulation with R

Phil Spector

This book presents a wide array of methods applicable for reading data into R, and efficiently manipulating that data. In addition to the built-in functions, a number of readily available packages from CRAN (the Comprehensive R Archive Network) are also covered. All of the methods presented take advantage of the core features of R: vectorization, efficient use of subscripting, and the proper use of the varied functions in R that are provided for common data management tasks.

2008. 164 pp. (Use R) Softcover
ISBN 978-0-387-74730-9

Lattice:
Multivariate Data Visualization with R

Deepayan Sarkar

The book contains close to 150 figures produced with lattice. Many of the examples emphasize principles of good graphical design; almost all use real data sets that are publicly available in various R packages. All code and figures in the book are also available online, along with supplementary material covering more advanced topics.

2008. Approx. 290 pp. (Use R!) Softcover
ISBN 978-0-387-75968-5

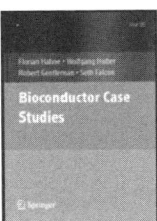

Bioconductor Case Studies

Huber F. Hahne, R.W. Gentleman and S.Falcon

In this book, the authors present a collection of cases to apply Bioconductor tools in the analysis of microarray gene expression data. Each chapter of this book describes an analysis of real data using hands-on example driven approaches. Short exercises help in the learning process and invite more advanced considerations of key topics. The book is a dynamic document. All the code shown can be executed on a local computer, and readers are able to reproduce every computation, figure, and table.

2008. 292 pp. (Use R!) Softcover
ISBN 978-0-387-77239-4

Printed in the United States